Signals and Communication Technology

For further volumes:
http://www.springer.com/series/4748

Björn W. Schuller

Intelligent Audio Analysis

Springer

Björn W. Schuller
LS für Mensch-Maschine-Kommunikation
TU München
München
Germany

ISSN 1860-4862
ISBN 978-3-642-36805-9 ISBN 978-3-642-36806-6 (eBook)
DOI 10.1007/978-3-642-36806-6
Springer Heidelberg New York Dordrecht London

Library of Congress Control Number: 2013933575

Printed on acid-free paper

Springer is part of Springer Science+Business Media (www.springer.com)

For Thorben Amadeus Bryan &
Benno Olav Sylvain

Foreword

Intelligent Audio Analysis unites methods of audio signal processing and machine learning. Other terms exist for this field or sub-fields and might have been used instead, such as *Computer Audition* or *Machine Listening*—each of which is being used by partly different research communities with slightly different understanding of the core application field and the inventory of methods.

Besides Automatic Speech Recognition being researched since more than half a century, recently an increasing number of further speech and speaker characterisation tasks have been pursued in the literature. In addition, the younger field of Music Information Retrieval is growing and there is emerging interest in the computationally 'intelligent' analysis of general sound events. Fields of application comprise audio coding, edition, interaction, search, surveillance as well as coaching and entertainment applications.

This book first propagates a unified view on the multiplicity of resulting tasks. It further provides a broad overview of the field enriched by extensive recent research application examples mostly based on the author's latest work. The focus thereby lies on realistic conditions and standardisation by open-source software implementations and comparative evaluations. The main goal is to increase robustness by temporary and innovative methods such as automated data-acquisition by semi-supervised learning, audio signal enhancement by non-negative matrix factorisation, systematic feature brute-forcing and application of memory-enhanced learning algorithms—for example in combination with graphical model structures. Machine-based recognition of speech, non-linguistic vocalisations and para-linguistic speaker states and traits serve as examples of application in the domain of speech processing. As for music processing, examples include blind separation of instruments, determination of tempo, metre and ballroom dance style, as well as analysis of musical key, chord progression and structure, next to estimation of music mood and singer traits. Finally, examples are complemented by the recognition of general sound events along with their emotional connotation.

In the outlook, avenues towards evolutionary, unsupervised and holistic audio-signal analysis are shown.

It is thus hoped that the book may find interest by the very broad and interdisciplinary range of researchers and practitioners in academia and industry reaching from engineering and computer science to the fields of speech, language, music and general audio science with their manifold sub-fields. It further addresses levels from early to very advanced level—obviously, though, not all details can be provided at any time, and further reading will be of help where the reader finds it most helpful for oneself.

Preface

This book is based on my habilitation thesis and by that on selected essential research application examples added by explanatory chapters made during the period of my habilitation at the Institute for Human–Machine Communication of the Technische Universität München (TUM) in Munich, Germany, to obtain the German state doctorate (*fakultas docendi*) and private lectureship (*venia legendi*, German PD) in the subject area of Signal Processing and Machine Intelligence. A representative selection of application examples was made basing on coverage of the broader field, scientific relevance and recency. The book further includes knowledge and findings of research conducted and lectures held during this period at TUM, the CNRS LIMSI's Spoken Language Processing Group in Orsay, France, the Imperial College London's Department of Computing in London, UK, the Università Politecinicà delle Marche in Ancona, Italy and the National ICT Australia in Sydney, Australia.

The aim is to provide a handbook that can be read from the beginning to the end, structured into methods and examples of their application. Reference to the original research is repeatedly made throughout, such that the interested reader is referred to these, as well as to further reading from myself and my colleagues or further research in the field. By that, the book introduces a broader view on and new avenues towards the computational and 'intelligent' analysis of audio aiming at the higher goal of lending machines the ability to listen to and understand arbitrary and complex compounds of speech, music and sound.

Gilching, December 2012 Björn W. Schuller

Acknowledgments

> *Great discoveries and improvements invariably involve the co-operation of many minds. I may be given credit for having blazed the trail, but when I look at the subsequent developments I feel the credit is due to others rather than to myself.*
>
> —Alexander Graham Bell

I would first like to express my deepest gratitude to my habilitation mentor and year-long advisor Prof. Gerhard Rigoll of Technische Universität München in Munich, Germany, for his immeasurable technical inspiration and guidance and consultancy along the academic path. Next, for their utmost valuable preceptorship and chance of common scientific work my further habilitation mentors Prof. Kristian Kroschel of the Karlsruhe Institute of Technology and the Fraunhofer IOSB and Prof. Andreas Wendemuth of the Otto-von-Guericke-University in Magdeburg.

For guidance along this path up to the doctoral level, I thank my former advisor Prof. Manfred K. Lang again at this point.

Next, I would particularly like to thank my precursor Prof. Günther Ruske for yielding his lectures on Digital Processing of Speech Signals, Automatic Pattern Recognition in Speech Processing and Data Analysis and Information Reduction to myself and of course for his countless advice.

A special thank you is also dedicated to my superiors at the CNRS-LIMSI and its Spoken Language Processing Group in Orsay, France, Prof. Laurence Devillers, Dr. Jean-Luc Gauvain and his wife Dr. Lori Lamel and Dr. Joseph-Jean Mariani, as well as to my colleagues at the time, Dr. Matthias Brendel and Dr. Riccardo Zaccarelli for the common work and to Clement Chastagnol, Agnes Delaborde, Nicolas Foucault, Marie Tahon and Christophe Vaudable for the manifold discussions—merci à tous.

For welcoming me at the Imperial College London's Department of Computing in London, UK, and for our constant co-operation at the time of writing, I would also like to particularly thank Prof. Maja Pantic and her excellent team of the Intelligent Behaviour Understanding group (i-bug), in particular Dr. Hatice Gunes

(Queen Mary University of London) and Dr. Michel Valstar (University of Nottingham) for our common research.

For hosting me as guest researcher at the National ICT Australia (NICTA) in Sydney, Australia, and for their infinite hospitality, I would next like to thank Dr. Julien Epps and Dr. Fang Chen.

For hosting me as guest lecturer and researcher at the Harbin Institute of Technology in Harbin, China, and endless care during these days, I want to further express my thanks to Prof. Haifeng Li.

Further, for inviting me as lecturer in Intelligent Audio Analysis and hosting me at the Università Politecnica delle Marche in Ancona, Italy, as well as for manifold common work and organisation with the A3Lab, I would like to thank Dr. Stefano Squartini and Prof. Francesco Piazza and from their team in particular Dr. Emanuele Principi and Dr. Rudy Rotili—grazie mille tutti.

For their outstanding contributions, I would further like to thank my team of post doctoral and doctoral students at the Institute for Human–Machine Communication, namely Dr. Silvia Monica Feraru, Dr. Cyril Joder, Dr. Fabien Ringeval and Jun Deng, Florian Eyben, Erik Marchi, Felix Weninger, Martin Wöllmer and Zixing Zhang.

Next, I would like to thank the external doctoral student Raymond Brückner, and those who visited my working group in 'sandwich doctoral programs', namely, Wenjing Han (Harbin Institute of Technology, China) and Masao Yamagichi (Tokyo Institute of Technology, Japan).

Also, I would like to thank the research assistants for their contributions in this period in temporal order from the beginning: Thomas Hook, Claudia Tiddia, Fabian Bross, Erik Marchi, Jordi Feliu, Simone antke, Thomas Knauer, Otto-Thadäus Jandl and Youssef Adel, as well as the student trainees Sun Yang, Judith Köppe and Salma Zouari.

A very warm and special thank you is also dedicated to all the master's students at the Institute for Human–Machine Communication of this period, who pursued their theses under my guidance, in temporal order Thomas Mikschl, Ricardo Minguez, Naijiang Lu, Florian Dibiasi, Hermann Anton Karl, Jin Yao, Martin Wöllmer, Florian Eyben, Yuan Ma, Josef Zwack, Sebastian Kopp, Xiaodi Wang, Cong Hui, Peter Grosche, Christoph Scheuermann, Tobias Knaup, Fabian Heinemann, Yang Sun, Haroon Jacob, Yunpeng Sun, Felix Weninger, Benedikt Gollan, Xiaohua Zhang, Christian Landsiedel, Erik Marchi and Duc Bui Tran, as well as the bachelor students—also in temporal order—Moritz Dausinger, Florian Eyben, Qianqian Xu, Benedikt Gollan, Xiaohua Zhang, Ugur Özcan, Kayssar Chaabane, Chi He, Haitao Liu, Jian Li, Clemens Hage, Johannes Dorfner, Jürgen Glatz, Christoph Kozielski, Judith Köppe and Felix Friedmann, and the interdisciplinary project students Damir Ismailovic, Kevin Tunstall, Maximilian Wöhrl, Christian Kern, Lorenz Mösenlechner, Tayfur Coskun, Alexander Lehmann, Felix Weninger and Marcel Knapp.

Among the many colleagues at the Institute in Munich, I will mention only a few from these years, but thank them all for unforgettable years and exchanges:

Dr. Dejan Arsić for manifold common publications and research, as well as Dr. Marc Al-Hames, Dr. Benedikt Hörnler and Dr. Joachim Schenk.

Next, for the constant collaboration including organisation of the first official INTER SPEECH 2009 Emotion Challenge, the INTER SPEECH 2010 Paralinguistic Challenge, the INTER SPEECH 2011 Speaker State Challenge, the INTER SPEECH 2012 Speaker Trait Challenge and the manifold commonly written articles in this time I would like to thank my colleagues Dr. Anton Batliner of the Friedrich-Alexander-University (FAU) of Erlangen-Nürnberg and Dr. Stefan Steidl of the International Computer Science Institute in Berkeley, as well as Prof. Shrikanth Narayanan of the University of Southern California, Dr. Felix Burkhardt of the Deutsche Telekom, Prof. Elmar Nöth of the FAU, Prof. Jarek Krajewski of the University of Wuppertal, Dr. Chrisitian Müller of the DfKI, Rob van Son of the University of Amsterdam and Florian Schiel of the University of Munich.

Further, for commonly conducted research in the framework of the European Community's Seventh Framework Programme's awarded SEMAINE project and for inviting me as expert to contribute to the EmotionML standard within the W3C, I would like to thank Dr. Marc Schröder of the DfKI in Saarbrücken, Germany, as well as the further partners of SEMAINE, Prof. Roddy Cowie and his wife Prof. Ellen Douglas-Cowie of the Queen's University Belfast, Prof. Catherine Pelachaud of the CNRS and Prof. Dirk Heylen of the University of Twente including their teams, as well as Prof. Gaël Richard of Telecom ParisTech and Prof. Hubertus Feussner of Technische Universität München for common research.

For their great collaboration in the European ASC-Inclusion project I coordinated at the time, I would further like to thank Prof. Simon Barron-Cohen and Prof. Peter Robinson of the University of Cambridge, Prof. Ofer Golan of the Bar-Ilan University, Shai Newman of Compedia, Antonio Camurri of the University of Genova, Prof. Sven Bölte of the Karolinska Institutet and Aurélie Barranger of Autism-Europe.

For the highly synergistic collaboration and funding in common projects, I would also like to thank Anja Höthker and Dr. Hitoshi Konosu of the Toyota Motor Corporation, Dr. Nhu Nguyen Thien of the Continental Corporation and Elle Baese and David Virette of Huawei.

For manifold fruitful discussions I would further like to thank Dr. Olivier Pietquin of Supélec, Prof. Anna Esposito of the IIASS, Dr. Bogdan Vlasenko of the Otto-von-Guericke University and Dr. Noam Amir of the Tel Aviv University, and for their utmost kind invitations and discussions Prof. Klaus Scherer of the Swiss Center for Affective Sciences in Geneva, Switzerland, Dr. Mohamed Chetouani of the Sorbonne Universities in Paris, France, Dr. Florian Metze of the Carnegie Mellon University in Pittsburgh, U.S.A., Dr. Meinard Müller of the Max-Planck Institute in Saarbrücken, Germany, Prof. Mohan Trivedi of the University of California San Diego, U.S.A., Prof. Elisabeth André of the University of Augsburg, Germany, Dr. Friedhelm Schwenker of the University of Ulm, Germany, Prof. Monika Rathert of the University of Wuppertal, Germany, Prof. Berthold Färber of the Universität der Bundeswehr München in Munich, Germany,

Prof. Jianhua Tao of the Chinese Academy of Sciences in Beijing, Germany and Prof. Christian Kaernbach of the Christian Albrecht University in Kiel, Germany.

Then, I would like to thank all of my 200 co-authors up to this date as follows—partly including persons already named above and all without academic titles for the sake of better readability: Markus Ablaßmeier, Vered Aharonson, Marc Al-Hames, Frank Althoff, Noam Amir, Ofer Amir, Dejan Arsić, Paolo Baggia, Anton Batliner, Elisabetta Bevacqua, Christoph Blaschke, Tobias Bocklet, Sebastian Böck, Simon Bozonnet, Artur Braun, Mátyás Brendel, Bernardo José Brüning Schmitt, Felix Burkhardt, Erik Cambria, Salman Can, Sydney D'Mello, Ellen-Douglas Cowie, Roddy Cowie, Cate Cox, Etienne de Sevin, Jun Deng, Laurence Devillers, Florian Dibiasi, Johannes Dorfner, Jean-Louis Durrieu, Jens Edlund, Julien Epps, Nicholas Evans, Sebastian Ewert, Florian Eyben, Berthold Färber, Jordi Feliu, Hubertus Feussner, Shimrit Fridenzon, Jürgen Gast, Jürgen Geiger, Michael Geiger, Karla Geiss, Jort Gemmeke, Matteo Gerosa, Ofer Golan, Benedikt Gollan, Masataka Goto, Art Graesser, Alex Graves, Michael Grimm, Peter Grosche, Hatice Gunes, Benedikt Hörnler, Anja Höthker, Clemens Hage, Wenjing Han, Alan Hanjalic, Simone Hantke, Helen Harris, Catherine Havasi, Michael Hawellek, Dirk Heylen, Martin Hofmann, Encho Hristov, Antti Hurmalainen, Amir Hussain, Raquel Jiménez Villar, Cyril Joder, Filip Jurcicek, Florian Kaiser, Moritz Kaiser, Kostas Karpouzis, Athanasios Katsamanis, Nassos Katsamanis, Christian Kern, Joseph Keshet, Loic Kessous, Nikolaj Klebert, Tobias Knaup, Niels Köhler, Hitoshi Konosu, Christoph Kozielski, Jarek Krajewski, Matthias Kranz, Michael Kranzfelder, Kristian Kroschel, Frank Kurth, Mohamed Anouar Lakhal, Christian Landsiedel, Manfred Lang, Kornel Laskowski, Alexander Lehmann, Nicolas Lehment, Haifeng Li, Cynthia Liem, Bing Liu, Atanas Lyutskanov, Lin Ma, Andreas Maier, Erik Marchi, Jean-Claude Martin, Matthias Mauch, Stefan Mayer, Gregor McGlaun, Gary McKeown, Guenter Meier, Kinfe Tadesse Mengistu, Angeliki Metallinou, Florian Metze, Christine Meyer, Ricardo Minguez, Lorenz Mösenlechner, Gelareh Mohammadi, Tobias Moosmayr, Peter Morguet, Christian Müller, Meinard Müller, Ronald Müller, Shrikanth Narayanan, Clifford Nass, Daniel Neiberg, Robert Neuss, Nhu Nguyen-Thien, Ralf Nieschulz, Elmar Nöth, Sathish Pammi, Maja Pantic, Geoffroy Peeters, Catherine Pelachaud, Christian Peter, Stavros Petridis, Hannes Pirker, Tony Poitschke, Tim Polzehl, Emanuele Principi, Dmytro Prylipko, Bernd Radig, Stefan Reifinger, Stephan Reiter, Gaël Richard, Gerhard Rigoll, Luis Roalter, Nicolas Rollet, Irit Ronen, Paolo Rosso, Rudy Rotili, Günther Ruske, Sonja Schaeffer, Joachim Schenk, Christoph Scheuermann, Florian Schiel, Thomas Schindl, Sebastian Schnieder, Sascha Schreiber, Marc Schröder, Dagmar Schuller, Dino Seppi, David Sommer, Stefano Squartini, Andre Störmer, Jan Stadermann, Stefan Steidl, André Stuhlsatz, Yang Sun, Shahar Tal, Jianhua Tao, Mark ter Maat, Benjamin Treffich, George Tzimiropoulos, Stefan Ungruh, Michel Valstar, Rob van Son, Laurence Vidrascu, Alessandro Vinciarelli, Ravichander Vipperla, David Virette, Tuomas Virtanen, Bogdan Vlasenko, Thurid Vogt, Johannes Wagner, Frank Wallhoff, Dong Wang, Haixun Wang, Benjamin Weiss, Andreas Wendemuth, Felix Weninger, Ian Wilson, Matthias Wimmer, Martin Wöllmer, Kazuyoshi Yoshii, Riccardo Zaccarelli,

Stefanos Zafeiriou, Stefan Zettl, Xiaohua Zhang, Xiaopeng Zhang, Zixing Zhang, Thomas Zielke, Martin Zobl and Enrico Zovato.

As for public funding, the research leading to these results has received funding from the European Community's Seventh Framework Programme (FP7/ 2007–2013) under grant agreement nos. 211486 (SEMAINE) and 289021 (ASC-Inclusion) and the ERASMUS Teaching Mobility Programme. This work was further supported by the Federal Republic of Germany through the German Research Foundation (DFG) under grant nos.SCHU2508/2-1 ("Non- Negative Matrix Factorization for Robust Feature Extraction in Speech Processing") and SCHU 2508/4-1 ("Context-Sensitive Automatic Recognition of Spontaneous Speech by BLSTM Networks"), the Chinese Research Council and the National ICT Australia (NICTA).

In particular for their outstanding help with the preparation of the final version of the book starting from my habilitation theses, I would like to thank Florian Eyben and Felix Weninger.

For valuable discussions around the topics of the book I further express my thanks to Dr. Cyril Joder and Martin Wöllmer.

For her constant support and understanding, I particularly thank my wife Dagmar Maria.

Finally, I thank the publisher for the excellent co-operation during the finalisation of this book.

Contents

Acronyms

AAA	Average Absolute Amplitude
ACF	Auto Correlation Function
AEC	Acoustic Event Classification
AED	Acoustic Event Detection
AF	Analytic Feature
AFE	Advanced Frontend
ALC	Alcohol Language Corpus
AM	Acoustic Model
AMC	Audio Mood Classification
AMDF	Average Magnitude Difference Function
ANN	Artificial Neural Network
API	Application Programming Interface
AR	Autoregressive
ARFF	Attribute Relation File Format
ARMA	Autoregressive Moving Average
ASE	Audio Sub-Event
ASR	Automatic Speech Recognition
AUC	Area Under Curve
AVIC	Audiovisual Interest Corpus
AWGN	Additive White Gaussian Noise
BAC	Blood Alcohol Concentration
BLSTM	Bi-directional Long Short-Term Memory
BN	Bayesian Network
BoW	Bag of Words
BoNG	Bag of N-Grams
BoCNG	Bag of Character N-Grams
BPM	Beats per Minute
BPTT	Back Propagation Through Time
BRAC	Breath Alcohol Concentration
BRD	Ballroom Dance Style
BRNN	Bi-directional Recurrent Neural Network
BASS	Blind Audio Source Separation
CASA	Computational Auditory Scene Analysis

CAN	Controller Area Network
CC	Correlation Coefficient
CD	Compact Disc
CENS	Chroma Energy-Distribution Normalised Statistics
CI	Computational Intelligence
CMS	Cepstral Mean Subtraction
COSINE	Conversational Speech In Noisy Environments
CPF	Conditional Probability Function
CPT	Conditional Probability Table
CPU	Central Processing Unit
CRF	Conditional Random Field
CSV	Comma Separated Value
DAG	Directed Acyclic Graph
DBN	Dynamic Bayesian Network
DCT	Discrete Cosine Transformation
DDR RAM	Double Data Rate Random-Access Memory
DES	Danish Emotional Speech Database
DF	Document Frequency
DFT	Discrete Fourier Transformation
DIN	German Standardisation Institution (German: Deutsches Institut für Normung)
DJ	Disc Jockey
DT	Determiner
Dom	Dominant
DSR	Distributed Speech Recognition
DT	Decision Tree
DTW	Dynamic Time Warping
EC	Error Carousel or Expectation Correction
ED	Euclidean Distance
EER	Equal Error Rate
EM	Expectation Maximisation
EMMA	Extensible Multi Modal Annotation markup language
ETSI	European Telecommunications Standards Institute
EWE	Evaluator Weighted Estimator
F0	Fundamental Frequency
F1–7	Formant 1–7
FFT	Fast Fourier Transformation
FMLLR	Feature space Maximum Likelihood Linear Regression
FNN	Feed-forward Neural Network
FPR	False Positive Rate
FPS	Frames per Second
GMM	Gaussian Mixture Model
GPB	Generalised Pseudo-Bayesian
GT	Ground Truth
HCRF	Hidden Conditional Random Field

HCS	Hierarchical Classification System
HEQ	Histogram Equalisation
HFC	High Frequency Content
HFCC	Human Factor Cepstral Coefficients
HMM	Hidden Markov Model
HNR	Harmonics-to-Noise Ratio
HTK	Hidden Markov Model Toolkit
HU-ASA	Humboldt University Animal Sound Archive
ICA	Independent Component Analysis
ID3	Iterative Dichotomiser 3
IDCT	Inverse Discrete Cosine Transformation
IDFT	Inverse Discrete Fourier Transformation
IDF	Inverse Document Frequency
IDSF	International Dance Sport Federation
IEC	International Electrotechnical Commission
IEEE	Institute of Electrical and Electronics Engineers, Inc
IG	Information Gain
IGR	Information Gain Ratio
IIR	Infinite Impulse Response
IOI	Inter-Onset Interval
IP	Internet Protocol
IS	Itakura-Saito
ISA	Independent Subspace Analysis
ISO	International Organisation for Standardisation
ITU	International Telecommunication Union
JJ	Adjective
JPD	Joint Probability Distribution
KL	Kullback-Leibler
kNN	k Nearest Neighbour
KSS	Karolinska Sleepiness Scale
LDA	Linear Discriminant Analysis
LDM	Linear Dynamic Model
LLD	Low-Level Descriptor
LM	Language Model
LOO	Leave One Out
LOSO	Leave One Song/Speaker Out
LP	Linear Prediction
LPC	Linear Predictive Coding
LPCC	Linear Prediction Cepstral Coefficient
LSP	Line Spectral Pairs
LSTM	Long Short-Term Memory
LoI	Level of Interest
LPC	Linear Prediction Coding
LVCSR	Large Vocabulary Continuous Speech Recognition
MAP	Maximum A-Posteriori

MAE	Mean Absolute Error
MC	Matched Condition
MCELR	Minimum Classification Error Linear Regression
MFB	Mel Frequency Bands
MFCC	Mel Frequency Cepstral Coefficient
MIDI	Musical Instrument Digital Interface
MIML	Multimodal Interaction Markup Language
MIR	Music Information Retrieval
MIREX	Music Information Retrieval Evaluation eXchange
ML	Maximum Likelihood
MLE	Maximum Likelihood Estimation or Mean Linear Error
MLLR	Maximum Likelihood Linear Regression
MLP	Multi-Layer Perceptron
MLR	Multiple Linear Regression
MMC	Mismatched Condition
MMI	Man–Machine Interface
MMSE	Minimum Mean Square Error
MP3	ISO MPEG 1 Audio-Layer-3
MPEG	Motion Picture Expert Group
MSE	Mean Square Error
MTV	Music Television
MVN	Mean and Variance Normalisation
NHR	Noise-to-Harmonics Ratio
NIST	National Institute of Standards and Technology
NMD	Non-Negative Matrix Deconvolution
NMF	Non-Negative Matrix Factorisation
NN	Noun
NLL	Negative Log-Likelihood
NP	Noun Phrase or Non-deterministic Polynomial-time
NPP	Non-pitched Percussive
NTWICM	Now That's What I Call Music
NWPD	Normalised Weighted Phase Deviation
OKS	Online Knowledge Source
openBliSSART	open Blind Source Separation for Audio Retrieval Tasks
openEAR	open Emotion and Affect Recognition toolkit
openSMILE	open Speech and Music Interpretation by Large space Extraction
OOV	Out of Vocabulary
PC	Principal Component
PCA	Principal Component Analysis
PCM	Pulse Code Modulation
PCP	Pitch Class Profiles
PD	Phase Deviation
PDA	Pitch Detection Algorithm
PDF	Probability Density Function

PESQ	Perceptual Evaluation of Speech Quality
PLP	Perceptual Linear Prediction
PMI	Pointwise Mutual Information
PNP	Pitched Non-Percussive
POS	Part of Speech
PP	Prepositional Phrase or Pitched Percussive
PSD	Power Spectral Density
PTR	Probe Tone Rating
RASTA	RelAtive SpecTrA
RB	Adverb
RBF	Radial Basis Function
RCD	Rectified Complex Domain
RF	Random Forest
RIR	Room Impulse Response
RMS	Root Mean Square
RMSE	Root Mean Square Error
RNN	Recurrent Neural Network
ROC	Receiver Operating Characteristic
ROI	Region of Interest
RTF	Real-Time Factor
SACF	Summarised Auto Correlation Function
SAD	Speech Activity Detection
SAR	Switching Autoregressive
SCV	Stratified Cross Validation
SD	Spectral Difference
SF	Spectral Flux
SFFS	Sequential Floating Forward Search
SFS	Speech Filing System
SHS	Sub-Harmonic Summation
SI	International System of Units (French: Système international d'unités)
SIFT	Simplified Inverse Filtering Technique
SLC	Sleepy Language Corpus
SLDM	Switching Linear Dynamic Model
SLDS	Switching Linear Dynamic System
SMA	Simple Moving Average
SMI	Self Mutual Information
SMO	Sequential Minimal Optimisation
SMOTE	Synthetic Minority Oversampling TEchnique
SNR	Signal-to-Noise Ratio
SPL	Sound Pressure Level
SSK	String Subsequence Kernel
STFT	Short-Time Fourier Transformation
Sub	Sub-Dominant
SVM	Support Vector Machine

SVR	Support Vector Regression
TF	Term-Frequency, also Time-Frequency
TPR	True Positive Rate
TT	Total Time
TUM	Technische Universität München
UA	Unweighted Accuracy
USS	Unsupervised Spectral Subtraction
VAD	Voice Activity Detection
VB	Verb
VP	Verb Phrase
WA	Weighted Accuracy
WA	Weighted Accuracy or Word Accuracy
WDC	World Dance Council
WGN	White Gaussian Noise
XML	eXtensible Markup Language
ZCR	Zero Crossing Rate

Part I
Introduction

The first part of this book will motivate research in the field of Intelligent Audio Analysis and set the aims of this book. Further, besides the definition of Audio Analysis in the sense of this book, a solution statement will be provided and the structure of the book will be presented.

Chapter 1
Intelligent Audio Analysis: A Definition

> *Joy, sorrow, tears, lamentation, laughter—to all these music gives voice, but in such a way that we are transported from the world of unrest to a world of peace, and see reality in a new way, as if we were sitting by a mountain lake and contemplating hills and woods and clouds in the tranquil and fathomless water.*
> —Albert Schweitzer

For a start, a short definition of Intelligent Audio Analysis shall be given. This will be followed by an explanation and clarification of the focus chosen for this book: real-life conditions.

1.1 Intelligent Audio Analysis

In general, *audio* is understood as a representation of sound. In this book, this representation is in first given as analogue electrical signal, usually by voltage, then numerically by digitalisation, i.e., transformation to a pulse-code modulated (PCM) stream by regularly sampling at uniform intervals in time and quantising to the nearest value in given digital steps. By that, and in the first place, we deal with mechanical waves, i.e., a complex series of changes in or oscillation of pressure as compound of frequencies within the acoustic range available to humans and at sufficiently intense level to be perceived, i.e., *audible* by them. These waves may be transmitted by solid, liquid, or gas—however, in this book practical examples are limited to air transmission. This goes, however, without general limitation of the methods presented in other cases. Further, in this book, sound is broken down into speech, music, and general sound. The latter—general sound—may from now on also be referred to as 'sound', omitting 'general' for the sake of simplification.

The *analysis* of audio aims at the extraction of information and—on a higher level—attachment of semantic meaning to audio signals.

B. W. Schuller, *Intelligent Audio Analysis*, Signals and Communication Technology,
DOI: 10.1007/978-3-642-36806-6_1, © Springer-Verlag Berlin Heidelberg 2013

Finally, *intelligent* audio analysis refers to the involvement of computational intelligence (CI) algorithms as provided by the means and methods of machine learning going beyond mere signal processing. Such algorithms are often nature-inspired such as neural networks and genetic algorithms or of statistical nature and aim at the ability of reasoning and decision-making, usually in the form of generalisation from exemplary learning material. This is also referred to as *recognition*. In a further development, *evolving* intelligence tries to imitate self-learning abilities from experience including self-made models and clustering and unsupervised and semi-supervised adaptation. Besides parameters, also structures and even the learning algorithm may be adapted by such systems, ideally on-line. While this field is still at its very beginning, first attempts into this direction are given in this book.

1.2 In Real-life Conditions

With a focus on "real-life conditions", lower recognition rates are accepted in order to foster realism and allow for a realistic estimate of system performances as to be expected for a running system in 'real life' rather than under lab conditions. The according requirements made are as listed below:

Non-prototypical test data: An Intelligent Audio Analysis system 'in the wild' is usually confronted with subtle nuances of audio phenomena. For speech, this means non-prompted, but spontaneous speech. As for speaker states, these should ideally not be acted, but realistic. In music, as an example, the down-beat may be played in a rather subtle way than in an exaggerated one. In a similar manner, when producing sounds such as a door shut, a door slam could be overly prototypical. Also in real-life data, obviously, such more prototypical data may occur, but the test data should consist of a representative collection of different facets and nuances.

Non-preselected test data: In automatic speech recognition (ASR) normally, all data have been employed apart from, for instance, non-linguistic vocalisations, etc., which are treated as 'garbage'; but they are still treated and not removed from the signal before processing. This is different in speaker state and trait classification, when it comes to unreliable gold standard, such as emotion or personality analysis: Often a subset of the full database is taken consisting of somehow clear, i.e., more or less decided cases [1]. Using 'realistic data', thus means as well using all data. The first, qualitative aspect, has been taken into account by several studies, yet, the second, sort of 'quantitative' aspect, has still been neglected by and large. In research challenges organised by the author of this book, it was dealt with this second aspect by employing the full database independent of the inter-labeller agreement as is the situation in real applications. In ASR, a rough estimate for the difference between read and spontaneous data was that, at least to start with, one could expect an error rate for spontaneous data twice the size than the one for read data [2]. One cannot simply transfer this empirically obtained estimate onto other audio analysis problems—still, we definitely will have to deal with a plainly lower classification performance. In music analysis, an example could be to use a whole CD collection and not pick some 'friendly cases' where a system works well without transparency on how these examples were selected. In a similar way, a whole archive of sounds should be processed.

Independent test data: For intelligent speech analysis tasks, this usually refers to speaker or speaker group independence, i.e., a speaker for testing of an intelligent speech analysis system is 'seen' for the first time by it. In a similar way, music should not be previously known. This holds in multiple senses, such as no music by the same artist or group was seen before, or no variation of the same musical piece, etc. Finally, as for sound, it needs to be produced by an independent sound source. For example, when recognising a door shut, it should be produced by another door in another building, best by another person in test condition.

No optimisation on test data: Besides the test data being independent from learning data as described above, no optimisation whatsoever should ideally have taken place on this data, and repeated measurement is only made for the sake of illustration of system behaviour. To this end, (a partition of) disjoint learning material is used for all kinds of system optimisation.

Meta-data is retrieved from the Internet: Whenever non-audio meta information such as lyrics or genre tags for a piece of music are used in a system's decision making, this information needs to be retrieved automatically from the Internet. As a consequence, the algorithms have to be able to deal with erroneous and missing information also in this sense.

Monaural audio capture: In this book, no use is made of stereophonic or multichannel audio recording by intention. Using such information can of course be highly beneficial in a real-life use-case, however, in order not to limit applicability of the methods, this decision was made. In many situations, recordings are not given by multiple channels, such as in telephone transmission or older monophonic music recordings, etc.

These requirements are enforced wherever reasonably applicable in the results presented in this book. Further, high emphasis is laid throughout experiments and results on reproducibility of the findings by availability of data and transparent configuration. One further requirement of a real-world system, however, has been partly ignored for better focus on the task at hand:

Fully automatic chunking: In a working system, audio can be expected to be recorded in a continuous stream. Thus, as opposed to most off-line test-beds for Intelligent Audio Analysis, where audio is already end-pointed or 'cut' in the sense of pre-segmented, this step needs to be carried out in a real system and can easily become challenging when audio is blended or noisy, as for example when labelling chunks of speech in the presence of background music, etc. This is not always given in the exemplary results presented in this book, as this can make evaluation considerably more difficult. Methods for chunking are, however, presented.

Certainly, other more or less significant limitations of realism are ever-present in the respective literature and the results presented in this book for sheer practicability and feasibility of research. An example is the additive superposition of noise for robustness analysis. Such superposition simply allows controlled noise overlay in different levels or the evaluation of audio separation quality as the clean original is available. Yet, is not realistic, as different reverberation occurs for the signals, and for example speakers are not influenced by the overlaid noise such as in the Lombard effect. To ease this fact, additional results with originally noisy data are presented in parallel where appropriate.

References

1. Seppi, D., Batliner, A., Schuller, B., Steidl, S., Vogt, T., Wagner, J., Devillers, L., Vidrascu, L., Amir, N., Aharonson, V.: Patterns, prototypes, performance: classifying emotional user states. In: Proceedings INTERSPEECH 2008, 9th Annual Conference of the International Speech Communication Association, Incorporating 12th Australasian International Conference on Speech Science and Technology, SST 2008, pp. 601–604. Brisbane, Australia, Sept 2008. ISCA/ASSTA, ISCA.
2. Lippmann, R.: Speech recognition by machines and humans. Speech. Commun **22**(1), 1–16 (1997)

Chapter 2
Motivation, Aims, and Solutions

It is not knowledge, but the act of learning, not possession but the act of getting there, which grants the greatest enjoyment.
—Carl Friedrich Gauss

2.1 Motivation of Intelligent Audio Analysis

There are numerous scenarios and fields for potential application of Intelligent Audio Analysis that are commercially interesting and may help us in our daily lives. These are detailed out in the application part of this book (Part III) that aims to give some practical examples, but a more general perspective on use-cases of the whole field is given for a motivational introduction at this point. Intelligent Audio Analysis is currently used and holds future promises in particular for

Audio Encoding: Obviously, in an acoustic representation, highest bitrates are required, which can be eased step-wise by going to partly or fully parametric representation [1], and partly or fully symbolic representation (cf. Fig. 2.1). As for speech, 'symbolic' could thereby be phones as acoustic realisations of phonemes, which are "the smallest segmental unit of sound employed to form meaningful contrasts between utterances" [2]. In the case of music, 'symbolic' could refer to note events or chords, etc. However, highest bit rate reduction is only reached by semantic encoding—though obviously at the highest loss factor as, rather than preserving the original audio, only its semantics are kept for storage or transmission via highly band limited channels. This then requires to synthesise audio at the moment of decoding based on these semantics. In music, an example would be note events and instrumentation saved in symbolic representation for storage and later synthesis for play-back. However, compromises can be made also at this level by combination with (few) parameters or even highly compressed acoustics—the semantics can then touch certain aspects of the audio signal for good reproduction at the moment of decoding and regeneration.

B. W. Schuller, *Intelligent Audio Analysis*, Signals and Communication Technology, DOI: 10.1007/978-3-642-36806-6_2, © Springer-Verlag Berlin Heidelberg 2013

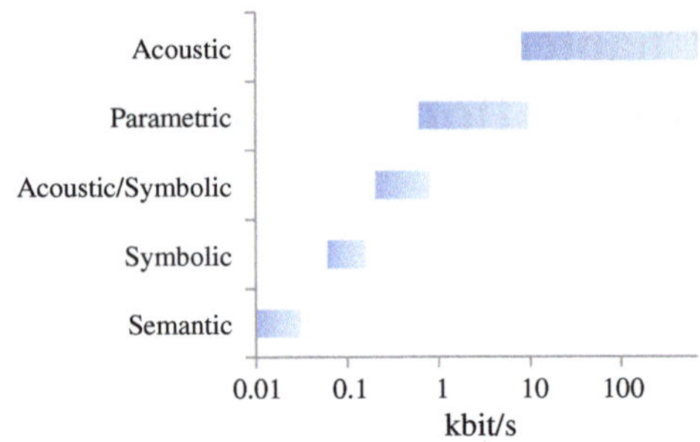

Fig. 2.1 A rough overview on obtainable audio bit-rates by partly lossy compression depending on the representation type.

Audio Alteration: In a chain of analysis, edition, and synthesis, audio can be modified and altered. Examples include voice transformation [3] including for example the change of the emotional tone of a voice, and music alteration such as combining drum tracks from one musical piece with the singer of another, etc.
Audio Retrieval: In audio search, manifold search tags are used today and can be used in the future such as by speaker identity or emotion, music artist or genre and positions of the chorus, sound type, etc. However, such information needs to be provided at first and additionally stored. As this may involve considerable human labelling effort and labelling may easily be erroneous if larger user groups of laymen are involved, Intelligent Audio Analysis may help to assess such information fully automatically off-line or even on-line.
Audio-based Interaction: In Human-Machine and Human-Robot communication, machine listening and understanding capabilities beyond speech and sound recognition and interpretation can allow for injection of 'social competence'. For example, speaker state and trait analysis allows for improved socio-emotional contextual comprehension of a machine. In music analysis, powerful user-interfaces can be provided to musicians, that allow for example for user input by clapping, singing, humming or playing of real musical instruments for interaction with the machine.
Monitoring and Surveillance: In this domain, speaker states can be of interest, such as sleepiness or intoxication of responsible persons in steering and control tasks [4]. Another example in this respect is monitoring of a customer's interest in sales presentations [5]. Also terrorism and vandalism alert systems may be realised by such systems—potentially combining speech and sound analysis [6]. An example of a hardware product is the WhyCry®—a device that aims to indicate a new-born's annoyance, boredom, hunger, sleepiness, and stress to less experienced parents. In music analysis, monitoring can for example be used for on-line auto mixing and balancing. Sound monitoring can for example be used to ensure proper functionality of bearings, pipelines, etc.
Coaching: Voice coaching includes training for public speeches or help in foreign language acquisition [4], but also holds promises for empowerment and inclusion. In the European ASC-Inclusion project,[1] children with autism spectrum condition shall

[1] http://www.asc-inclusion.eu

acquire improved socio-emotional skills by digital gaming including appropriate interpretation and expression of emotion. This example also includes monitoring and alteration, as their vocal expression is monitored in the game and the voice is altered for exemplification. In the music domain, a learning software can notify a student of an instrument if mistakes occur as by Fraunhofer's "Songs2See", or help in training vibrato singing [7], etc.

Entertainment: As the entertainment sector can often be more forgiving if accuracies are not at perfection level, this domain has seen many products make it to the market by now. Such software includes a console game around speech-based deception recognition ("Truth or Lies—Someone Will Get Caught", THQ® Entertainment) already appeared on the market. Software centred around singing intonation in Karaoke-style games such as "SingStar" and "RockBand" by Harmonix or more recently Ubisoft®'s guitar learning game "Rocksmith®" based on real guitar audio analysis are examples of huge market success.

Despite the appearance of first commercial and non-commercial usage of Intelligent Audio Analysis products and solutions, the state-of-the-art today is often not sufficient for the often very high requirements given by several of the above use-cases. According research work is thus still urgently needed. In addition, standard references in the literature that provide a broader perspective are just to appear given the rather young age of the field and its more recent emergence on a broader level.

2.2 Aims of the Book

It is the aim of this book to help allow for improved and extended exploitation of Intelligent Audio Analysis in the illustrated and further application scenarios. In particular and by that, the goals are as follows:

1. To provide a unified perspective on audio analysis tasks and a broad overview on recent advancements in the field exemplified primarily by work of the author and his colleagues. The intention is to stimulate synergies arising from transfer of methods and lead to a holistic audio analysis [8]—audio is usually highly complex and blended in the real world, but research is usually focused on isolated aspects at the present day.

2. To help approach improved robustness and reliability of today's Intelligent Audio Analysis systems by suited and innovative methods.

3. To stimulate extension of the range of Intelligent Audio Analysis applications by showing its potential in new tasks that were not or hardly touched in the literature so far, which, however, can be of broad commercial and technical interest.

4. To provide the reader with benchmark results and standardised test-beds for a broader range of audio analysis tasks. The main focus thereby lies on the parallel advancement of realism in audio analysis, as too often today's results are overly optimistic owing to idealised testing conditions.

5. To show deficiencies in current approaches and future perspectives in and for the field.

2.3 Solutions

From a technical point of view, the discussed solutions to the described ends foremostly consist of the inventory provided by the methods of pattern recognition. This includes advanced and recent methods of signal processing and machine learning. In more detail, these are:

Audio enhancement and source separation as needed for emphasising the characteristics and isolation of the signal part of interest.

Brute-forcing of large heterogenous audio feature spaces to provide a broad feature basis for the space initialisation in the approach of new audio tasks.

Careful design of new audio feature types as systematic brute-forcing may have its limitations.

Combination, adaptation, and application of recent learning methods to profit from synergies and inject new paradigms such as graphical modelling aspects and long short-term memory into the machine learning process and enable partly supervised self-learning.

As for the non-technical side, practical solutions include in the first place:

Establishment of unified test-beds and transparent benchmarks as this invites the research community to compare results in a well-defined way and by that may help to advance on the state-of-the-art. This includes or partly requires the following two aspects worth mentioning in isolation.

Collection and annotation of suited audio data to consider new tasks of Intelligent Audio Analysis or enrich the ever sparse data-base in the field.

Provision of standardised (open-source) software implementations where such are currently missing to allow for comparability of findings and potentially code additions by others.

References

1. Ruske, G.: Automatische Spracherkennung, 2nd edn. Methoden der Klassifikation und Merkmalsextraktion. Oldenbourg, Munich, Germany (1993)
2. I. P. Association: Phonetic Description and the IPA Chart, Chapter 2, pp. 3–17. Cambridge University Press, Cambridge (1999)
3. Stylianou, Y.: Voice transformation: A survey. In: Proceedings of IEEE International Conference on Acoustics, Speech, and Signal Processing (ICASSP), pp 3585–3588. Taipei, Taiwan (2009)
4. Schuller, B., Steidl, S., Batliner, A., Burkhardt, F., Devillers, L., Müller, C., Narayanan, S.: Paralinguistics in speech and language–state-of-the-art and the challenge. Comp. Speech Lang. Special Issue Paralinguistics Naturalistic Speech Lang. 27(1), 4–39 (2013)
5. Schuller, B., Müller, R., Eyben, F., Gast, J., Hörnler, B., Wöllmer, M., Rigoll, G., Höthker, A., Konosu, H.: Being bored? recognising natural interest by extensive audiovisual integration for real-life application. Image Vis. Comput. Special Issue Vis. Multimodal Anal. Hum. Spontaneous Behavior 27(12), 1760–1774 (2009)
6. Schuller, B., Wimmer, M., Arsić, D., Moosmayr, T., Rigoll, G.: Detection of security related affect and behaviour in passenger transport. In: Proceedings INTERSPEECH 2008, 9th Annual Conference of the International Speech Communication Association, incorporating

12th Australasian International Conference on Speech Science and Technology, SST 2008, pp. 265–268, Brisbane, Australia, ISCA/ASSTA, ISCA (2008)
7. Weninger, F., Amir, N., Amir, O., Ronen, I., Eyben, F., Schuller, B.: Robust feature extraction for automatic recognition of vibrato singing in recorded polyphonic music. In: Proceedings 37th IEEE International Conference on Acoustics, Speech, and Signal Processing, ICASSP 2012, pp. 85–88, Kyoto, Japan, IEEE, IEEE (2012)
8. Weninger, F., Schuller, B., Liem, C., Kurth, F., Hanjalic, A.: Music information retrieval: An inspirational guide to transfer from related disciplines. In: Müller, M., Goto, M. (eds.) Multimodal Music Processing, Seminar, vol. 69, pp. 195–215. 1041 of Dagstuhl Follow-UpsSchloss Dagstuhl, Germany (2012)

Chapter 3
Structure of the Book

Some people do best studying in structured, linear way, while others do best jumping around, surrounding a subject rather than traversing it.

—William "Bill" Henry Gates III

To provide a good overview for the reader, the book is divided into four main parts as will follow.

Introduction (Part I): This part provides an introductory short motivation with general aims and solutions towards reaching of these (Chap. 2), definition of audio analysis per se (Chap. 1), and the current overview on the book (Chap. 3).

Audio Analysis Methods (Part II): The aim of this part is to provide the reader with the knowledge necessary for comprehension of the application part thereafter. Basic knowledge in information technology and in particular in the field of signal processing is assumed. Additional knowledge in machine learning is of help, yet, not mandatory. In order to keep focus on more recent advances in the field, well known and well formulated standard techniques may be introduced only in short and recommended reading reference is provided. The presentation line of methods follows that of the typical Intelligent Audio Analysis chain as will be discussed first in this part (Chap. 4) by going from audio data (Chap. 5) to audio features (Chap. 6), and audio recognition (Chap. 7). Further, audio source separation (Chap. 8), and enhancement and robustness (Chap. 9) will be discussed.

Audio Analysis Applications (Part III): In this part, selected applications will be shown for the three types of audio: speech (Chap. 10), music (Chap. 11), and general sound (Chap. 12). Each task will be shortly introduced, usually followed by a description of the specific data and methods applied for illustration, experiments and results, and a conclusion for this specific task. The idea is to illustrate the applicability of the inventory of methods previously introduced in a broader range of tasks. The transfer to further such tasks can often follow highly similar patterns. At the same time, however, these tasks also demonstrate that unification will always find its limits and 'slightly' task specific modifications will often be advantageous. Examples from

B. W. Schuller, *Intelligent Audio Analysis*, Signals and Communication Technology,
DOI: 10.1007/978-3-642-36806-6_3, © Springer-Verlag Berlin Heidelberg 2013

many authors and publications could have been chosen—merely for the sake of unification, it was, however, decided to pick all examples from research of the author of the book and his colleagues. Obviously, the state-of-the-art is advancing quickly in this young field and other solutions exist and may or do lead to partly better results. Reference to such work is made throughout—while these cannot be complete, they aim to provide good entry points for further reading on more specific questions.

For speech, these tasks include the recognition of speech and non-linguistic vocalisation such as laughter and sighs as well as classification of speakers by their states and traits including sentiment in text and emotion, interest, age, gender, height, intoxication, and sleepiness, from the acoustic properties of the speech signal. These are partly grouped by research challenges the author had organised in the field.

As for music, separation of drum beats, localisation of onsets, and tempo, metre and ballroom dance style recognition deal fore mostly with the rhythmic section in music. This is followed by the tonal analysis of the musical key, the chord progression, and structural analysis in particular aiming at localisation of the chorus section. Many of this information is then combined in the higher level analysis of mood in music that bases on higher level features from the above listed extracted information of a musical piece including linguistic cues from the lyrics of a song. Then, analysis of singer traits is presented. To this end, the singing voice is enhanced before applying similar methods as are presented in the speaker analysis. In the examples, the type of music is limited to western music but covers different facets such as Classical, Jazz or Popular and Rock music. However, most if not all methods should be directly applicable or easily transferable to other genres of music.

As for sound, three illustrative application examples have been selected: First, the recognition of animal sounds in up to five broader groups is presented, followed by the classification of general sound events including animals into seven groups. For this application, semi-supervised learning is exemplified. Finally, emotion in sound is classified.

Conclusion (Part IV): At this point, goals of the book set at its beginning will be discussed in light of the content presented (Chap. 13). This is followed by a general discussion and vision for future research and application (Chap. 14).

Appendix:

The Appendix summarises the acoustic feature sets as were used throughout most of the speech and sound analysis in a single table.

Part II
Intelligent Audio Analysis Methods

The aim of this part is to provide a deep insight into the methods of audio analysis following the chain of processing.

Chapter 4
Chain of Audio Processing

A complex system that works is invariably found to have evolved from a simple system that works.

—John Gaule

In the following, let us have a look at the overall process of Intelligent Audio Analysis as introduced in [1].

In Fig. 4.1, a unified overview on a typical Intelligent Audio Analysis system is given. Its chain of processing is followed in the ongoing, and each component is described in detail.

Preprocessing: Subsequently to capturing the audio by a single microphone or array of microphones and digitalising it, the audio is preprocessed. This step usually aims at enhancement of the audio signal of interest or (blind) separation of individual sources which are mixed in the captured audio stream. Usually, de-noising is dealt with in the literature more frequently than de-reverberation that aims at reducing the influence of varying room impulse responses. Popular (blind) source separation methods comprise Independent Component Analysis (ICA) [2] in the case of multiple microphones/arrays, and Non-Negative Matrix Factorisation (NMF) [3] in the case of single microphones (cf. Chap. 8). Popular audio enhancement algorithms include Wiener filtering and unsupervised spectral subtraction (cf. Chap. 9).

Low Level Descriptor extraction: After the components of interest of the digital signal have been extracted, parameters must be extracted from the signal which contain—ideally only—information for a given analysis task but discard other information. Such parameters are, e.g., the signal energy and the pitch. Instead of the term 'parameters' we also find the names 'features'. Since audio analysis is mostly based on short-time analysis, i.e., analysis of short frames of audio, in which we can assume the signal to be stationary, the specific set of parameters that are extracted at this stage are called the Low-Level Descriptors (LLDs). This is detailed in Sect. 6.1.2.

LLDs are extracted at approximately 100 frames per second with typical frame sizes of 10–30 ms. Typically multiple LLD are extracted per frame; we refer to an LLD (feature) vector in this case. Windowing functions are usually rectangular

B. W. Schuller, *Intelligent Audio Analysis*, Signals and Communication Technology, DOI: 10.1007/978-3-642-36806-6_4, © Springer-Verlag Berlin Heidelberg 2013

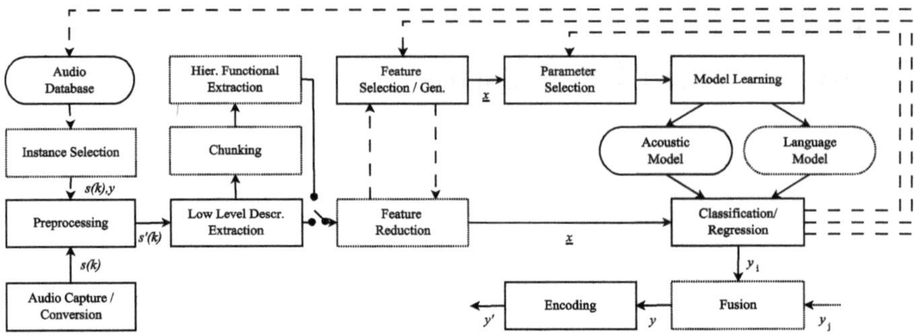

Fig. 4.1 Unified overview of typical Intelligent Audio Analysis systems. *Dotted boxes* indicate optional components. *Dashed lines* indicate steps carried out only during system training or adaptation phases, where $s(k)$, x, y are the audio signal, feature vector and target (vector), respectively, high comma indicates altered versions and subscripts indicate diverse vectors. The connection from classification or regression back to the audio database indicates active and semi-supervised or unsupervised learning. The fusion block allows for integration of other signals by late 'semantic' fusion

for extraction of a Low Level Descriptor (LLD) in the time domain and smooth (e.g., Hamming or Hann) for extraction in the frequency or time-frequency (TF, e.g., Gaussian or general wavelets) domains. To compensate artefacts introduced by the windowing function, typically a smoothing of the LLD with a moving average filter of 3 frames length is done.

Many systems process features on the LLD level (also referred to as frame level) directly, either to provide a frame-by-frame estimate, or by sliding windows of feature vectors of fixed length, or by dynamical approaches that provide some sort of temporal alignment and warping such as Hidden Markov Models (HMMs) or general Dynamic Bayesian Networks (DBNs).

Typical audio LLDs cover: intonation (pitch, etc.), intensity (energy, etc.), Linear Prediction Cepstral Coefficients (LPCCs), Perceptual Linear Prediction (PLP), Cepstral Coefficients (MFCCs, etc.), formants (amplitude, position, width, etc.), spectrum (Mel Frequency Bands (MFBs), NMF-based components, MPEG-7 audio, roll-off, etc.), harmonicity (Harmonics-to-Noise Ratio (HNR), Noise-to-Harmonics Ratio (NHR), etc.), perturbation (jitter, shimmer, etc.), pitch class profiles, etc.

Note that one can also introduce string-type LLDs to describe, e.g., linguistic content. Their extraction usually requires chunking and speech recognition or similar.

Chunking: In most applications intelligent audio analysis algorithms have to consider longer segments of audio, as attributes such as emotion, speaking style, music mood, instruments, musical chord progression, or general sound events are characterised by the dynamics of the signal over time. Depending on the task, the right segment of analysis, i.e., the chunking, has to be found. Methods for chunking comprise: choosing a fixed number of frames, acoustic chunking (e.g., by Bayesian Information Criterion), voiced/unvoiced parts, and for speech units such as phonemes, syllables, words, or sub-turns in the sense of syntactically or semantically motivated chunkings below the turn level or complete turns, etc. [4]. For music, these can be beats, single or multiple consecutive bars, and parts such as chorus or bridge, etc. Obviously, higher level chunking requires suited pre-analysis such as audio activity detection, voicing analysis, or complex structural analysis (see Sect. 6.1.3 for a discussion).

Supra segmental analysis and (hierarchical) functional extraction: Next, the method of segment level analysis has to be defined. If—as mentioned in the previous section—a classifier operates directly on the LLD frames, either dynamic approaches have to be used, or the frame-wise results have to be combined to a single segment level result (late fusion, cf. below). Alternatively, or additionally, LLD feature vectors can be combined into a single feature vector per segment, and then only a single classification result is obtained. We refer to this method as 'supra-segmental' analysis. In case that the length of all segments is constant, we can concatenate all LLD feature vectors within the segment to a single, higher-dimensional feature vector. If the length varies (e.g., for sentences, beats or bars in music, etc.), this approach is not feasible, as the dimensionality of the resulting high-dimensional vector will not be constant—which is usually required by classifiers. In this case, it is common practice to summarise the LLD feature vectors by applying 'functionals' to them. These can be statistical descriptors such as mean or standard deviation; in this case, information from a pre-trained Gaussian (mixture) model of the features can be used to obtain more robust estimates ('universal background model' approach). Other commonly used statistics of the feature distribution comprise percentiles and higher moments. Furthermore, one can compute descriptors related to the temporal evolution of the LLDs, such as statistics of peaks (number, distances, etc.), spectrum (e.g., DCT coefficients) or autoregressive coefficients. The result is a feature vector per segment with a constant dimensionality $d = N_{\mathrm{LLD}} \cdot N_{\mathrm{func}}$. Thereby N_{LLD} and N_{func} are the numbers of LLDs and functionals, respectively. This method of summarisation can also be repeated on higher levels, i.e., 'functionals of functionals' can be computed, etc. This leads to a hierarchical representation, referred to as analytical features [5] and feature brute-forcing [6, 7].

Feature reduction: As in any other pattern recognition task, the reduction of the parameter space to those parameters which are most highly correlated with the classification problem of interest, is beneficial in terms of classification accuracy, model complexity, and speed.

In this step the the feature space is transformed in order to reduce the covariance between features in the new space—usually by a translation into the origin of the original feature space and a rotation to reduce covariances outside the main diagonal of the covariance matrix. This is typically achieved by the Principal Component Analysis (PCA) [8]. Linear Discriminant Analysis (LDA) additionally employs target information (usually discrete class labels) to maximise the distance between class centres and minimise dispersion of classes. Next, a reduction by selecting a limited number of features in the new space takes place—in the case of PCA and LDA, by choosing the components with the highest according eigenvalues. These features still require extraction of all features in the original space—in the case of principle components, this comes as the features in the new space are linear combinations of all original ones.

Feature selection/generation: To further reduce the feature space dimensionality, in this step it is decided which features to keep in the feature space and which to discard. This may be of interest if a new task—e.g., estimation of a speaker's weight, body surface, race or heart rate, playing effects on a Cajon or Blues harp or mal-function of a technical system from acoustic properties—is not well known. In such a case, a multiplicity of features can be 'brute-forced'. From these, the ones well suited for the task at hand can be kept. Typically, a target function is defined first. In the case of 'open loop' selection, typical target functions are of information theoretic nature such as IG or statistical nature such as correlation among features and of features with the target of the task at hand. In the case of 'closed loop', the target function is the learning algorithm's accuracy to be maximised. Usually a search function is needed in addition as an exhaustive search in the feature space is computationally hardly feasible. Such a search may start with an empty set adding features in 'forward' direction, with the full set deleting features in 'backward' direction or bi-directional starting 'somewhere in the middle'. Often random is injected or the search is based entirely on random selection guided by principles such as evolutionary, i.e., genetic algorithms. As the search is usually based on accepting a sub-optimal solution but reducing computation effort, 'floating' is often added to overcome nesting effects [9, 10]. That is, in the case of forward search, (limited) backward steps are added to avoid a too 'greedy' search. This 'Sequential Forward Floating Search' is among the most popular in the field, as one typically searches a small number of final features out of a large set. In addition, generation of further feature variants can be considered within the selection of features, e.g., by applying single feature or multiple feature mathematical operations such as logarithm or division which can lead to better representation in the feature space.

Parameter selection: Parameter selection 'fine tunes' the learning algorithm. This can comprise optimisation of a learning algorithm's topology, initialisation, the type of functions, or step sizes in the learning phase, etc. Indeed, the performance of a machine learning algorithm can be significantly influenced by optimal or sub-optimal parametrisation. While this step is seldom carried out systematically apart from varying expert-picked 'typical' values, the most popular approach is likely grid search. As for the feature selection, it is crucial not to 'tune' on instances used for evaluation as obviously this would lead to overestimation of performance.

Model learning: This is the actual training phase in which the classifier or regressor model is built based on labelled data. There are classifiers or regressors that do not need this phase (so-called 'lazy learners') as they only decide at run-time by training instances' properties which class to choose, e.g., by the training instance with shortest distance in the feature space to the testing ones. However, these are seldom used, as they typically do not lead to sufficient accuracy in the rather complex tasks of Intelligent Audio Analysis and are usually slow and memory consuming at run-time.

Classification/regression: This step assigns the actual target to an unknown test instance. In the case of classification, these are discrete labels. In the case of regression, the output is a continuous value. In general, a high diversity exists in the field of

Intelligent Audio Analysis on which types of classifiers or regressors are used, partly owing to the diverse requirements arising from the variety of tasks (cf. Chap. 7).

Fusion: (optional): This stage exists if information is fused on the 'late semantic' level rather than on early feature level (cf., e.g., [11]).

Encoding: (optional): Once the final decision is made, the information needs to be represented in an optimal way for system integration such as a music or sound search or spoken language dialogue system [12]. Here, standards may be employed to ensure utmost re-usability such as VoiceXML, Extensible MultiModal Annotation markup language (EMMA) [13], Emotion Markup Language (EmotionML) [14], Multimodal Interaction Markup Language (MIML) [15], ID3 tags, etc. Additional information such as confidences can reasonably be added to allow for disambiguation strategies or similar.

Audio databases: They comprise the stored audio of exemplary speech, music, and sound for model learning and evaluation. In addition, a transcription of the spoken content or note events, etc., may be given and/or the labelling of further target tasks.

Acoustic model (AM): consists of the learnt dependencies between acoustic observations and classes, or continuous values in the case of regression.

Language model (LM): stores the learnt dependencies of linguistic observations and according assignments.

In the following, all these steps (except for fusion and encoding) will be explained in detail (remaining Part II), then practical applications are shown (Part III).

References

1. Schuller, B.: Voice and speech analysis in search of states and traits. In: Salah, A.A., Gevers, T. (eds.) Computer Analysis of Human Behavior, Advances in Pattern Recognition, chapter 9, pp. 227–253. Springer, Berlin (2011)
2. Hyvärinen, A., Karhunen, J., Oja, E.: Independent Component Analysis. Wiley, New York (2001)
3. Cichocki, A., Zdunek, R., Phan, A.H., Amari, S.-I.: Nonnegative Matrix and Tensor Factorizations. Wiley, Chichester (2009)
4. Batliner, A., Seppi, D., Steidl, S., Schuller, B.: Segmenting into adequate units for automatic recognition of emotion-related episodes: a speech-based approach. Advances in Human Computer Interaction, Special Issue on Emotion-Aware Natural Interaction, 2010(Article ID 782802), 1–15 (2010)
5. Pachet, F., Roy, P.: Analytical features: a knowledge-based approach to audio feature generation. EURASIP J. Audio Speech Music Process. **1**, 1–23 (2009)
6. Schuller, B., Wimmer, M., Mösenlechner, L., Kern, C., Arsić, D., Rigoll, G.: Brute-forcing hierarchical functionals for paralinguistics: a waste of feature space? In: Proceedings 33rd IEEE International Conference on Acoustics, Speech, and Signal Processing, ICASSP 2008, pp. 4501–4504. IEEE, Las Vegas (2008)
7. Eyben, F., Wöllmer, M., Schuller, B.: Opensmile—the munich versatile and fast open-source audio feature extractor. In: Proceedings of the 9th ACM International Conference on Multimedia, MM 2010, pp. 1459–1462. ACM, Florence (2010)
8. Jolliffe, I.T.: Principal Component Analysis. Springer, Berlin (2002)
9. Pudil, P., Novovicova, J., Kittler, J.: Floating search methods in feature selection. Pattern Recogn. Lett. **15**, 1119–1125 (1994)

10. Ververidis, D., Kotropoulos, C.: Fast sequential floating forward selection applied to emotional speech features estimated on des and susas data collection. In: Proceedings of European Signal Processing Conference (EUSIPCO 2006), Florence (2006)
11. Bocklet, T., Stemmer, G., Zeissler, V., Nöth, E.: Age and gender recognition based on multiple systems—early versus late fusion. In: Proceedings of Interspeech, pp. 2830–2833. Makuhari, Japan (2010)
12. De Melo, C., Paiva, A.: Expression of emotions in virtual humans using lights, shadows, composition and filters, volume 4738 LNCS of Lecture Notes in Computer Science (including subseries Lecture Notes in Artificial Intelligence and Lecture Notes in Bioinformatics). Springer, Heidelberg (2007)
13. Baggia, P., Burnett, D.C., Carter, J., Dahl, D.A., McCobb, G., Raggett, D.: EMMA: Extensible MultiModal Annotation markup language (2007)
14. Schröder, M., Devillers, L., Karpouzis, K., Martin, J.-C., Pelachaud, C., Peter, C., Pirker, H., Schuller, B., Tao, J., Wilson, I.: What should a generic emotion markup language be able to represent? In: Paiva, A., Picard, R.W., Prada, R. (eds.) Affective Computing and Intelligent Interaction: Second International Conference, ACII 2007, Lisbon, Portugal. Proceedings, volume 4738/2007 of Lecture Notes on Computer Science (LNCS), pp. 440–451. Springer, Berlin, 12–14 Sept 2007
15. Mao, X., Li, Z., Bao, H.: An extension of MPML with emotion recognition functions attached, volume 5208 LNAI of Lecture Notes in Computer Science (including subseries Lecture Notes in Artificial Intelligence and Lecture Notes in Bioinformatics). Springer, Berlin (2008)

Chapter 5
Audio Data

It is a capital mistake to theorize before one has data. Insensibly one begins to twist facts to suit theories, instead of theories to suit facts.

—Sir Arthur Conan Doyle

5.1 Audio Data Requirements

In order to train and test intelligent audio systems, audio data is needed. In fact, this is often considered as one of the main bottle necks and the common opinion is that there is "no data like more data". However, there are several pre-requisites apart from the sheer quantity of the data, and in fact, obtaining considerable amounts of data can be difficult and laboursome [1], also, as data usually also needs to be labelled. Table 5.1 provides an overview on the most relevant of these requirements when building an (audio) database for learning and testing of classifiers and regressors.

To reach annotations with labels y_n for instance n of the Intelligent Audio Analysis task of interest with reduced cost, new methods for community or distributed annotation such as crowd sourcing, e.g., by Amazon Mechanical Turk[1] will be of interest. If one further wants to reduce the amount of audio data prior to the labelling to those instances that will likely result in the best gain for the system, the field of active learning provides solutions to this end [2]. In addition, to obtain even larger amounts of data without typically involved efforts in annotation, uniting of databases for training [3] and semi-supervised learning techniques have recently been shown beneficial [4, 5]. In particular the latter allows for exploitation of practically infinite amounts of data, such as on-line available audio and audiovisual video streams. A more complex, yet also very promising alternative was shown in [6], where

[1] https://www.mturk.com/mturk/

DOI: 10.1007/978-3-642-36806-6_5,

Table 5.1 Requirements for database building

Requirement	Example
Quantity	"There's no data like more data"
	High diversity with respect to manifold influence factors
	Reasonably balanced distribution of instances among classes/range
	Knowledge of natural distribution among classes/range ('priors')
Quality	Adequate data
	Realistic data
	Ideal capture conditions
	Intended corruption
Modelling	Reasonable categorisation
	Well-defined mappings between models
Labelling	Unique and additional labelling (text+events, labeller tracks, context, etc.)
	High number of labellers
	Provision of gold standard's reliability
Release	Documentation of side conditions
	Additional perception tests
	Free release of the data with high accessibility
	Defined partitioning

synthesised training material was shown to be highly beneficial in cross-corpus testing, i.e., using a different database for training then for testing.

5.2 Ground Truth and Gold Standard

Often in Intelligent Audio Analysis, the gold standard is not reliable, i.e., the training and testing labels themselves may be erroneous. This highly depends on the task: For example, the age of a speaker is usually known, but the emotion of a speaker is usually difficult to assess. Similarly, the tempo of a musical piece can be determined somewhat reliably by human annotators, while the ballroom dance style may be ambiguous for a pop or rock song, as often several can fit, etc.

The terms 'ground truth' and 'gold standard' are often used more or less as synonyms in the literature—here, we want to define 'ground truth' as the actual truth as measured on the ground as compared to the 'gold standard' that might ideally be identical with the ground truth, however, it might also be the (slightly) error-prone labelling as seen from the 'sky above'.[2] When interpreting results, one thus has to bear in mind that the reference is usually the gold standard and not necessarily the ground truth. This has a double impact: On the one hand side the learnt models are error-prone—on the other hand side, the test results might be over- or under-interpretations.

[2] The term ground truth indeed originated in the fields of aerial photographs and satellite imagery.

Thus, in order to achieve a reliable gold standard close to the ground truth, usually several annotators (or labellers, raters) are used—the less certain the task is, the more. There are a couple of measures to identify the agreement among labellers. If the task is modelled continuously, such as likability of a speaker on a continuous scale or tempo in beats per minute (BPM), correlation or mean linear/absolute error (MLE, MAE) among labellers are frequently used.

Further, labellers can be weighted individually in order to reach highest consent among these with the gold standard. The justification is that labellers may lack in concentration if they have to label huge amounts of data, or do not take labelling seriously at any time. The evaluator weighted estimator (EWE) as described in [7] provides an elegant model to reach a weighted gold standard $y_{EWE,n}$:

$$y_{EWE,n} = \frac{1}{\sum_{k=1}^{K} r_k} \sum_{k=1}^{K} r_k y_{n,k}, \tag{5.1}$$

where the subscript k represents the rater with $k = 1, \ldots, K$, $y_{n,k}$ is the label of rater k for the instance n, and r_k is an evaluator-dependent weight. The EWE's average of the individual evaluators' responses thus takes the fact that each evaluator is subject to an individual amount of disturbance during evaluation into account:

$$r_k = \frac{\sum_{n=1}^{N} \left(y_{n,k} - \frac{1}{N}\sum_{n'=1}^{N} y_{n',k}\right)\left(\bar{y}_n - \frac{1}{N}\sum_{n'=1}^{N} \bar{y}_{n'}\right)}{\sqrt{\sum_{n=1}^{N} \left(y_{n,k} - \frac{1}{N}\sum_{n'=1}^{N} y_{n',k}\right)^2} \sqrt{\sum_{n=1}^{N} \left(\bar{y}_n - \frac{1}{N}\sum_{n'=1}^{N} \bar{y}_{n'}\right)^2}}. \tag{5.2}$$

These weights measure the correlation between the listener's estimations $y_{n,k}$ and the average ratings of all evaluators, $\bar{y}_n$, where

$$\bar{y}_n = \frac{1}{K} \sum_{k=1}^{K} y_{n,k}. \tag{5.3}$$

The inter-evaluator agreement can be described by the correlation coefficients (CCs) r_k using Eq. (5.2) and by the standard deviations σ_n of the assessments,

$$\sigma_n = \sqrt{\frac{1}{K-1} \sum_{k=1}^{K} \left(y_{n,k} - y_{EWE,n}\right)^2}. \tag{5.4}$$

The standard deviation indicates how similar an audio instance is perceived by the human listeners. The inter-evaluator correlation measures the agreement among the individual evaluators and thus focuses on the more general evaluation performance [7]. If the weights are chosen constant among raters, the gold standard is the simple mean of the raters' continuous labels $y_{n,k}$.

In the case of categorical modelling, usually majority votes among the individual ratings $y_{n,k}$ of the raters are used. A variety of measures can be employed for agreement evaluation such as Krippendorff's alpha [8], or (Cohen's) kappa [9]. As a continuum can be discretised, the latter statistics can also be used in this case—often with a linear or quadratical weighting. In the ongoing, we will consider exclusively kappa, which is defined as follows:

$$\kappa = \frac{p_O - p_c}{1 - p_c}, \tag{5.5}$$

where p_O is the measured agreement among two labellers and p_c is the chance-level of agreement. If labellers agree throughout, κ equals 1. If they agree only on the same level as chance would, then κ equals 0. Negative values indicate systematic disagreement. According to [10], values of 0.4–0.6 indicate moderate agreement, such above are considered as good to excellent agreement. This is known as Cohen's kappa [9]—the extension to several raters is known as Fleiss's kappa, and linear and quadratic weighting are commonly used in the case of ordinal-scaled class properties [11].

In order to demonstrate typical data collection for Intelligent Audio Analysis, three examples are picked in the ongoing: One from speech, music, and general sound data, each.

5.3 Exemplary Databases

5.3.1 Example in Speech: TUM AVIC

Let us first exemplify the collection of speech data in the context of determining speaker interest. This task particularly demonstrates the difficulty of collecting diverse data: Various levels of interest need to be captured in a realistic setting.

In TUM's Audiovisual Interest Corpus (TUM AVIC) as described in detail in [12], an experimenter and a subject are sitting on opposite sides of a desk. The experimenter plays the role of a product presenter and leads the subject through a commercial presentation. The subject's role is to listen to explanations and topic presentations of the experimenter, ask several questions of her/his interest, and actively interact with the experimenter considering his/her interest in the addressed topics. The subject was explicitly asked not to worry about being polite to the experimenter, e.g., by always showing a certain level of 'polite' attention. Voice data was recorded by two microphones—one headset and one far-field microphone. Recordings were stored with 44.1 kHz, 16 bit. 21 subjects took part in the recordings, three of them Asian, the remaining European. The language throughout experiments is English, and all subjects are non-native, yet very experienced English speakers.

More details on the subjects are summarised in Table 5.2.

Table 5.2 Details on subjects contained in the TUM AVIC database

Group	Subjects [#]	Mean age (years)	Rec. time (h)
All	21	29.9	10:22:30
Male	11	29.7	5:14:30
Female	10	30.1	5:08:00
Age <30	11	23.4	5:13:10
Age 30–40	7	32.0	3:37:50
Age >40	3	47.7	1:31:30

To acquire reliable labels of a subject's 'Level of Interest' (LoI), the collected material was first segmented into speaker- and sub-speaker-turns. Then, it was labelled by four male annotators, independently from each other. The annotators were undergraduate students of psychology. The intention was to annotate observed interest in the 'common sense'. A speaker-turn thereby was defined as a continuous speech segment produced solely by one speaker. Back channel interjections ("*mhm*", etc.) were ignored, i.e., every time there is a speaker change, a new speaker-turn begins. This is in accordance with the common understanding of 'turn-taking'. By that, speaker-turns can contain multiple and partially long sentences. In order to provide LoI analysis on a finer time scale, the speaker-turns were additionally segmented at grammatical phrase boundaries. A turn lasting longer than 2 s is split by punctuation and syntactical and grammatical rules, until each remaining segment is shorter than 2 s. The segments resulting from this 'chunking' are referred to as 'sub-speaker-turns'.

The LoI is annotated per such sub-speaker-turn. To familiarise the annotators with a subject's character and behaviour patterns prior to the actual annotation task, the annotators had to watch approximately five minutes of a subject's video at first. Each sub-speaker-turn had to be viewed at least once to label the LoI displayed by the subject. Five LoI were distinguished as follows:

- LoI $-$ 2: *Disinterest* (the subject is tired of listening and talking about the topic, is totally passive, and does not follow)
- LoI $-$ 1: *Indifference* (the subject is passive, does not give much feedback to the experimenter's explanations, and asks unmotivated questions, if any)
- LoI0: *Neutrality* (the subject follows and participates in the discourse; it cannot be recognised if she/he is interested or indifferent in the topic)
- LoI $+$ 1: *Interest* (the subject wants to discuss the topic, closely follows the explanations, and asks questions)
- LoI $+$ 2: *Curiosity* (there is a strong wish of the subject to talk and learn more about the topic).

In addition to the LoI annotation, the spoken content has been transcribed by one annotator and counter-checked by another. In this process, *long pauses*, *short pauses*, and additionally various types of non-linguistic vocalisations have been labelled. These vocalisations are *breathing* (452), *consent* (325), *hesitation* (1 147),

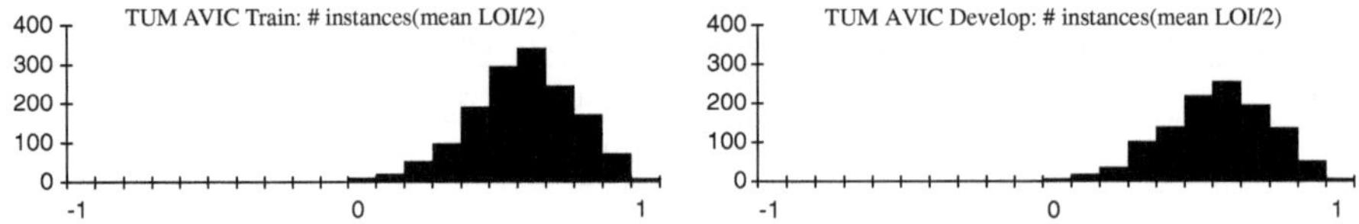

Fig. 5.1 Mean Level of Interest (LoI, divided by 2) histograms for the train and develop partitions of TUM AVIC [12]

laughter (261), and *coughing, other human noise* (716). There is a total of 18 581 spoken words, and 23 084 word-like units including 2 901 non-linguistic vocalisations (19.5 %). The overall annotation thus contains per sub-speaker-turn information on the spoken content, non-linguistic vocalisations, individual LoI annotator tracks, and the mean LoI across annotators.

The gold standard is established either by majority vote on discrete ordinal classes or by shifting to a continuous scale obtained by averaging over the single annotators' LoI. The histogram for this mean LoI is shown in Fig. 5.1. As can be seen in the figure, the subjects had a tendency to be rather polite: Almost no negative average LoI was annotated. Note that here the original LoI scale reaching from LoI $-$ 2 to LoI $+$ 2 is mapped to $[-1, 1]$ by division by 2 in accordance with the scaling as is adopted in other corpora in this field, e.g., [13]. Apart from a higher resolution of LoI, the continuous representation form allows for subtraction of a subject's long-term interest profile to adapt to the mood or personality of the individual.

The overall 21 speakers (and 3 880 sub-speaker-turns) were partitioned speaker-independently in the best achievable balance with priority on gender, next age, and then ethnicity into three partitions: Train (1 512 sub-speaker-turns in 51:44 min of speech of 4 female, 4 male speakers), Develop (1 161 sub-speaker-turns in 43:07 min of speech of 3 female, 3 male speakers), and Test (1 207 sub-speaker-turns in 42:44 min of speech of 3 female, 4 male speakers).

5.3.2 Example in Music: NTWICM

In the second example, we emphasise more on the problem of choosing an appropriate model and measuring reliability of labellers. A particularly ambiguous task was chosen for illustration—the mood in music. The data set was introduced in [14] for a classification task, which was later extended to fully continuous modelling [15].

For building a music database annotated by mood, the compilation *"Now That's What I Call Music!"* (U.K. series, volumes 1–69, double CDs, each) was selected for the following reasons: No audio needed to be recorded—only the process of its annotation was needed. The choice of a commercially available series allows reproducibility by other researchers at a reasonable cost—the annotation can be distributed freely. Further, the decision to include a complete series ensures transparent 'non-prototypicality', i.e., no music pieces were pre-selected for example by choosing the 'easy cases'—this reflects a realistic database management setting.

The overall series contains 2 648 titles—roughly a week of continuous play time. It covers the time span from 1983 until 2008 and represents music styles popular today ranging from Pop and Rock music over Rap, R&B to electronic dance music such as Techno music or House music. The original stereo sound files were 'ripped' from CD and MPEG-1 Audio Layer 3 (MP3) encoded using a sampling rate of 44.1 kHz and a variable bit rate of at least 128 kBit/s. This simulates universal use-cases of an automatic mood classification system, as these are likely faced with compressed music if stored in large digital archives.

For training and testing, a suitable mood representation needs to be decided on, next. Two different approaches are currently utilised predominantly in this field: a discrete [14] and a dimensional description [15].

A *discrete* model relies on a list of adjectives with each describing a state of mood such as *happy*, *sad* or *depressed*. Hevner [16] was the first to suggest a collection of eight word clusters overall consisting of 68 words. Later, Farnsworth [17] regrouped these into ten labelled groups which were used and expanded to 13 groups more recently [18]. Also the popular Music Information Retrieval Evaluation eXchange (MIREX) uses word clusters for its Audio Mood Classification (AMC) task [19]. However, the labelling by adjective groups can easily suffer from being overly ambiguous for a concise estimation of mood in music. In addition, one runs the risk that different adjective groups are increasingly correlated with each other when increasing their number [20]. This implies that a less redundant representation of mood can be found.

Dimensional mood models assume that different mood states are composed by linear combinations of a low number (i.e., two or three) of basic moods. Likely the best known model is the circumplex model of affect presented by Russell [21] consisting of a "two-dimensional space of pleasure-displeasure and degree of arousal". It allows to identify emotional tags as points in the 'mood space' as shown in Fig. 5.2. Thayer [22] divided this mood space into four quadrants as depicted in Fig. 5.2. This model is frequently encountered [23–25], probably because it leads to two binary classification problems with comparably low complexity.

A mood model based on the two dimensions valence ($=: \nu$) and arousal ($=: \alpha$) was used to annotate the music in the NTWICM set. Thayer's mood model is slightly extended, as only four possible values $(\nu, \alpha) \in (1, 1), (-1, 1), (-1, -1), (1, -1)$ seem not to be capable to cover musical mood satisfyingly [24]. To refine the model, first, a pseudo-continuous annotation was considered, i.e., $(\nu, \alpha) \in [-1, 1] \times [-1, 1]$. However, after the annotation of 250 songs this approach was considered to be too complex to achieve a coherent rating. The final model during annotation thus uses five discrete values per dimension. With $D := \{-2, -1, 0, 1, 2\}$ all songs receive a rating $(\nu, \alpha) \in D^2$ as is visualised in Fig. 5.3.

Some implementations have used excerpts of songs to investigate characteristic song parts. This requires an algorithm to locate relevant parts as presented, e.g., in [26–29] and later in this book. Instead of performing any selection, the songs are considered in full length in this section. Mood may well change within a song, such as change of more and less lively passages or change from a sad song to a positive resolution, etc. Annotation in such detail is particularly time-intensive. It was thus

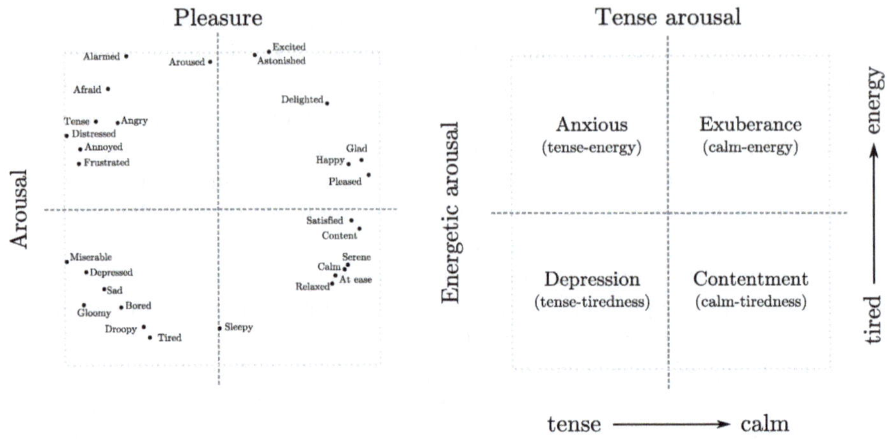

Fig. 5.2 Dimensional mood model development: multidimensional scaling of emotion-related tags as by Russell (*left*) and Thayer's model with four mood clusters (*right*) [14]

Fig. 5.3 Dimensional mood model with five discrete values for arousal and valence [14]

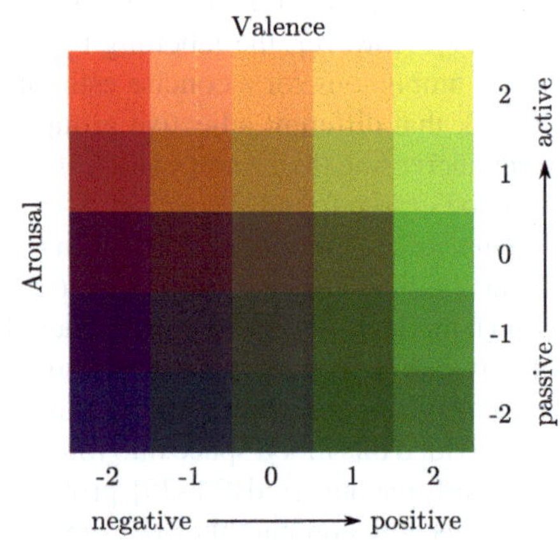

decided in favour of a large database where changes in mood during a song are 'averaged out' in the annotation process, i.e., assignment of the *connotative mood* one would overall have on mind. In fact, this can be sufficient in many applications, such as for automatic music suggestion by the mood that best fits a listener's mood. A different question is whether a learning model would benefit from a 'cleaner' representation without change of mood over the length of a musical piece. For NTWICM, one can assume the contained mainstream popular and commercially oriented music to be less affected by such variation as might be found, e.g., in longer arrangements of classical music. In fact, an analogon can be found in human emotion recognition: Up to less than half of the duration of a spoken utterance may portray the perceived emotion when annotated on isolated word level [30]. Yet, state-of-the-art emotion recognition from speech usually ignores this fact by using turn-level labels rather than word-level based labels [31].

Table 5.3 Overview on the raters (A–D) by age, gender, ethnicity, professional relation to music, instruments played, and ballroom dance abilities, as well as CC between arousal (*A*) and valence (*V*) for each rater's annotations

Rater	Age (years)	Gender	Ethnicity	Prof. Relation	Instruments	Dancing	CC(V, A)
A	34	m	European	club DJ	guitar, drums	Standard/Latin	0.34
B	23	m	European	–	piano	Standard	0.08
C	26	m	European	–	piano	Latin	0.09
D	32	f	Asian	–	–	–	0.43

As mood perception is generally known to be highly subjective [19], it was decided for four labellers. Details on these (three male, one female, aged between 23 and 34 years, average: 29 years) and their relation to music are provided in Table 5.3. Raters A–C stated that they listen to music several hours per day and have no distinct preference of musical style, while rater D stated to listen to music every second day on average and prefers Pop music.

As can be seen, they were picked to form a well-balanced set. They were asked to make a forced decision assigning values in $\{-2, -1, 0, 1, 2\}$ for arousal and valence. They annotated by the perceived mood, i.e., the 'represented' mood, not by the induced mood, i.e., the 'felt' one, which could have resulted in too high labelling ambiguity: One may know the represented mood, but it is not mandatory that the intended or equal mood is actually felt by the raters. Indeed, depending on perceived arousal and valence, different behavioural, physiological, and psychological mechanisms are involved and contextual associations are often highly decisive [32].

The labellers listened via external sound proof headphones in an isolated and silent laboratory environment. Labelling was carried out independently of the other raters within a period of maximum 20 consecutive working days. Each session took a maximum time of two hours. Each song was fully listened to with a maximum of three times forward skipping by 30 s, followed by a short break. Playback of songs was allowed, and the annotation could be reviewed. For the annotation a plugin[3] to the open-source audio player Foobar[4] was provided. It displays the valence-arousal plane in colour code as is shown in Fig. 5.3 and allows for selecting a class by clicking.

Based on each rater's labelling, Table 5.3 depicts the CC of valence and arousal (rightmost column).[5] Clear differences are indicated looking at the variance among these correlations. The distribution of labels per rater as depicted in Fig. 5.4 further visualises these differences in individual perception of music mood.

To establish a gold standard that considers also songs that do not possess a majority agreement in label, a new strategy has to be found: In the literature such instances are usually discarded, which does not reflect a real world usage where any musical

[3] Available at http://www.openaudio.eu.

[4] http://www.foobar2000.org

[5] The complete annotation by the four individuals is available at http://www.openaudio.eu to ensure reproducibility by others.

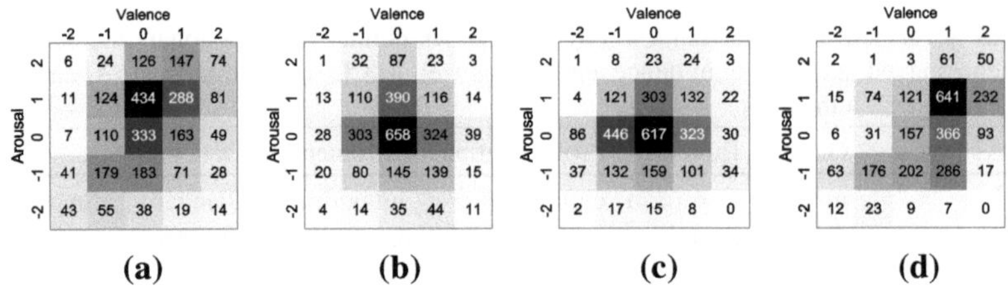

Fig. 5.4 5×5 class distributions of the music database (2,648 total instances) for the annotation of each rater (A–D) [14]. **a** Rater A, **b** Rater B, **c** Rater C, **d** Rater D

piece needs to be handled. Introduction of an additional 'garbage' class [33] was found unsuitable in this case, as the perception among the raters differs considerably, and a learnt model may be affected too strongly by such a garbage class that may easily 'consume' the majority of instances due to the lack of a sharp definition. Two strategies that both benefit from the fact that these 'classes' are ordinal in nature may help: Usage of the *mean* of each rater's label or the *median*, which is known to better cope with outliers. To match from mean or median back to classes, a binning is needed, if one does not want to introduce novel classes 'in between' (for example, if two raters judge '0' and '1', one would obtain the new class '0.5'). A simple round operation was thus chosen to preserve the original five 'classes'. To find the better suited representation among these two types of gold standard calculation, Table 5.4 shows mean kappa values with none, linear, and quadratic weighting over all raters. In addition to the five classes (in the ongoing abbreviated as A5 for arousal and V5 for valence), it considers a clustering of the positive and negative values resulting in three classes per dimension (A3 and V3, respectively). The observed increase in kappa going from no weighting to linear weighting to quadratic weighting indicates that confusions between a rater and the established gold standard favourably occur more frequently between neighbouring classes. As stated, kappa values larger 0.4 are considered as moderate agreement, while such larger 0.7 can be considered as good agreement [34]. As can be seen, the median is the better choice for NTWICM. Further, three classes show better agreement except when considering quadratic weighting—this comes as less confusions with far spread classes can occur in the case of less classes. The choice of gold standard for NTWICM thus is the (rounded) median after clustering to three classes. The preference of three over five classes is further motivated by the lack of sufficient instances for the 'extreme' classes in the case of five classes. This becomes obvious looking at the resulting distribution of instances by the rounded median gold standard for the original five classes per dimension as provided in Fig. 5.5.

For partitioning, all 2 648 songs in the NTWICM database are used in a dataset named AllInst. For evaluation of 'true' learning success, training, development, and test partitions are constructed. As stated, a transparent definition allows easy reproducibility and is not optimised in any respect: Training and development sets are obtained by selecting all songs from odd years, whereby development is assigned by choosing every second odd year. By that, the test set is defined using every even

Table 5.4 Mean kappa values over the raters (A–D) for four different calculations of gold standard obtained either by employing rounded mean or median of the labels per song

#Classes	Gold Standard	κ	κ^1	κ^2
Arousal				
5	mean	0.328	0.477	0.634
5	median	0.415	0.518	0.626
3	mean	0.475	0.496	0.533
3	median	0.526	0.545	0.578
Valence				
5	mean	0.307	0.453	0.602
5	median	0.411	0.510	0.604
3	mean	0.440	0.461	0.498
3	median	0.519	0.535	0.561

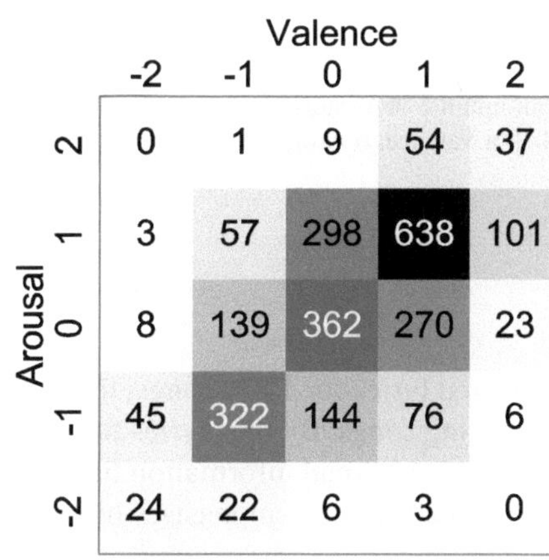

Fig. 5.5 5×5 class distribution of the music database (2,648 total instances) after annotation based on rounded median of all raters [14]

year. The distributions of instances per partition are displayed in Fig. 5.6 following the three degrees per dimension.

To reveal the impact of limiting to musical pieces with clear agreement by a majority of raters, one can additionally consider the sets Min2/4 for the case of agreement of two out of four raters, while the other two have to disagree among each other, resembling unity among two and draw between the others, and the set Min3/4, where three out of four raters have to agree. Note that the minimum agreement is based on the original five degrees per dimension and that this sub-set is only used for the testing instances, to keep training conditions fixed for better transparency of effects of such 'prototypisation'. The numbers of instances per degree of agreement are shown in 5.7.

For 1 937 of 2 648 songs in the NTWICM database lyrics can automatically be collected from two on-line databases: In a first run LyricsDB[6] was applied, which

[6] LyricsDB (http://lyrics.mirkforce.net)

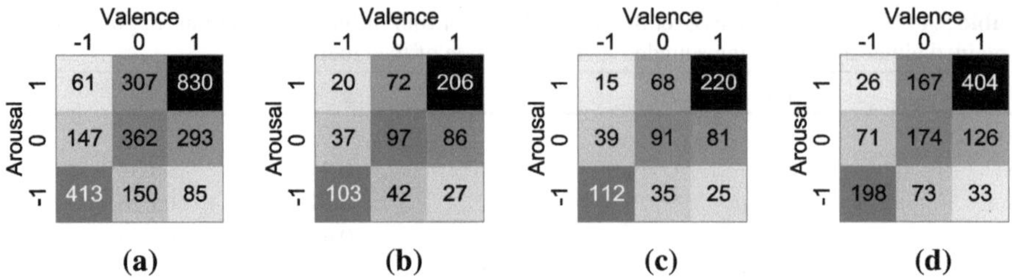

Fig. 5.6 3×3 class distribution of the music database (2 648 total instances) after annotation based on rounded median of all raters and clustering of positive and negative instances. Shown are all, train, development, and test instances [14]. **a** All, **b** Train, **c** Development, **d** Test

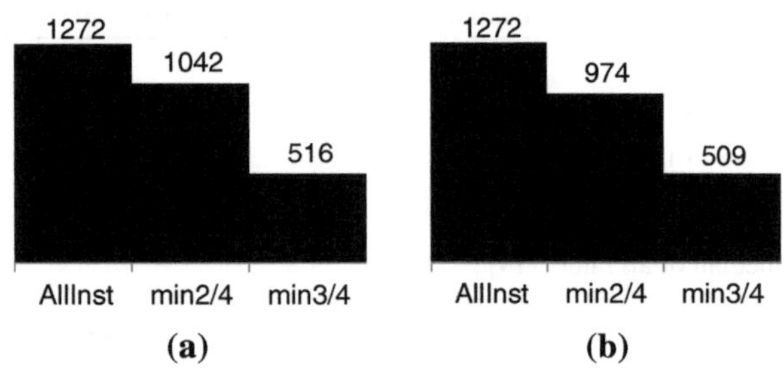

Fig. 5.7 Distributions of test instances in dependence of prototypicality: AllInst, Min2/4 (minimum 2 of 4 raters agree), and Min3/4 (minimum 3 of 4 raters agree) [14]. **a** Valence, **b** Arousal

delivered lyrics for 1 779 songs, then LyricWiki[7] was searched automatically for the remaining songs. By this, lyrics for 158 additional songs could be retrieved. Retrieving such additional information fully automatically—even at the risk of incompleteness or faultiness—emphasises high realism.

5.3.3 *Example in Sound: FindSounds Database*

The final example of the collection and annotation of audio data stems from the domain of general sound. It includes another example related to affect—just as in the two examples above. Further, audio is retrieved from the web rather than recorded or based on a commercial series and annotator weighting is applied.

5.3.3.1 FindSounds Database

For the modelling and recognition of sound events, audio data was first retrieved from the web via the FindSounds site.[8] This site provides a large amount and variety of sound events from real life recordings. These sounds are readily categorised into 16 cover classes and 365 sub-categories [35]. For the creation of the FindSounds

[7] http://www.lyricwiki.org

[8] http://www.findsounds.com/types.html, accessed 25 July 2011.

Database, it is generally sticked with this schema. However, categories without sufficient audio instances were discarded, or, in the case of the sound type 'birds', clustered (with 'animals'). This procedure leaves the following seven common categories out of 16 original cover classes [5]:

- *People*: 45 different human behaviours, such as biting, baby's crying, coughing, laughing, moaning, kissing, etc.
- *Animals* (including birds): 69 different non-bird animals (such as cat, frog, bear, lamb, etc.) and 16 kinds of birds (such as blackbird, etc.)
- *Nature*: 19 kinds of sounds from nature environment, for instance, earthquake, ocean waves, flame, rain, wind, etc.
- *Vehicles*: 34 different types of vehicles and their behaviours, such as motorcycling, braking, helicopter, closing (vehicle) door, etc.
- *Noisemakers*: 13 various events in this domain such as alarm, bell, whistle, horn, etc.
- *Office*: office space sound events including keyboard typing, printing, telephoning, mouse clicking, etc.
- *Musical Instruments*: 62 various musical instruments such as bass, drum, synthesiser, etc.

All audio files were converted into raw 16 bit encoding, mono-channel, at 16 kHz sampling rate. This was needed to unify the various formats and rates used in the original version as retrieved from the web. Each of the sound clips lasts between 1 s to 10 s. Roughly 15 hours of recording time and 16 937 instances were obtained in total, covering 276 sub-categories of real-life sound events. This set will be referred to as FindSounds database in the ongoing. Details on the distribution of FindSound's instances and total play time per category are summarised in Table 5.5. Note that, owing to the sheer size of the database, categorisation was not counter-checked, i.e., the gold standard is based on the categorisation found on the web which has been created by experts according to [35].

5.3.3.2 Emotional FindSounds Database

As was shown in the last section, the FindSounds database is well suited for sound event classification. If one additionally aims at recognising the emotion evoked in a listener of a sound, an additional annotation is needed, as described in [36]. As we had seen for the annotation for music mood above, a typical problem in general emotion recognition is the selection of a suited emotion representation model [37, 38]. For the recognition of emotion evoked in a human listener by sound, Thayer's frequently encountered 2-D model [22] with valence and arousal as dimensions is again adopted. Respecting the divergence between individual labellers, the EWE as gold standard can improve the robustness of sound emotion recognition (here regression) results by making the gold standard more consistent.

To build the 'Emotional FindSounds Database', instances were chosen from the rather huge FindSounds database as was described above. 390 sound files were

Table 5.5 Quantitative description of the FindSounds database

Category	#Subsets	#Clips	Duration (h)
People	45	2540	2h 9min
Animals+Birds	85	2841	2h 42min
Nature	19	937	1h 17min
Vehicles	34	2166	2h 47min
Noisemakers	13	2010	1h 56min
Office	18	1769	1h 01min
Musical Instruments	62	4674	3h 49min
Total	276	16937	15h 41min

Table 5.6 Details on the Emotional FindSounds database

Class	Clips	Duration		Arousal				Valence			
		total	mean	CC	κ	κ^1	κ^2	CC	κ	κ^1	κ^2
All	390	24:53.55	3.50	0.584	0.386	0.411	0.436	0.796	0.490	0.601	0.699
Animals	90	6:06.53	4.05	0.524	0.350	0.364	0.378	0.685	0.448	0.507	0.569
Musical Instruments	75	3:41.17	2.57	0.659	0.392	0.458	0.529	0.712	0.435	0.505	0.592
Nature	30	2:43.65	5.29	0.541	0.355	0.360	0.356	0.759	0.430	0.511	0.575
Noisemakers	30	1:58.12	3.56	0 .569	0.409	0.406	0.415	0.869	0.522	0.650	0.747
People	60	3:20.55	3.21	0.629	0.344	0.386	0.414	0.823	0.495	0.622	0.722
Sports	30	1:37.63	3.17	0.550	0.389	0.390	0.396	0.607	0.347	0.363	0.384
Tools	30	2:09.48	4.20	0.621	0.435	0.454	0.474	0.738	0.480	0.543	0.607
Vehicles	45	3:16.43	4.22	0.473	0.357	0.322	0.281	0.688	0.414	0.459	0.518

Times in (minutes:) seconds. milliseconds. Human agreement: mean CC and majority kappa values over the labellers

selected out of the overall 16937 different sound clips. It was decided to use the following eight categories and sub-categories from FindSounds' taxonomy: *Animals*, *Musical Instruments*, *Nature*, *Noisemaker*, *People*, *Sports*, *Tools* and *Vehicles*. With this choice of cover classes, the database represents a broad variety of frequently occurring sounds in every day environments. More details on the used part of the FindSounds database are given in Table 5.6. The corpus size of this 'Emotional FindSounds' database is well in line with first datasets of emotional speech (such as the Berlin or Danish emotional speech databases) or music (such as the first MIREX mood classification task set).

The Emotional FindSounds database was annotated by four labellers, just as NTWICM (by ID: A: male, 25 years; B: female, 28 years; C: male, 27 years, plays guitar; D: male, 26 years, plays Chinese DiZi flute). They were all post graduate students working in the field of Intelligent Audio Analysis. All labellers are of Southeast-Asian origin (Chinese and Japanese) in order not to introduce strong cross-cultural differences—such questions need to be left for further investigations. For the annotation, these four listeners were asked to make their decision again assigning values on a five-point scale in $\{-2, -1, 0, 1, 2\}$ for arousal and valence using the same tool as was introduced for the NTWICM corpus's annotation. In further

Table 5.7 Overview on the labellers' (ID A–D) agreement: CC of the individual labellers and mean, and κ and weighted κ for labellers' agreement with the majority vote for arousal and valence

ID	CC		κ		κ^1		κ^2	
	Arousal	Valence	Arousal	Valence	Arousal	Valence	Arousal	Valence
A	0.343	0.769	0.265	0.442	0.186	0.544	0.099	0.635
B	0.701	0.869	0.445	0.590	0.505	0.702	0.566	0.794
C	0.542	0.744	0.399	0.477	0.435	0.582	0.474	0.683
D	0.749	0.800	0.435	0.454	0.519	0.575	0.604	0.684

analogy to NTWICM, they were instructed to annotate the perceived emotion and could repeatedly listen to the sounds that were presented in random order across categories. Annotation was also carried out individually and independently by each of the labellers.

Due to the novelty of the task, it has to be investigated whether it is well-defined, or, how to deduce a gold standard from the individual human labels that is to be used as target for learning algorithms. Taking into account the ordinal scale nature of the dimensional emotion ratings, weighted kappa (κ) statistics are of particular interest. Further, CC is considered for the measurement of inter-labeller agreement. It is computed by the mean rating for each instance, followed by calculating the CC of each labeller with this mean. Inter-labeller agreement in terms of κ is calculated per labeller with the majority vote of the labellers. The results of this agreement analysis are shown in Table 5.7. Interestingly, the agreement is considerably higher ($\kappa^2 = 0.699$) for valence than for arousal ($\kappa^2 = 0.436$). Furthermore, a more detailed analysis by sound category reveals that the human agreement—particularly, on valence—strongly dependends on the sound category. For instance, the valence of noisemakers is highly agreed upon ($\kappa^2 = 0.747$) while sounds from sports are not agreed as consistently upon ($\kappa^2 = 0.384$). For arousal, the strongest agreement is found for the group of musical instruments ($\kappa^2 = 0.529$), and vehicles ($\kappa^2 = 0.281$) are observed at the lower end of the agreement scale. Self agreement in a complete repetition (in shuffled order) of the labeller's original annotation was also taken into account to measure the consistency of labelling. This was done after one full week of pause. It was highest for labeller B ($\kappa^2 = 0.554$ for arousal, $\kappa^2 = 0.772$ for valence) who also displayed highest agreement with the gold standard (cf. Table 5.7). Considering the 'reliability' of individual labellers, i.e., their agreement with the 'consensus', one can observe differences especially for arousal: Here, CC ranges from 0.343 (labeller A) to 0.749 (labeller D). This is also reflected in the κ statistics ($\kappa^2 = 0.099$ for labeller A, $\kappa^2 = 0.604$ for labeller D). For valence, less pronounced differences can be observed. Labeller B shows the strongest agreement with the 'consensus'. Overall, the EWE provides a robust estimate of the desired labeller-independent emotion rating in addition to the arithmetic mean. This finding is backed up by regression results on the corpus [36].

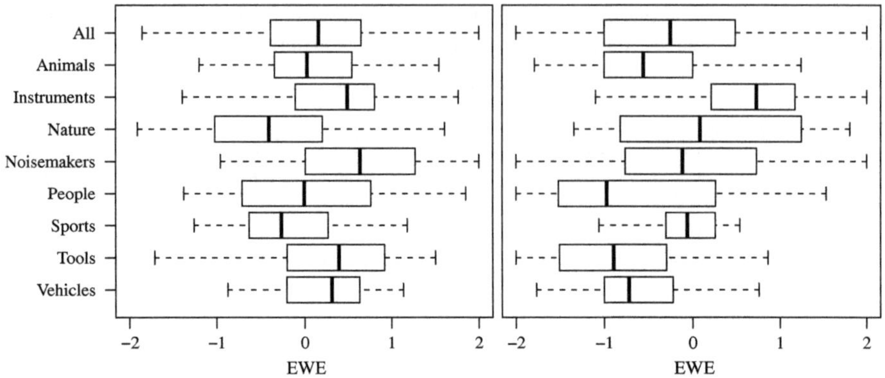

Fig. 5.8 Boxplots of the EWE per sound category: arousal (*left*) and valence (*right*) [36]

The distribution of the EWE for each sound category is shown in Fig. 5.8 as box-and-whisker plot.

References

1. Nieschulz, R., Schuller, B., Geiger, M., Neuss, R.: Aspects of efficient usability engineering. Inf. Technol. Spec. Issue Usability Eng **44**(1), 23–30 (2002)
2. Riccardi, G., Hakkani-Tur, D.: Active learning: theory and applications to automatic speech recognition. IEEE Trans. Speech Audio Process. **13**(4), 504–511 (2005)
3. Schuller, B., Zhang, Z., Weninger, F., Rigoll, G.: Using multiple databases for training in emotion recognition: To unite or to vote? In: Proceedings INTERSPEECH 2011, 12th Annual Conference of the International Speech Communication Association, pp. 1553–1556, Florence, Italy, August 2011 (ISCA, ISCA)
4. Zhang, Z., Weninger, F., Wöllmer, M., Schuller, B.: Unsupervised learning in cross-corpus acoustic emotion recognition. In: Proceedings 12th Biannual IEEE Automatic Speech Recognition and Understanding Workshop, ASRU 2011, pp. 523–528, Big Island, HY, December 2011 (IEEE, IEEE)
5. Zhang, Z., Schuller, B.: Semi-supervised learning helps in sound event classification. In: Proceedings 37th IEEE International Conference on Acoustics, Speech, and Signal Processing, ICASSP 2012, pp. 333–336, Kyoto, Japan, March 2012 (IEEE, IEEE)
6. Schuller, B., Burkhardt, F.: Learning with synthesized speech for automatic emotion recognition. In: Proceedings 35th IEEE International Conference on Acoustics, Speech, and Signal Processing, ICASSP 2010, pp. 5150–515, Dallas, TX, March 2010 (IEEE, IEEE)
7. Grimm, M., Kroschel, K.: Evaluation of natural emotions using self assessment manikins. In: Proceedings of ASRU, pp. 381–385 (2005) (IEEE)
8. Krippendorff, K.: Content Analysis, An Introduction to Its Methodology, 2nd edn. Sage Publications, Thousand Oaks, CA, U. S. A. (2004)
9. Cohen, J.: A coefficient of agreement for nominal scales. Educ. Psychol. Meas. **20**, 37–46 (1960)
10. Landis, J., Koch, G.: The measurement of observer agreement for categorical data. Biometrics **33**, 159–174 (1977)
11. Fleiss, J.: The measurement of interrater agreement. In: Statistical Methods for Rates and Proportions, Chapter 13, pp. 212–236, 2nd edn. John Wiley & Sons, New York (1981)

12. Schuller, B., Müller, R., Eyben, F., Gast, J., Hörnler, B., Wöllmer, M., Rigoll, G., Höthker, A., Konosu, H.: Being bored? recognising natural interest by extensive audiovisual integration for real-life application. Imag. Vis. Comput. Special Issue on Visual and Multimodal Analysis of Human Spontaneous Behavior **27**(12), 1760–1774 (2009)
13. Grimm, M., Kroschel, K., Narayanan, S.: The Vera am Mittag German Audio-Visual Emotional Speech Database. In: Proceedings of the IEEE International Conference on Multimedia and Expo (ICME), pp. 865–868, Hannover, Germany (2008)
14. Schuller, B., Dorfner, J., Rigoll, G.: Determination of non-prototypical valence and arousal in popular music: Features and performances. EURASIP J. Audio Speech Music Process. Special Issue on Scalable Audio-Content Analysis, 2010(Article ID 735854), 19 (2010)
15. Schuller, B., Weninger, F., Dorfner, J.: Multi-modal non-prototypical music mood analysis in continuous space: reliability and performances. In: Proceedings 12th International Society for Music Information Retrieval Conference, ISMIR 2011, pp. 759–764, Miami, FL, October 2011 (ISMIR, ISMIR)
16. Hevner, K.: Experimental studies of the elements of expression in music. Am. J. Psychol. **48**, 246–268 (1936)
17. Farnsworth, P.R.: The Social Psychology of Music. The Dryden Press, New York (1958)
18. Li, T., Ogihara, M.: Detecting emotion in music. In: Proceedings of ISMIR, pp. 239–240, Baltimore, MD (2003)
19. Hu, X., Downie, J.S., Laurier, C., Bay, M., Ehmann, A.F.: The 2007 MIREX audio mood classification task: lessons learned. In: Proceedings 9th International Conference on Music Information Retrieval (ISMIR), pp. 462–467, Philadelphia, PA (2008)
20. Russell, J.A.: The Measurement of Emotions, Volume 4 of Emotion, Theory, Research, and Experience, Chapter Measures of Emotion, pp. 83–111. Academic Press, San Diego (1989)
21. Russell, J.A.: A circumplex model of affect. J. Pers. Soc. Psychol. **39**(6), 1161–1178 (1980)
22. Thayer, R.E.: The Biopsychology of Mood and Arousal. Oxford University Press, New York (1990)
23. Liu, D.: Automatic mood detection from acoustic music data. In: Proceedings International Conference on Music, Information Retrieval, pp. 13–17 (2003)
24. Lu, L., Liu, D., Zhang, H.: Automatic mood detection and tracking of music audio signals. IEEE Trans. Audio Speech Lang. Process. **14**(1), 5–18 (2006)
25. Xiao, Z., Dellandréa, E., Dou, W., Chen, L.: What is the best segment duration for music mood analysis? In: Proceedings of International Workshop on Content-Based Multimedia Indexing (CBMI), pp. 17–24, (2008)
26. Bartsch, M.A., Wakefield, G.H.: To Catch a Chorus: Using Chroma-Based Representations for Audio Thumbnailing. In: IEEE Workshop on Applications of Signal Processing to Audio and Acoustics 2001, pp. 15–18, New Paltz, NY, October 2001
27. Schuller, B., Rigoll, G., Lang, M.: Hmm-based music retrieval using stereophonic feature information and framelength adaptation. In: Proceedings 4th IEEE International Conference on Multimedia and Expo, ICME 2003, vol. II, pp. 713–716, Baltimore, MD, July 2003 (IEEE, IEEE)
28. Goto, M.: A chorus section detection method for musical audio signals and its application to a music listening station. IEEE Trans. Audio Speech Lang. Process. **14**(5), 1783–1794 (2006)
29. Müller, M., Kurth, F.: Towards structural analysis of audio recordings in the presence of mucical variations. EURASIP J. Adv. Signal Process. ID 89686, (2007)
30. S. Steidl, A. Batliner, D. Seppi, and B. Schuller. On the impact of children's emotional speech on acoustic and language models. EURASIP J. Audio Speech Music Process. Special Issue on Atypical Speech, 2010(Article ID 783954), 2010. pp. 14
31. Zeng, Z., Pantic, M., Roisman, G.I., Huang, T.S.: A survey of affect recognition methods audio, visual, and spontaneous expressions. IEEE Trans. Pattern Anal. Mach. Intell. **31**(1), 39–58 (2009)
32. Gabrielsson, A.: Emotion perceived and emotion felt: same or different? Musicae Scientiae, pp. 123–147 (2002)

33. Steidl, S., Schuller, B., Seppi, D., Batliner, A.: The hinterland of emotions: facing the open-microphone challenge. In: Proceedings 3rd International Conference on Affective Computing and Intelligent Interaction and Workshops, ACII 2009, vol. I, pp. 690–697, Amsterdam, The Netherlands, September 2009 (HUMAINE Association, IEEE)
34. Carletta, J.: Assessing agreement on classification tasks: the kappa statistic. Comput. Linguist. **22**(2), 249–254 (June 1996)
35. Rice, S.V., Bailey, S.M.: A web search engine for sound effects. In: Proceedings of 119th AES, New York (2005)
36. Schuller, B., Hantke, S., Weninger, F., Han, W., Zhang, Z., Narayanan, S.: Automatic recognition of emotion evoked by general sound events. In: Proceedings 37th IEEE International Conference on Acoustics, Speech, and Signal Processing, ICASSP 2012, pp. 341–344, Kyoto, Japan, March 2012 (IEEE, IEEE)
37. Gunes, H., Schuller, B., Pantic, M., Cowie, R.: Emotion representation, analysis and synthesis in continuous space: A survey. In: Proceedings International Workshop on Emotion Synthesis, rePresentation, and Analysis in Continuous spacE, EmoSPACE 2011, held in conjunction with the 9th IEEE International Conference on Automatic Face & Gesture Recognition and Workshops, FG 2011, pp. 827–834, Santa Barbara, CA, March 2011 (IEEE, IEEE)
38. Kim, Y., Schmidt, E., Migneco, R., Morton, B., Richardson, P., Scott, J., Speck, J., Turnbull, D.: Music emotion recognition: a state of the art review. In: Proceedings of ISMIR, pp. 255–266, Utrecht, The Netherlands (2010)

Chapter 6
Audio Features

The ability to focus attention on important things is a defining characteristic of intelligence.

—Robert J. Shiller

To represent the information contained in the audio (stream) in a compact way focussing on the task of interest, a parametrised form is usually chosen.These parameters describe properties of the audio usually in a highly information reduced form and typically at a considerably lower rate, such as the mean energy or pitch over a longer period of time. As different Intelligent audio analysis tasks are often best represented by different such 'features', a broad selection of the most typical ones will be presented in the ongoing—these will be the ones that are later also used in the application examples in this book. The determination of the features will include the digitalisation and segmentation of the audio prior to their actual calculation or extraction.

6.1 Audio Chunking

This section describes the digitalisation of audio and subsequent chunking in order to go from an analogue stream to digitised chunks as 'units of analysis' that can be processed computationally.

6.1.1 Digital Audio

In order to process the audio signal in a digital way, the analogue signal $s_{ana}(t)$ with t representing continuous time is represented by a sequence of equidistant (interval Δt) with index k at the times $t = f(k\Delta t)$ [1]. The area of these impulses is proportional to

B. W. Schuller, *Intelligent Audio Analysis*, Signals and Communication Technology,
DOI: 10.1007/978-3-642-36806-6_6, © Springer-Verlag Berlin Heidelberg 2013

the analogue value $s_{ana}(k\Delta t)$. If the sample impulse $a(t)$ is chosen very narrow and with the area equalling one, the sampling can be described by a discrete convolution of $s_{ana}(t)$ and $a(t)$ in time steps Δt [2]:

$$s_{ana,T}(t) = \sum_{k=-\infty}^{+\infty} \Delta t \cdot s_{ana}(k\Delta t) \cdot a(t - k\Delta t). \tag{6.1}$$

For ideally short sampling impulses the function $a(t)$ can be approximated by a Dirac impulse [1]:

$$a(t) = \begin{cases} 0 & \text{for } |t| > \frac{\tau}{2} \\ \frac{1}{\tau} & \text{for } |t| \leq \frac{\tau}{2}, \end{cases} \tag{6.2}$$

$$\lim_{\tau \to 0} a(t) = \delta(t).$$

The Dirac impulses at the positions $t = k\Delta t$ cause that $s_{ana}(t)$ is only analysed at the positions $t = k\Delta t$. In the discrete convolution one can thus exchange the function $s_{ana}(k\Delta t)$ by $s_{ana}(t)$ and change the order:

$$s_{ana,T}(t) = s_{ana}(t) \cdot \sum_{k=-\infty}^{+\infty} \Delta t\, \delta(t - k\Delta t). \tag{6.3}$$

The process of sampling can then be represented as a product of the function $s_{ana}(t)$ with a sampling function. This sampling function is an infinite series of Dirac impulses multiplied with the constant Δt [1].

The Fourier transform $S_{ana,T}(f)$ of the sampled signal with the continuous frequency f is an overlap of single spectra that result from the original spectrum $S_{ana}(f)$ by shift of integer multiples of $S_{ana,0}$ [3]. This may result in aliasing spectra [4]. If, however, the signal $s_{ana}(t)$ was band-limited to $-B < f < B$, the spectra of the analogue signal $s_{ana}(t)$ and the sampled signal $s_{ana,T}(t)$ are identical in the region $[-B, +B]$. In other words, the sampling theorem applies for the choice of the sampling frequency f_{sample}:

$$f_{sample} = \frac{1}{\Delta t} \geq 2B. \tag{6.4}$$

The perfect inverse transformation and reconstruction of the original signal is possible if the periodic spectral parts are cut by an ideal Küpfmüller low-pass filter [3]. The according convolution in the time domain can be interpreted as an interpolation with sinc-functions, which in their sum reconstruct the orignial signal:

$$s_{ana}(t) = \sum_{k=-\infty}^{+\infty} s_{ana,T}(k\Delta t)\, \mathrm{si}(\pi \frac{t - k\Delta t}{\Delta t}). \tag{6.5}$$

To ease the requirements on the edge steepness of the band-limiting filters, the sampling frequency can be chosen accordingly higher than the cutoff frequency. Taking the highest frequency audible to the human ear and the requirement of doubling this frequency given by the sampling theorem into consideration, one arrives at the typical value of 44.1 kHz as used in CD audio. For speech digitisation, lower values of 16 kHz (broad-band telephony) or even 8 kHz (narrow-band 'standard' telephony) are typically chosen.

In addition to the time discretisation by sampling, the continuous analogue values need to be discretised to digital (binary) values [5]. The word length w of the binary number is usually limited (mostly 16 bit as in CD audio, or 8 bit as in narrow-band telephone speech). In the case of binary representation the number Q of quantisation steps is:

$$Q = 2^w. \tag{6.6}$$

This limited number of steps results in a quantisation error which is the deviation between the original value and its quantised counter part. This error leads to quantisation noise. For linear quantisation, i.e., in case of equally sized quantisation intervals, this quantisation noise can be estimated in terms of signal-to-noise ratio (SNR) r_q as:

$$r_q = 10\,lg\frac{P_S}{P_N} \simeq 20\,lgQ = 20\,lg2^w \quad [\text{dB}], \tag{6.7}$$

where P_S is is the standard power of the preferred signal, and P_N is the according power of the unwanted noise. For longer word lengths this means [2]:

$$r_q \simeq 6\,\text{dB/bit}. \tag{6.8}$$

Better values can be reached by adapting the quantisation steps to the signal characteristics such as by the ITU's A-law as primarily used in Europe or the μ-law as primarily used in Northern America and Japan for telephony in the ITU G.711 standard for Pulse Code Modulation (http://www.itu.int/rec/T-REC-G.711/e).

6.1.2 Short Time Analysis

Audio signals change over time, i.e., they are time variant [6]. Thus, the main parameters are also time-dependent. However, one can make the assumption that these parameters change relatively slower than the signal frequencies. The evolution over time of the parameters are thus new signals which are, however, sampled with a considerably lower frequency as compared to the original signal [7]. Their sampling frequency will be referred to as parameter sampling frequency in the ongoing in contrast to the signal sampling frequency as was introduced above.

The short time analysis considers the signal in a given short interval within which the audio signal is considered to be stationary [6]. To this end, a weighting of the signal in the time domain by a weighting 'window' function $w(\tau)$ is carried out. The window emphasises the audio signal's values around the time instant t and suppresses distant values [5]. The faded signal part at time t can be described by a multiplication with the window as:

$$s_{ana}(\tau)w(t-\tau). \tag{6.9}$$

In particular two opposing considerations influence the choice of the window length T: Most importantly, the window needs to be sufficiently long in order to allow for reliable determination of the parameter of interest. At the same time, however, it needs to be short enough to ensure that the measurement is still valid, i.e., the audio signal is 'quasi-stationary' within the window. As a result, a compromise has to be made that leads to a certain uncertainty in analogy to Heisenberg's uncertainty principle. For spectral transformation, for example, holds the Heisenberg-Gabor limit that a function cannot be both time limited and band limited:

$$\Delta\tau \cdot \Delta f \geq \frac{1}{4\pi} \tag{6.10}$$

Thereby, $\Delta\tau$ and Δf are the uncertainty in time and frequency.

Further, the sampling theorem (Eq. 6.4) holds for the choice of the parameter sampling instants t. Typical window lengths for speech analysis are 20–40 ms. In music analysis the length is sometimes chosen longer, around 50–80 ms. However, the windows are usually chosen to be overlapping if the window function is a soft function and not a rectangular one. Typical audio parameter sampling frequencies are thus around 100 Hz, i.e., the window shift (also referred to as step size) is typically around 10 ms. As the windowed audio signal is usually referred to as 'audio frame', the parameter sampling frequency is often measured in frames per second (FPS). Usually, the audio signal values outside the window are set to zero. This is known as the 'stationary' approach. The non-stationary approach assumes the signal outside of the window as undefined [2]. Other approaches consider the window's content as periodic, i.e., the signal is continued periodically outside of the window and the period length is usually equal to the window length T. This is known as the 'periodic' approach [2]. In addition, for periodic signals one could attempt to synchronise the window T with the audio signal's fundamental period T_0 in order to benefit from the signal's inherent periodicity and reduce windowing distortions to a minimum.

A crucial factor thus is the choice of the optimal window function. In fact, this choice depends on the audio parameter to be determined. For parameters in the time domain, rectangular windows are often sufficient. For time frequency transformation, one desires narrow and rectangular windows in the frequency domain, which do, however, also decay rapidly in the time domain (cf. the Heisenberg-Gabor limit Eq. (6.10)). A compromise are thus comparably 'soft' window functions which rise and fall slowly in time and by that also in frequency. In addition, one wishes to avoid side maxima in the respectively other domain. Consider the rectangular window, for example: In the frequency domain it corresponds to the wavy sinc-function. The Gaussian function at the other extreme has no side maxima, neither in time nor frequency domain, but is of infinite length. The reduction of side maxima in general comes at the cost of a wider main maximum. Common window functions (represented for the 'open' region $[-\frac{T}{2}, +\frac{T}{2}]$) include:

- **The rectangular window:** It is characterised by having the narrowest main maximum in the frequency domain, i.e., the smallest bandwidth. However, this comes at the cost of large side maxima—the first one still having an amplitude of −16 dB. The rectangular window is given by:

$$w_{Rect}(\tau) = \begin{cases} 1 & \text{for } \tau = -\frac{T}{2}, \ldots, +\frac{T}{2} \\ 0 & \text{otherwise.} \end{cases} \tag{6.11}$$

- **The Hamming window:** This window is most frequently encountered in audio signal analysis for parameters in the frequency domain. Its side maxima are the smallest at −42 dB almost independent of their frequency [2]. It is given by:

$$w_{Ham}(\tau) = 0.54 + 0.46 \cos\left(2\pi \frac{\tau}{T}\right) \quad \text{and} \quad \tau = -\frac{T}{2}, \ldots, +\frac{T}{2}. \tag{6.12}$$

- **The Hanning window:** It can be represented as a $\cos^2$ window or as a Hamming window with different constants. In comparison to the Hamming window, it reaches zero at the side ends in the time domain. It is further the narrowest in the time domain of these three windows and thus often preferred for analysis in this domain, in particular for pitch or harmonic analysis. Moreover, due to its symmetries it is preferred for audio signal processing where a time domain signal is transformed into the frequency domain, modified, and transformed back to the time domain. It is given by:

$$w_{Han}(\tau) = \cos^2\left(\pi \frac{\tau}{T}\right) \quad \text{and} \quad \tau = -\frac{T}{2}, \ldots, +\frac{T}{2}. \tag{6.13}$$

These window functions are shown in Fig. 6.1 within the interval $[-\frac{T}{2}, +\frac{T}{2}]$ – outside, they are set to zero following the stationary approach. Other window functions include the family of Kaiser windows based on Bessel functions. Their advantage is that side maxima can be further lowered, but again by broadening the main maximum [2]. However, these types are hardly encountered in practical solutions.

6.1.3 Audio Activity Detection

For many applications, one can analyse a continuous audio stream directly frame-by-frame, i.e., make decisions on the frame level. If information that is contained in the dynamics of frame-level parameters is of interest, so-called 'supra-segmental' features can be used. These are summaries of the frame-level features over a given

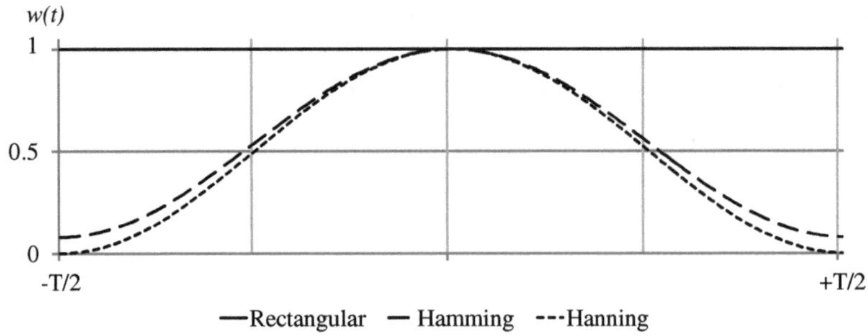

Fig. 6.1 Commonly used audio windowing functions: rectangular, Hamming, and Hanning. Shown is the value over time in the window region $t \in [-\frac{T}{2} + \frac{T}{2}]$

time span of frames, i.e., a segment. The choice of these segments or chunks, also known as the 'chunking', is important. In the case of speech analysis, such chunks may be voiced or unvoiced segments, syllables, words, or larger entities, such as sentences or paragraphs. In the case of music analysis, the unit of analysis may be beats, bars, sections such as verse, bridge, or chorus. A simple alternative is fixed length chunks in analogy to the short time windows [8, 9].

In many cases—in particular for speech or general sound analysis—audio events in between 'silences' (= pauses) are analysed. These silences may be filled with background noise, and the acoustic events of interest could be words or other events, such as animal sounds, for example.

We will call this discrimination between pauses and audio events audio activity detection in the ongoing as a short form for detection of activity of the 'audio signal of interest'. E.g., when searching for speech or singing voice activity, background noise or music are also present, but not of interest. In the specific case of speech, one generally speaks of voice activity detection (VAD) or—more recently—speech activity detection (SAD). The simplest method is the use of a threshold for the audio signal energy. Usually, a hysteresis is used with two thresholds. Once the first threshold is exceeded, a second, lower threshold may not be under-run during a given time length in order to detect a speech, music, or sound onset. If one can assume the background noise to change less quickly than the audio signal of interest, one can use an adaptive algorithm based on histograms: One determines the histogram of the signal level. In the described case, this results in a significant maximum at the level of the background noise. This level plus a certain delta can then be used for audio onset determination. The histogram then needs to be updated on-line. In addition, the histogram of the derivative of the signal level can be used analogously.

More complex solutions are based on multi-dimensional feature information such as spectral analysis with a trained classifier. Such approaches can be trained very well to the signal of interest and thus usually allow for better results—however, at the cost of higher effort. In addition, these two approaches can be efficiently combined: Only when an audio onset is expected based on signal level characteristics, the classifier-trained decision is made to assure that the audio onset belongs to the type of signal one is interested in. Some standard methods are found in [10–14].

6.2 Audio Low Level Descriptors

This section introduces a variety of important acoustic low-level descriptors (LLDs) which are commonly used in the fields of speech, music, and general sound analysis.

The following description of audio LLDs is based on the assumption of digitised audio. By that, the signal is represented as $s(k)$ in the discrete time domain with the discrete time index k as the index of the k-*th* sample. Further, the sampling of parameters by windowing of the signal requires the use of a second time variable: a time n for the instant of measurement of parameters over a window of length N (see Sect. 6.1 for details on digital audio signal representations and windowing).

6.2.1 Speech Descriptors

Among the most important descriptors for speech signals are the intensity, the fundamental frequency F_0 together with the probability of voiced/unvoiced speech, the formants, i.e., resonance frequencies F_X of the vocal tract, with X typically between 1 and 7, together with anti-formants. Further, the voice quality parameters jitter and shimmer are often of interest—these are micro perturbations of the fundamental frequency period lengths and intensities, respectively. Parameters describing the structure of the spectrogram are thereby particularly coined by the characteristics of the vocal tract.

6.2.1.1 Intensity

Rather than modelling the psycho-acoustically perceived intensity which usually depends on the energy, pitch, duration, and the spectral shape of a stimulus [15], just the physical energy E of the signal $s(k)$ is used as a approximate measure of intensity. It is defined as [16]:

$$E = \sum_{k=-\infty}^{+\infty} s^2(k). \tag{6.14}$$

With short time analysis, the energy $E(n)$ at time n is determined as

$$E(n) = \sum_{k=n-\frac{N}{2}}^{n+\frac{N}{2}-1} [s(k)w(n-k-1)]^2, \tag{6.15}$$

assuming a window function being different from zero for $k = n - \frac{N}{2}, \ldots, n + \frac{N}{2} - 1$. This same assumption is made for all the following LLDs.

The square operation emphasises differences between softer and louder parts in the signal and is physically motivated. A more commonly used alternative of E is the root means square (RMS) amplitude, signal power, or RMS energy E_{rms}:

$$E_{rms}(n) = \sqrt{\frac{1}{N} \sum_{k=n-\frac{N}{2}}^{n+\frac{N}{2}-1} [s(k)w(n-k-1)]^2}, \tag{6.16}$$

A linear alternative is the average absolute amplitude (AAA, also average magnitude) $A(n)$ at time instant n (assuming a time limited window) [2]:

$$A(n) = \frac{1}{N} \sum_{k=n-\frac{N}{2}}^{n+\frac{N}{2}-1} |s(k)w(n-k-1)|. \tag{6.17}$$

The typical window function $w(\tau)$ in the time domain is a rectangular window $w_{Rect}(\tau)$ (Eq. 6.11).

6.2.1.2 Zero Crossings

The number of zero crossings per frame, i.e., the Zero Crossing Rate (ZCR) [6], is defined as:

$$ZCR(n) = \sum_{k=n-\frac{N}{2}}^{n+\frac{N}{2}-1} s_0(k) \quad \text{with} \quad s_0(k) = \begin{cases} 0 & \text{if } sgn[s(k)] = sgn[s(k-1)] \\ 1 & \text{if } sgn[s(k)] \neq sgn[s(k-1)]. \end{cases} \tag{6.18}$$

While the signal energy does not provide any information on the frequency distribution, the zero crossing rate does [17]. For a pure sine tone, for example, the number of zero crossings is twice the tone's frequency. Since the general audio signal is usually a complex compound of different frequency components, the ZCR only roughly indicates whether the signal contains high frequency components—in this case ZCR would also be high—or not. This is very useful to see if a speech signal is voiced or not, for example. A voiced/harmonic signal usually has a low ZCR, as it is periodic at a lower frequency, whereas unvoiced speech signal parts or noisy parts are characterised by high frequency components. Together, energy and ZCR are therefore suited to realise a basic speech/pause detector [2].

6.2.1.3 Auto Correlation

Another important basic descriptor is the auto correlation function (ACF) $R(d)$, here the short time ACF [5]. For signals that are infinite in time it is defined as:

$$R(d) = \sum_{k=-\infty}^{+\infty} s(k)s(k+d) \tag{6.19}$$

or normalised as:

$$r(d) = \frac{R(d)}{R(0)}, \tag{6.20}$$

where d is the delay parameter. An integration is performed over the product of the function and the function shifted by d. The ACF is axis-symmetric:

$$R(-d) = R(d). \tag{6.21}$$

For time variant signals, the short time ACF can be defined in two ways: First, by the stationary approach using a weighted part of the signal at time instant n [2]:

$$R(n,d) = \sum_{k=-\infty}^{+\infty} s(k)w(n-k-1)s(k+d)w(n-k-1-d). \tag{6.22}$$

This definition is in accordance with the ACF for signals infinite in time. The finite limits result from setting values outside the window to zero. Some important characteristics of the ACF $R(d)$ are as follows [7]:

- At the origin, i.e., $R(0)$, there is a global maximum identical to the energy of the analysed signal.
- The ACF of a periodic signal is periodic itself.
- Scaling of the amplitude by x results in a scaling of $R(d)$ by x^2.
- In the case of a (quasi-)periodic signal structure, a shift of the window has a comparably mild influence on the ACF, i.e., a certain phasing invariance is given.

The disadvantage, however, of the stationary approach is, that with increasing delay d less values are available for the ACF's computation: for $d \rightarrow T$ the ACF approaches zero, i.e., fades out. The non-stationary approach overcomes this problem. In this case, the time signal is weighted only 'once', i.e., one computes the product of the weighted time signal with the non-weighted infinite delayed version of the time signal [2]:

$$R(n,d) = \sum_{k=-\infty}^{+\infty} s(k)s(k+d)w(n-k-1). \tag{6.23}$$

Strictly speaking, this is a cross-correlation and the result is not axis-symmetrical. Further, negative values may result, as opposed to the normal ACF. It is, however, better suited in the case of short analysis windows as in this case the effect of fading is particularly significant for the stationary approach. Overall, however, the stationary approach is preferred.

6.2.1.4 Spectrum and Cepstrum

With the speech and most audio signals generally being a non-stationary process that can be considered 'quasi-stationary' only for short time periods, one determines short time spectra instead of transforming the whole signal into the spectral domain [2, 18]. From the time signal $s(k)$ with a suitable window function $w(k)$ we can determine the short time spectrum at time k with n as variable for the Fourier transformation. The short time spectrum by that is a function of time n and frequency m.

With the DFT given as

$$S(m) = \sum_{k=0}^{N-1} s(k) e^{\frac{-j2\pi mk}{N}}, \tag{6.24}$$

the complex short time spectrum $S(m, n)$ is obtained by [3]:

$$S(m, n) = \sum_{k=n-\frac{N}{2}}^{n+\frac{N}{2}-1} s(k) w(n-k-1) e^{\frac{-j2\pi mk}{N}}. \tag{6.25}$$

Note that, implementation wise the Fast Fourier Transform (FFT)—is commonly used for DFT calculation.

To improve readability, in the following consideration we switch back to an analogue frequency description with f as the continuous frequency; still, the described concept is valid also for the discrete time and frequency domain.

According to the simplified linear source filter model of speech production, the speech signal can be modelled by the convolution of the excitation/source signal $E(f)$ with,

- the excitation transfer function $G(f)$,
- the transfer function of the vocal tract $H(f)$,
- and a transfer function $R(f)$, which describes the sound wave propagation into the space outside the human body

weighted by an amplitude factor A[6, 19].

If the influence of the source is to be eliminated, a deconvolution of the source and the transfer functions is required. This can be easily achieved in the frequency domain where the convolution is expressed as product of the signal and all transfer functions:

$$S(f) = E(f) \cdot G(f) \cdot H(f) \cdot R(f) \cdot A. \tag{6.26}$$

In the logarithmic domain, this product turns into a summation. The signal part that is owed to $E(f)$ can be eliminated by high- or band-pass filtering. In the case of high-pass filtering this requires that these parts are indeed low-frequent, in order not to cut away formants (cf. Sect. 6.2.1.8). This high-pass can be best realised on the back-transformation of the logarithmised powers of the spectrum into the time-domain. This leads to the so-called cepstrum, with the independent variable d, the 'quefrency' [7]. These names have been artificially created from the terms 'spectrum' and 'frequency' by re-ordering of characters. The variable d is a unit of time that corresponds to the delay in the ACF, which is the reason for the choice of the same identifier. By applying the logarithm to the power spectrum, the product relationship of the source signal and the transfer functions turns into a sum relationship. After the back-transformation to the time domain (i.e., in the cepstrum) the additive concatenation of the linear source filter model components remains [2]:

$$x(d) = IDFT[log|S(f)|^2] \tag{6.27}$$

$$= IDFT[log|E(f)|^2 + log|G(f)|^2 + log|H(f)|^2 + log|R(f)|^2 + log|A|^2] \tag{6.28}$$

$$= e(d) + g(d) + h(d) + r(d) + A, \tag{6.29}$$

where (I)DFT is the (Inverse) Discrete Fourier Transformation, and $e(d)$, $g(d)$, $h(d)$, and $r(d)$ are the equivalents of their capitalised frequency domain counterparts $E(f)$, $G(f)$, etc. The cepstrum is real valued, if computed from the amplitude or power spectrum, as these are both axis-symmetrical [6]. The desired high-pass can be obtained by trimming the cepstrum after the first fundamental period, i.e., at T_0.

Variations of the classical cepstrum use other back-transformations such as the Discrete Cosine Transformation (DCT) or PCA for de-correlation.

If one maps the power spectrum onto Mel-frequency scale bands, then takes the logarithms of the powers of each band, and applies a DCT transformation to the resulting values, one obtains the Mel-frequency cepstral coefficients (MFCCs). The mapping onto Mel-frequency scale bands is typically performed by triangular filters which are equidistantly spaced on the Mel-frequency scale. This scale takes the physiology of human hearing into account: the frequency resolution of the human ear is higher for low frequencies and lower for high frequencies; an approximately logarithmic relationship of the frequency resolution to the absolute frequency exists [5]. The Mel-frequency scale $Mel(f)$ is given by:

$$Mel(f) = 2595 \cdot \log\left(1 + \frac{f}{700}\right). \tag{6.30}$$

MFCCs are among the most popular audio features. Usually coefficients 0 up to 16 are used. For speech recognition in particular, coefficients 0–12 are applied most frequently.

6.2.1.5 Linear Prediction

A simple model for the production of speech bases on the assumption that voiced sounds—in particular vowels—can be well modelled by a few resonance frequencies, which are referred to as formants [6]. Therefore, one can assume that subsequent samples of a speech signal are not independent, but correlated to some degree, i.e., linear dependencies exist among consecutive frames [6]. By that, it should be possible to predict a sample value $s(k)$ by its predecessors [5].

Given a digital speech signal $s(k)$, with k from $-\infty \cdots +\infty$, we may assume the long term average to equal zero [2]. To estimate and model the linear dependencies, the method of Linear Predictive Coding (LPC) applies. The principle behind LPC is a linear system, which describes an output value $s(k)$ as a weighted sum, i.e., as linear combination of a limited number of preceding values $s(k-i)$ [17]:

$$\hat{s}(k) = -\sum_{i=1}^{p} a_i s(k-i). \tag{6.31}$$

The minus sign is chosen to simplify further calculations. In practice, one can only expect an error-prone estimation $\hat{s}(k)$ of the actual value $s(k)$. The error $e(k)$ between these two is:

$$e(k) = s(k) - \hat{s}(k). \tag{6.32}$$

With Eq. (6.31):

$$s(k) = -\sum_{i=1}^{p} a_i s(k-i) + e(k). \tag{6.33}$$

The weights a_i are the so-called predictor coefficients. The summation delimiter p is the order of the predictor. The predictor coefficients now have to be determined such that—within a given interval—the values k conform well with the actual values of $s(k)$, i.e., the prediction error is minimal. The optimisation criterion is the squared error. In addition, the order p should be minimal in order to require as few coefficients as possible [17]. Just like spectral parameters, the predictor coefficients need to be computed for short segments, as speech signals vary over time.

It can be seen that the predictor polynomial represents a digital filter of the order p which can be used either to produce the speech signal $s(k)$ or the error signal $e(k)$ by using $e(k)$ or $s(k)$ as input signal. The weights a_i completely describe the according linear system. If one uses the speech signal as input to the predictor, the system is a digital transversal filter and one obtains the error signal:

$$e(k) = s(k) + \sum_{i=1}^{p} a_i s(k-i). \tag{6.34}$$

In the following, we will use z-transformation for the mathematical derivation. The (two-sided) z-transformation is given by:

$$S(z) = \sum_{k=-\infty}^{+\infty} s(k)z^{-k}. \tag{6.35}$$

With the z-transformations $E(z)$ and $S(z)$ of the signals $e(k)$ and $s(k)$, respectively, and obeying the rule of the z-transformation that $s(k-i)$ corresponds to $S(z)z^{-i}$ in the z-domain, holds:

$$E(z) = S(z)(1 + \sum_{i=1}^{p} a_i z^{-i}), \tag{6.36}$$

and for the transfer function $H(z)$:

$$H(z) = \frac{E(z)}{S(z)} = 1 + \sum_{i=1}^{p} a_i z^{-i}. \tag{6.37}$$

In the inverse case the system is excited by the error signal and produces the speech signal—the filter then is a mere recursive filter and the transfer function the reciprocal. This is a simple model for speech production, where the vocal tract is seen as linear filter which is excited by regular pulses by the vocal chords. The excitation pulses are not linearly predictable at a low number of predictor coefficients within a short analysis interval and thus produce the prediction error. In the case of unvoiced sounds, excitation is given by white noise. The transfer function in this case has only poles and no zeros, i.e., the system is an all-pole model [6]. These poles can be determined directly from the predictor coefficients a_i. One now has to determine these for a given order p such that the deviation between the estimated signal and the real signal is minimal.

The squared error α within the interval of analysis (for the moment running from $k = -\infty$ to $+\infty$; later within the open window region) is:

$$\alpha = \sum_{k} e(k)^2 \tag{6.38}$$

$$\alpha = \sum_{k} \left[\sum_{i=0}^{p} a_i s(k-i) \right]^2. \tag{6.39}$$

Note that, for simplification a coefficient a_0 was introduced that equals one. In order to determine the minimum of this error, one differentiates the error partially per predictor coefficient and sets the derived error equal to zero:

$$\frac{d\alpha}{da_i} = \sum_k \left[2s(k-i) \sum_{j=0}^{p} a_j s(k-j) \right] \stackrel{!}{=} 0. \tag{6.40}$$

The order of summations may be exchanged:

$$\sum_{j=0}^{p} a_j \underbrace{\sum_k s(k-i)s(k-j)}_{r_{i,j}} \stackrel{!}{=} 0. \tag{6.41}$$

One can now substitute by the correlation coefficients $r_{i,j}$ as shown in the above Eq. (6.41). This results in a linear system of equations

$$\sum_{j=0}^{p} a_j r_{i,j} = 0 \quad \text{for } i = 1 \ldots p. \tag{6.42}$$

which can be solved for the p predictor coefficients a_j with diverse standard methods, which will not be detailed here. For more details please see [2].

It is interesting to note that the predictor error α_m, within an interval of analysis, monotonously decreases with increasing predictor length m because the estimated signal $\hat{s}(k)$ improves [2]:

$$\alpha_m \leq \alpha_{m-1}. \tag{6.43}$$

The linear prediction is also of relevance in the frequency domain and in fact, it is closely related to the ACF [2]. According to Parseval's theorem, the minimisation of the prediction error in the time domain results in an according minimisation in the frequency domain. It can be shown that by that, the filter has a tendency to result in the smooth envelope of the fine-grained spectrum [7]. At the same time the digital filter tends to whiten the spectrum of the error signal which means that its time signal either results more or less in a series of dirac pulses (such as the pulse train excitation in case of voiced sounds), or in white noise (e.g., in the case of unvoiced excitation) [19].

Let us first determine the LPC spectrum of the inverse filter denoted by the subscript 'inv'. As it is a mere transversal filter (cf. above), its impulse response is identical with the LPC coefficients a_i (extended by $a_0 = 1$):

$$h_{inv}(k) \equiv 1, a_1, a_2, \ldots, a_p. \tag{6.44}$$

The discrete complex spectrum is obtained directly by application of the DFT as:

$$H_{inv}(m) \equiv DFT(h_{inv}(k)) \quad \text{with:} \quad m = m\Delta f, \tag{6.45}$$

$$\Delta f = \frac{1}{N\Delta t} = \frac{f_{sample}}{N} \quad \text{and} \quad N = p + 1. \tag{6.46}$$

With the DFT:

$$H_{inv}(m) = \sum_{k=0}^{p} h_{inv}(k)e^{-j2\pi m\frac{k}{N}}. \tag{6.47}$$

The DFT thus has $\left[\frac{p+1}{2}\right] + 1$ significant real values and $\frac{p+1}{2}$ significant imaginary values. By computation of the absolute value and squaring of the complex spectrum the power spectrum with $\left[\frac{p+1}{2} + 1\right]$ values is obtained.

For the recursive all-pole model holds (denoted by the subscript'rec'):

$$H_{rec}(z) = \frac{1}{H_{inv}(z)} \tag{6.48}$$

and

$$H_{rec}(m) = \frac{1}{H_{inv}(m)}. \tag{6.49}$$

or logarithmised:

$$log[H_{rec}(m)] = -log[H_{inv}(m)]. \tag{6.50}$$

In logarithmic scaling, one thus only has to invert the sign in order to obtain the spectrum of the recursive filter from the inverse one.

As the LPC filter can only have poles, one can well model formants of vowels [2] but not zeros in the spectrum. The latter would be characteristic for nasal sounds. As an advantage over short time spectra obtained by Fourier transform, the spectra are very smooth and do not show the waviness due to the presence of the fundamental frequency. This comes, however, at the cost that for noisy patterns, such as fricative sounds, LPC modelling is not well suited because the spectrum is still approximated by only p poles.

As typical speech spectra fall by around 6 dB per octave [2], the efficiency of the LPC analysis can be improved by emphasising higher frequencies by a first order 'pre-emphasis' high pass with the following transfer function [7]:

$$H_{pre}(z) = 1 - \mu z^{-1}, \tag{6.51}$$

where the pre-emphasis factor μ is usually chosen between 0.9 and 0.95.

In order to best model the vocal tract transfer function $H(z)$, the formants need to be well captured. Per formant, a pole pair is needed [19], which results in a minimum order of $P_{min} = 2 \cdot$ number of the formants. Usually, one adds two to three additional poles to ensure that all formants are captured [2].

As stated above, the error in the analysis interval α falls monotonously with increasing predictor order p. For a small p it falls rapidly at first. Once all formants are captured, approximately around $p = 16$, it remains almost constant. Another significant decrease takes place once the fine-grained structure of the spectrum caused by the fundamental frequency and its harmonics is modelled. A small error always remains, however, due to non-linearities, zeros, etc.

The linear predictor coefficients a_n can be converted to a cepstrum representation x_n (see Sect. 6.2.1.4) by the following recursion starting at $n = 0$ (cf. [20]):

$$x_n = -a_n - \frac{1}{n} \sum i = 1n - 1\,(n - i)\, a_i x_{n-i} \tag{6.52}$$

6.2.1.6 Line Spectral Pairs

Line spectral pairs (LSP) or frequencies (LSF) are often employed for channel transmission of LPC due to their reduced sensitivity to quantisation noise, stability, and ability to be interpolated. The basic principle is the decomposition of the LP polynomial for $H(z)$ as given in Eq. (6.37) [21] into:

$$P(z) = H(z) + z^{-(p+1)} H(z^{-1}), \tag{6.53}$$

and

$$Q(z) = H(z) - z^{-(p+1)} H(z^{-1}), \tag{6.54}$$

where $P(z)$ and $Q(z)$ correspond to the vocal tract with the glottis closed and opened, respectively. These two have roots only on the unit circle as opposed to $H(z)$, which can have them anywhere in the z-plain. By that, they are palindromic and anti-palindromic polynomials, respectively [21]. For the determination of the LSP, one evaluates $P(e^{j\omega})$ and $Q(e^{j\omega})$ in a grid search for $\omega = 0, \ldots, \pi$, i.e., the roots of the two polynomials of order $p + 1$ need to be determined. These roots are all complex symmetrical pairs $\pm\omega$ – hence the name Line Spectral Pairs [17]. Two roots are found at 0 and p, and $p/2$ further roots need to be determined for $P(z)$ and $Q(z)$. Overall, the result is p roots, i.e., the same number as LPC coefficients.

6.2.1.7 Perceptual Linear Prediction

While the LP coefficients are well suited to focus on the phonetic content by good approximation of high-energy regions and filtering of the more speaker-specific fine harmonic structure of the speech spectrum owed to the source, they violate principles of human hearing. PLP thus extends LP by psychophysics of human hearing to derive an auditory spectrum estimate. The principles incorporated are

- **Critical-band spectral-resolution**: LP coefficients have equal treatment of all frequencies, whereas human spectral resolution is only roughly linear up to 800 or

1 000 Hz, but decreases thereafter. PLP overcomes this by remapping the frequency axis according to the Bark scale and integrating the energy in the critical bands for a critical-band spectrum approximation.
- **Equal-loudness hearing curve**: To simulate human hearing's higher sensitivity to the middle frequency range of the audible spectrum at normal conversational speech sound pressure levels, the critical-band spectrum is multiplied by an equal loudness curve that suppresses frequency ranges that are either relatively low or relatively high in comparison to the range from 400 to 1 200 Hz.
- **Intensity-loudness power law of hearing**: The non-linear relation of a sound's physical intensity and its human perceived loudness sensation is approximated by the power-law of hearing. A cube-root amplitude-compression of the loudness-equalised critical band spectrum estimate is applied.

The psychoacoustically derived spectrum shows less detail and is characterised by a smaller dynamic range. This allows for good modelling by a low-order all-pole model to weaken speaker characteristics: After estimation of the auditory-like spectrum it is converted to ACF values. Then, the autocorrelations are input to a standard LPC analysis, to output PLP coefficients [22]. These can be further be converted to cepstral coefficients by standard recursion (Eq. 6.52).

Interestingly, PLP allows for a smaller order as compared to LP coefficients. This reduces the number of features and by that the parameters needed in a learning algorithm.

A further variant are RASTA (RelAtive SpecTrA) PLP coefficients [23]. These aim at easing mismatches between training and testing data's recording conditions by linear filtering of the data. In the RASTA method, a bandpass filter is applied per spectral component in the critical band spectrum estimate to emphasise modulations in the range of the speech syllable rate. By that, frame-to-frame spectral changes between 1 and 10 Hz are emphasised by the following filter:

$$H(z) = 0.1 \cdot \frac{2 + z^{-1} - z^{-3} - 2z^{-4}}{z^{-4} \cdot (1 - 0.98z^{-1})}. \tag{6.55}$$

The authors in [23] stress, however that other filters could be used and that these could be adapted to the frequency.

The idea behind the RASTA method is that speech is modulated at a different rate as compared to channel effects, background noise, or non-linguistic vocalisations. Moreover, human hearing seems to be less sensitive to slowly varying stimuli [22].

In detail, the processing steps for RASTA PLP cepstral coefficients are: (1) DFT, (2) logarithm, (3) RASTA band-pass filtering, (4) equal loudness curve, (5) power-law of hearing, (6) inverse logarithm, (7) inverse DFT, (8) solving linear equations for LPC, (9) cepstral recursion.

6.2.1.8 Formants

The term 'formants' refers to resonance frequencies of the human vocal tract. In particular the lower resonance frequencies of the vocal tract, i.e., formants F_1 and F_2, are highly correlated with the phonetic content and allow for mapping of vowels and diphthongs (specific concatenation of two vowels) in the F_1, F_2 plane. In several languages such as Dutch F_3 also plays an important role for the spoken content, whereas the higher formants are usually more coined by speaker characteristics [2].

For vowels and non-nasal consonants the transfer function of the vocal tract $H(z)$ can be approximated as an all-pole transfer function (cf. Sect. 6.2.1.5). This corresponds to a mere recursive digital filter as realised by linear prediction. The poles of $H(z)$ are referred to as the formants of the speech signal. When determining the formants, one usually aims at—in order of relevance—the centre frequency, the bandwidth, and the amplitude. Formants are mostly assessed by linear prediction analysis. Alternative methods based on short time spectra also exist. Thereby, the formants are can be identified as dominant maxima, e.g., in the spectral envelope or even directly from the speech signal [24]. There are, however, a number of difficulties when using a spectral representation as starting point for formant determination—most dominantly single spikes may exist that exceed the vocal tract's resonance frequencies in amplitude—e.g., by the fundamental frequency or by noise. Next, these resonance frequencies or formants can be too close to each other, resulting in them being joined to a single spectral envelope maximum. These fundamental problems can be eased by LPC analysis.

Let us now consider formant analysis by linear prediction [2, 25]. The purely recursive filter of the linear prediction fits the smooth envelope of the short time spectra. Spectral maxima are modelled well—areas of low spectral energy are not. In the linear model, speech production is modelled by the chain of generation (cf. Sect. 6.2.1.4) starting with the excitation $E(z)$ (periodic or noise), excitation spectrum $G(z)$, vocal tract $H(z)$ and radiation $R(z)$ [1, 19]. However, we model the poles of the spectral function $S(z)$ of the speech signal. This means, we do not know of which of the components $G(z)$, $H(z)$, or $R(z)$ the poles found in the transfer function $H_{LP}(z)$ of the prediction filter do originate from. $H_{LP}(z)$ can thus not directly be assumed as the optimal approximation for $H(z)$. Rather, one has to determine which of the poles of $H_{LP}(z)$ belong to formants [26]. The poles of $H_{LP}(z)$ can first be determined by suitable algorithms such as the Newton-Raphson method: the algorithm is initiated by an estimate for the first pole and then calculates the polynomial value and its derivative. Then, in an iterative manner, improved estimates are calculated. This iteration terminates once the delta of subsequent solutions is smaller than a set threshold. The polynomial can then be divided by this pole and the algorithm starts over for the now reduced polynomial until all-poles are determined. A re-iteration per pole with the overall polynomial helps to ease limited precision in the first round. One can speed-up this process by using the poles from the last speech frame for initialisation, as the vocal tract position and by that the poles change comparably slowly over time. Now, a frequency range validation criterion could be applied to discard poles which do not belong to formants.

Another option to determine formants is by smoothing short time spectra based on DFT or FFT [2]. The idea is to obtain a smooth spectrum just as in the case of linear prediction, which is freed from the waviness caused by the fundamental frequency. This waviness results in maxima at a distance of F_0 apart due to the harmonics of the fundamental frequency. Obviously, these maxima can easily be confused with formants if the spectrum is not smoothed. Smearing of the maxima can be obtained by convolution with a smoothing function—however, this method is usually not very precise [2].

If analysis is based on the spectral appearance, peak-picking starting from a list of extreme values is needed to decide for the 'right' maxima. This holds for spectral smoothing or linear prediction spectra. Usually, candidates are first found per speech frame, then, the evolution over time is taken into consideration by also looking at neighbouring frames.

Overall, formant tracking is not solved to full satisfaction to the present day [27]. Among the main difficulties one can name unfavourable signal conditions, in particular insufficient spectral resolution in the case of neighbouring formants of similar amplitude. Further, formants are—strictly speaking—only defined for vowels radiated via the mouth. The shunt of the nasal cavity changes the frequency response of the vocal tract significantly, as novel nasal formants are added and formants may be compensated by anti-formants, i.e., zeros in the transfer function [6]. Such compensation may also occur due to zeros in the excitation spectrum $G(z)$. In addition, depending on the speaker and the phonemes—in particular dark vowels—the upper formants as of F_3 are too weak in comparison to surrounding noise. And finally, there exists no ground truth—only gold standards—if algorithms are tested with spontaneous human speech. There are, however, some sets as a partition of the TIMIT corpus—the MSR-UCLA VTR database—that are manually labelled by expert phoneticians [28]. Another standard approach to validity measurement is the usage of synthesised speech, where formant positions are known [29]. Obviously, this is less realistic than comparing performance on real human speech. In a similar way, this last problem of lacking ground truth also holds for fundamental frequency detection algorithms.

Due to these problems, formant tracking is still an active field of research, and new approaches are still introduced, such as biologically inspired algorithms basing on gammatone filter banks [27].

The tracking of anti-formants on the other hand is hardly pursued. As these are also not further considered in this book, we only refer to a few methods that aim to commonly describe formants and anti-formants. First is the autoregressive moving average (ARMA) method [30]: The auto regressive part deals with the recursive part of the filter to be determined, i.e., the poles, whereas the moving average part handles the non-recursive part, i.e., the zeros. A more common method, however, is to use the reciprocal or logarithmic transfer function and to apply the same methods as for the poles [31].

6.2.1.9 Fundamental Frequency and Voicing Probability

The fundamental frequency F_0 or the fundamental period length T_0 have a key role among speech parameters and the prosodic information. The human ear is considerably more sensitive to changes in the fundamental frequency as to changes in other parameters of the speech signal [15]. This makes it evident that high precision is required for its determination, and in fact, the correct determination of F_0 has significant influence on intelligent speech analysis as shown, e.g., in the author's work on emotion recognition in speech [32].

It may seem an easy task at first, as one has only to determine the period length of a quasi-periodic signal [33]. However, a number of factors makes it more challenging than that, and in fact one of the most difficult tasks of speech signal analysis [2]: As stated, in principle, speech production is a non-stationary process. The position of the vocal tract during articulation may change very quickly leading to significant changes of the structure of the time-signal of speech. This may occur already from one fundamental period to the following one [16]. Further, the multiplicity of used articulator positions of the human vocal tract in combination with the multiplicity of human voices result in huge variety of possible time structures of the speech signal. Then, narrow-band lower formants can easily be confused with the fundamental frequency. In particular the first formant can easily be confused with F_0 for female voices, where it is typically found around 200–1 400 Hz [2]. The excitation signal of the human voice itself is not always regular. This holds also in normal conditions, i.e., in the absence of pathological affects. The voice can further switch into the 'strohbass' register with a very low frequent and irregular excitation as low as 25 Hz. Across speakers, the fundamental frequency can further vary among almost four octaves (50–800 Hz). Finally, the transmission channel may lead to distortions or band limitations, such as in the case of (narrow-band) telephone speech (300–3 400 Hz).

This led to a considerable amount of Pitch Detection Algorithms (PDAs), of which none works to full satisfaction in arbitrary conditions [34]. Some of these aim at determination of the fundamental period T_0, which is equivalent to F_0 by:

$$F_0 = \frac{1}{T_0}. \tag{6.56}$$

If T_0 is to be determined, it is considered as momentary value, i.e., the time from the beginning of one period to the beginning of the subsequent one. If the speech signal was strictly periodic, both definitions would lead to the same result.

Each PDA can be sub-divided into three steps:

- the pre-processing that aims at a data reduction to focus on the problem at hand
- the actual extraction,
- and the post-processing that usually aims to smooth the overall pitch track and corrects minor errors, e.g., by Viterbi smoothing (cf. Sect. 7.3.2) [33].

Independent of these steps, PDAs can be parted into two families [7]: First are those operating in the short-time domain, i.e., windowing has taken place and a number of two to three consecutive fundamental periods are typically observed at a time. Second are the ones that operate in the time domain, i.e., input signal and extraction stage operate on the same time basis [2].

We will now first deal with the short-time PDAs. Direct determination of F_0 by localising the first spectral maximum is not sufficiently robust. Better results are obtained by looking at the sub-harmonic structure of the power spectrum [35]. This can be obtained by spectral compression: F_0 results as the largest common divisor of the frequencies of all harmonics. To this end, the power spectrum is compressed affinely along the frequency axis in the ratios 1:2, 1:3, etc. and then added to the original spectrum. By the coherent contribution of all higher order harmonics of F_0 a maximum at the actual frequency of F_0 is emphasised. This principle is known as sub-harmonic summation (SHS). Another approach is the measurement of neighbouring maxima in the power spectrum to determine the fundamental frequency. According to [15], these harmonics-based approaches are well justified by human pitch perception. They will even work well in cases where the actual fundamental frequency is missing due to band-limiting properties of the channel, for example. The human hearing is able to identify the correct pitch of a complex tone where the actual fundamental frequency is missing in the signal, but the structure of the higher harmonics of the fundamental is present. In such a case the actual perceived pitch is lower than the actual lowest frequency in the signal. This effect is known as virtual pitch [15].

A different PDA approach is based on the cepstrum [36]—the fundamental period T_0 can then be determined as significant maximum at the right end along the quefrency axis (cf. Sect. 6.2.1.4). At the left end close to the origin, the formants are located. Further, in the case of unvoiced sounds the excitation function is noise-like, i.e., non-periodic, such that no peaks occur in the cepstrum and the spectral energy is lower. By a simple threshold decision one can thus distinguish between voiced and unvoiced sounds. This decision can be added by the use of the ZCR as previously described. In general, the cepstral method can be considered as relatively robust.

Another method based on short-time analysis in the spectral domain makes use of the maximum likelihood (ML) principle [37]. For a limited segment in time, a periodic signal of unknown period length T_0 is by this method separated optimally from Gaussian-distributed noise. However, neither is the speech signal ideally periodic, nor the background noise Gaussian-distributed, which requires adjustments for the application to speech signals. As this method is not used in the ongoing, no further details are given at this point.

Let us now switch to PDAs based on correlation methods. The most straightforward approach is based on the ACF, as a periodic signal has a periodic ACF with distinct maxima at the beginning of each period. In order to ease the influence of the first formant, the spectrum can be flattened at first. This can be reached by LPC analysis: The signal is first band-limited to around 800 Hz. Next, it is inserted into an inverse filter with a low predictor order, e.g., of four. By that, computational effort remains small and a low order further ensures that the fundamental frequency is well preserved in the error or 'residual' output signal, whereas the first formant is already eliminated. This is also known as Simplified Inverse Filtering Technique (SIFT) [38]. Using the LPC error signal however gives best results only in the case of non-disturbed speech with sufficient presence of higher frequencies. For dark vowels, the error signal is usually rather weak. Noise will usually be forwarded without dampening into the error signal.

In the ACF-based PDAs, F_0 is then determined by the first peak after the one at the origin [39]. Erroneous period-values need to be eliminated or interpolated in this stage and potential changes in the period need to be foreseen. In addition, the ACF method can be used to determine the harmonicity of the speech signal—the Harmonics to Noise Ratio (HNR): The ACF's first peak at the origin reflects the overall signal's energy. HNR is then obtained by setting the peak at the origin in relation to the next occurring distinct peak. If this peak is considerably lower, one has a clear indication of a non-periodic signal, as self-similarity is low in case of higher delays in the ACF. The ratio can thus either be used as 'voicing probability' or by thresholding a hard decision between voiced / unvoiced can be made. Throughout this book, we calculate the logarithmic HNR by

$$HNR(n) = 10 \cdot \log \frac{ACF(T_0)}{ACF(0) - ACF(T_0)}, \tag{6.57}$$

where T_0 is the fundamental period.

A faster alternative to the ACF is the average magnitude difference function (AMDF) [40]. AMDF can be seen as anti correlation similar to comb-filtering where the resonance frequencies are determined by the delay d:

$$AMDF(d) = \sum_k |s(k) - s(k + d)|. \tag{6.58}$$

If the delay d equals the fundamental period T_0, a significant minimum is observed as opposed to the maximum in the case of ACF. AMDF's speed is based on the substitution of ACF's multiplications by subtractions. AMDF is usually applied either directly on the speech signal, the LPC error signal or in combination with a non-linear pre-processing and band pass filtering ('centre clipping') of the signal at first. Further, AMDF follows the principle of non-stationary analysis and thus allows for analysis of rather short segments [2]. These characteristics have helped to make AMDF-based PDAs popular.

Overall, PDAs based on short-time analysis are typically robust against noise, bandwidth limitation at the lower frequency end, and phase distortion. They do, however, not allow to provide a period-by-period determination of T_0 which is needed if one aims to measure the micro-perturbations of pitch and energy.

We thus now consider the PDAs that operate in the time domain. These can be differentiated by the amount of effort put on pre-processing. In fact, two extremes dominate which will both be exemplified by one PDA: First, to have all data reduction in the pre-processing stage. This may go as far as filtering all but the first partial oscillation. Second, in the other extreme, there is no pre-processing, but the extraction stage operates on the original signal directly.

The PDAs in the time domain analyse the signal period by period and set markers at periods' boundaries [33]. This makes them usually more susceptible to local deviations and by that less reliable than the majority of the short-time PDAs. In the case of highly non-periodic excitation signals, however, they usually provide the better results [2].

We will now first discuss the analysis of the time structure of a speech signal. The fundamental period is the response of the vocal tract to a single pulse of the excitation. Given the vocal tract to be a lossy and passive linear system, the impulse response is a sum of exponentially dampened oscillations. One can thus expect maxima and minima at the beginning of each period to be more significant as towards the end. This allows to search for maxima and minima in order to determine T_0. Problems in doing so include the momentary values of the formants which may change comparably quickly and dominate the frequencies of the relevant dampened oscillations. Further, F_1 is dampened only weakly whereas the signal envelope changes comparably faster. In case of a phase distorted signal the formants may appear as if excited at different moments in time. This makes analysis rather complex—however, computationally usually only comparisons and decisions are needed which makes these PDAs rather fast. The overall processing is as follows: The influence of higher order formants is eased by low-pass filtering. Then, all maxima and minima are determined. Such that are not significant are eliminated until a single extreme value per period remains. A final correction can take care of obviously error-prone candidates [2].

Low-pass filtering can go as far as attempting to preserve only the first partial oscillation. In this case, only zero-crossings need to be counted. Obviously, this is not trivial as we do not know a-priori in which range to expect T_0. Thus, to allow for less aggressive low-pass filtering, we can introduce a threshold above zero in the case of the preservation of more than one partial. This can be further extended by introduction of a hysteresis: A marker is set once an upper threshold is exceeded and only reset once a second lower threshold is under run. The requirement of severe low pass filtering demands for different frequency ranges of operation in any of these cases [2]. Further, the first partial needs to actually exist, which is, not the case in narrow-band telephone speech, for example.

6.2.1.10 Jitter and Shimmer

Jitter and shimmer are referred to as micro-prosodic descriptors as opposed to the prosodic descriptors intensity and intonation dealt with above. Like HNR, they describe the voice quality.

Jitter is the deviation of the fundamental period length from period to period. This information is, particularly suited in speaker age or pathology determination, for example. With increasing age or certain pathology the irregularity of the periodic excitation decreases. Further, the heart rate can have an influence on jitter [41]. One can distinguish between the cycle-to-cycle or local jitter J_{cc} – the deviation from one period to the next—given as:

$$J_{cc} = T_0(n) - T_0(n-1), \tag{6.59}$$

and the period or cycle jitter J_c of the deviation of the current fundamental period and the 'ideal' fundamental period $\overline{T}_0$ as obtained by averaging in the analysis interval:

$$J_c = T_0(n) - \overline{T}_0. \tag{6.60}$$

Jitter is known to be particularly high at the beginning and end of a sustained voiced sound.

In a similar way, shimmer is the deviation of amplitude—usually in dB—from period to period. A healthy speaker's shimmer is usually between 0.05 and 0.22 dB [42].

To end this section, Fig. 6.2 gives some example plots of speech LLDs as discussed above.

6.2.2 *Music Descriptors*

In this section we will deal with LLDs tailored in particular to the analysis of music. However, many of the previously discussed features are also used for music analysis. We will first look at basic Pitch Class Profiles (PCP)—in particular by CHROMA-type features. These are suited for the tonal analysis of music. We will then take a look at the music theoretic and human perception based variants as were first introduced in [44]. Rhythmic features are discussed at a later stage in Sect. 11.3.

6.2.2.1 Pitch Class Profiles

In music theory, 'notes' are characterised and named by their pitch class and their octave, where an octave is an 'interval' between two notes. An increase by one octave resembles a doubling of a note's frequency. It is further a special interval: Two notes played in different octaves sound nearly equal to human listeners and thus share the same name with different octave number. In western music, the octave interval is divided into twelve equally sized intervals with the tempered scale. These intervals are called semi-tones. Their names in western music are shown in Fig. 6.3 that also visualises the discussed principle.

PCP features are based on the principle of providing the spectral energy per semi-tone band. They are computed using a DFT with a suitable window length, window function, and a window overlap—typically around 0.5. Human loudness sensation can be taken into account, e.g., by applying the A-weighting according to DIN EN 61672-1:2003-10 to the DFT magnitudes. The weighting is given by:

$$H_A(f) = \frac{12200^2 \cdot f^4}{(f^2 + 20.6^2) \cdot (f^2 + 12200^2)} \cdot \frac{1}{\sqrt{f^2 + 107.7^2} \cdot \sqrt{f^2 + 737.9^2}}. \tag{6.61}$$

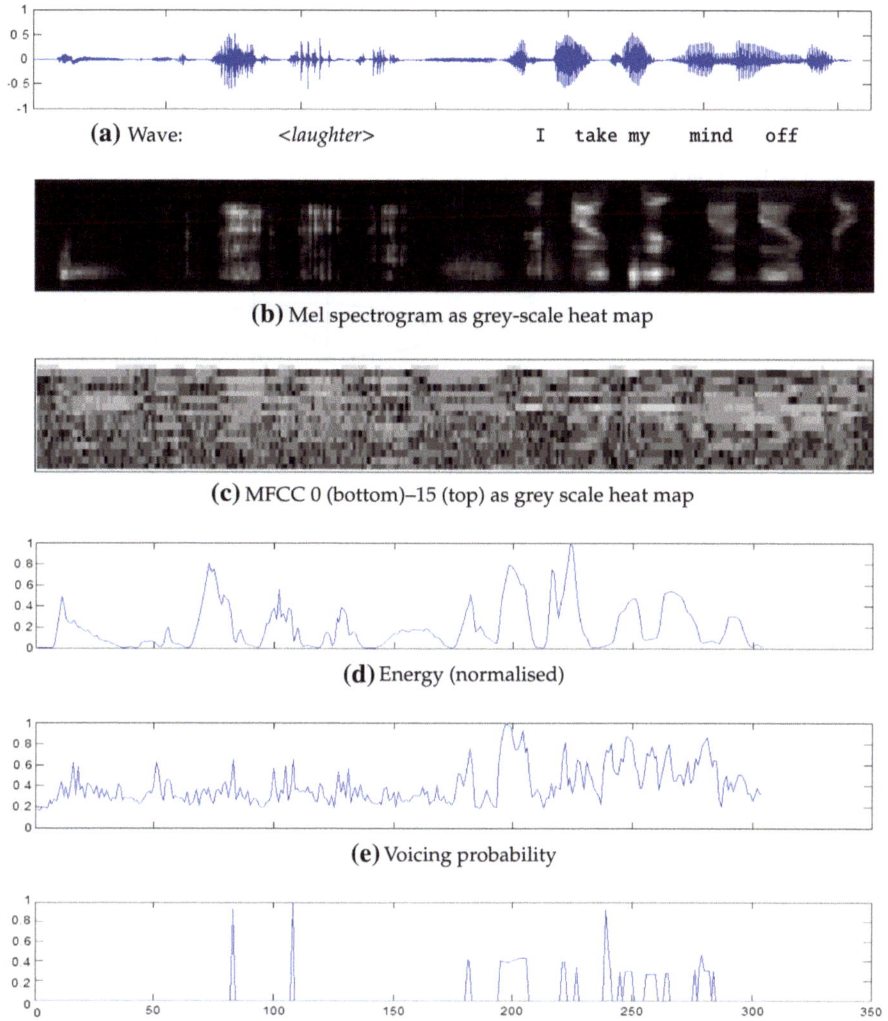

(a) Wave: *<laughter>* `I take my mind off`

(b) Mel spectrogram as grey-scale heat map

(c) MFCC 0 (bottom)–15 (top) as grey scale heat map

(d) Energy (normalised)

(e) Voicing probability

(f) Pitch (normalised)

Fig. 6.2 Exemplary speech wave form over time in ms: laughter (0.0–150 ms) followed by "I take my mind off" taken from the Sensitive Artificial Listener (SAL) database (male speaker) and selected LLDs [43]

Then, the audio signal is decomposed into frequency bands. These represent the semi-tones which are defined for equal temperament as

$$f_i = f_0 \cdot 2^{i/12} \qquad f_0 = f(\mathrm{A0}) = 27.5\,\mathrm{Hz}, \tag{6.62}$$

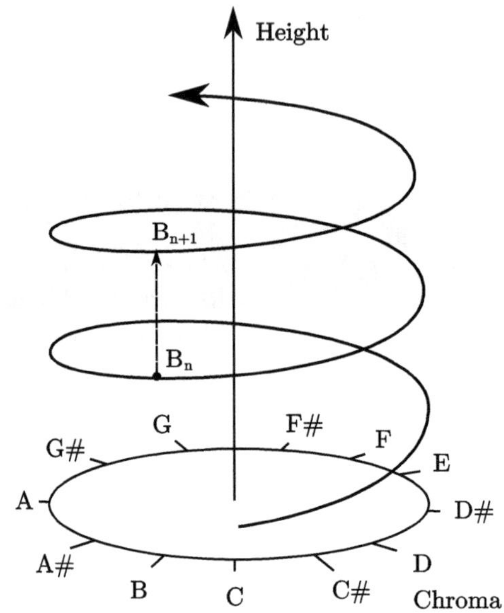

Fig. 6.3 The pitch helix as presented in [45]. The height axis is associated with a note's frequency and the rotation correspondsto the pitch class of a note. Here, B_n is one octave below B_{n+1} [46]

typically with $15 \leq i \leq 110$ (corresponding to the notes C2–B9) and therefore covering 96 semitones (8 octaves). In order to overcome 'tape speed variation' or intentionally different tunings, pitch correction can be applied as was suggested in [47]: a long term frequency analysis computes the prominent frequency f_p and determines a factor c

$$c = \frac{f_p}{f_r} \tag{6.63}$$

with

$$f_r = \arg\min_{f_i} \left\| \frac{f_p}{f_i} - 1 \right\| . \tag{6.64}$$

Next, all semitones f_i are multiplied with this correction factor c for pitch adjustment. For mapping of frequencies to the semitones, band-filters with Gaussians $g_i(x)$ centred at f_i given by

$$g_i(x) = \frac{1}{\sigma\sqrt{2\pi}} \cdot e^{-\frac{\left(\frac{x-f_i}{f_i-f_{i-1}}\right)^2}{2\sigma^2}} \qquad \sigma = 0.125 \tag{6.65}$$

can be used. The resulting sub-bands s_i are normalised by dividing each one belonging to the same octave O by the sum of these sub-bands according to

$$\hat{s}_i = \frac{s_{i,O}}{\sum s_{i,O}} \qquad s_{i,O} = s_i \in O. \tag{6.66}$$

These raw energies per semi-tone interval can then be re-grouped by summing up energies of octaves of a semi-tone to reduce the feature vector size to, e.g., 36, 24, or 12 bands. This will be now exemplified for the most frequently encountered choice of 12 dimensional 'CHROMA' features.

6.2.2.2 CHROMA

Rather than storing and analysing each individual musical semi-tone's energy for analysis of the chordal structure (a musical chord is defined as two or more simultaneously played notes) or the key, the feature vector $\underline{x}$ can be reduced to a limited number of octaves up to a single one, i.e., 12 features, as for CHROMA features [48]. This may be performed by addition of all bands belonging to the same semi-tone in different octaves. Finally the vector $\underline{x}$ is normalised by the number of merged bands. A 12 dimensional CHROMA vector $\underline{\mathrm{x}}$ thus provides the cumulative spectral energies per semitone $A, A\#, \ldots, G\#$ over all octaves:

$$\underline{x} = \left[A,\ \ A\#,\ B,\ \ C,\ \ C\#,\ D,\ \ D\#,\ E,\ \ F,\ \ F\#,\ G,\ G\#\right]^T \tag{6.67}$$

by adding up—as a final step to the previous PCP calculation—all sub-bands corresponding to the same relative pitch class.

In some implementations the length of the CHROMA vector is normalised to 1 in order to have energy independent CHROMA information. This is, however, problematic for low energy signals, as the noise (e.g., quantisation noise) present in this signal will dominate the CHROMA features instead of the desired harmonic information. To avoid this problem, the CHROMA values can be forced to 0, if the energy of the signal falls below a chosen threshold.

6.2.2.3 CENS

CHROMA-features provide only short-time information for an individual frame of analysis. CENS (CHROMA Energy-distribution Normalised Statistics) features are suggested in [49, 50] to provide a perspective beyond individual frames. The underlying principle resembles averaging CHROMA features over time. Yet, differing from a sheer prolongation of window-size, quantisation and temporal weighting of harmonic information are better modelled. As the local chroma features may be too sensitive concerning articulation effects and local tempo deviations, to each component of $\underline{x} = (x_1, \ldots, x_{12})$ a quantisation function Q defined as

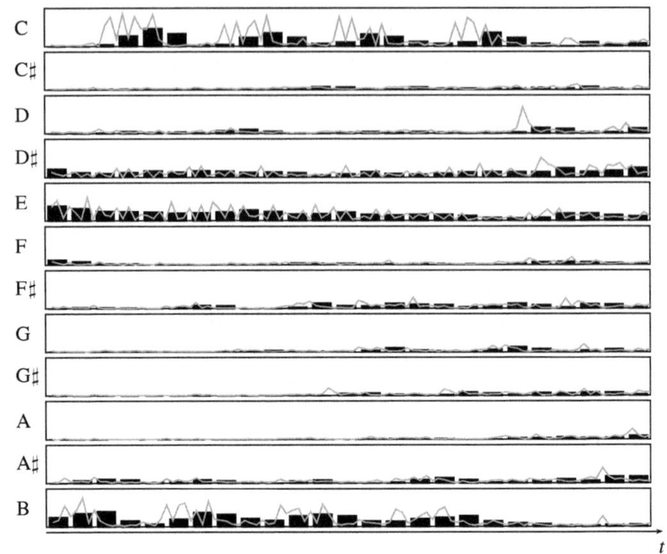

Fig. 6.4 Harmonic representation of the first 20 s of "Abba—Mamma Mia". The light *curves* illustrate the local CHROMA energy distribution, and the dark bars the CENS features [51]

$$Q(a) := \begin{cases} 4 & \text{for } 0.4 \le a \le 1 \\ 3 & \text{for } 0.2 \le a < 0.4 \\ 2 & \text{for } 0.1 \le a < 0.2 \\ 1 & \text{for } 0.05 \le a < 0.1 \\ 0 & \text{for } 0 \le a < 0.05 \end{cases} \tag{6.68}$$

is applied. In the next step, one convolves 11 consecutive quantised CHROMA vectors $Q(\underline{x})$ component-wise with a Hanning window. This results in a smoothing/low-pass filtering of the CHROMA components over time. Given the low-pass characteristics of the resulting components, temporal down-sampling by a factor of four is performed as the final step of CENS computation. A visual comparison of CENS and CHROMA features is provided in Fig. 6.4.

In the following, the creation of higher level features based on CHROMA features (CHROMA-based in the ongoing) will be described. The key 'C major' will be used as an example. These features are based on music theory and human perception. They were suggested in [44].

6.2.2.4 Scale-based

The first type of CHROMA-based features are scale-based descriptors that base on the principle of matching the pattern of a major key in all possible 12 combinations

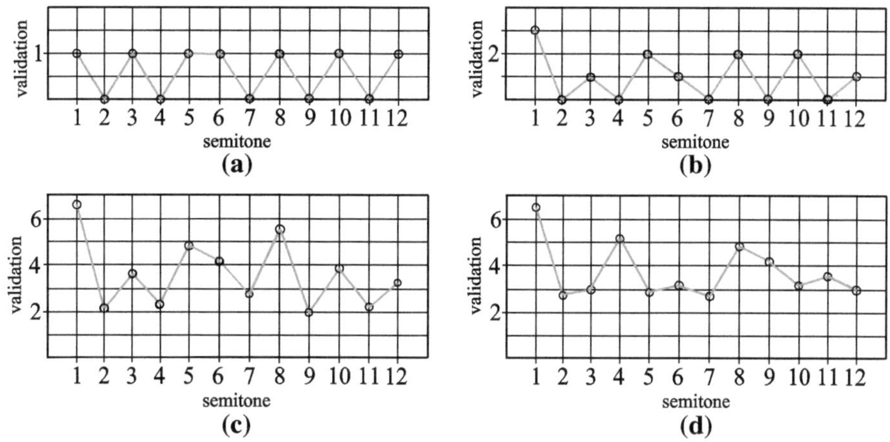

Fig. 6.5 Pattern for the basic feature types. The vertical axis label "validation" was taken over for all types from Krumhansl's Probe Tone Ratings, where human listeners rated how well a heard note fits a previously heard chord-progression and thus reflects the 'weight' of a semitone within a scale [44]. **a** Scale based pattern **b** chord based pattern **c** PTR-major pattern **d** PTR-minor pattern

to the CHROMA vector by correlation. In doing so, one obtains the pattern shown in Fig. 6.5a by labelling the notes appearing in the scale beginning with the root as semitone 1. In this 'hard' template the seven notes associated to the key are set to 1 and the remaining five notes outside the key are set to 0. Based on this pattern, one can construct an according template for each root note. For the feature type scale s and the root note C one obtains the following template $\underline{t}_s(C)$ (the index of the templates abbreviates the respective feature in the ongoing, and the root note is underlined in the vectors that always start with A as in (6.67)):

$$\underline{t}_s(C) = \left[1,\ 0,\ 1,\ \underline{1},\ 0,\ 1,\ 0,\ 1,\ 1,\ 0,\ 1,\ 0\right]^T. \tag{6.69}$$

To create corresponding templates for other keys, the same pattern is simply shifted (i.e., 'transposed') by multiplication with the shifting matrix $\underline{M}$ (6.70). The shifting matrix rotates the template vector and keeps the pattern but starts on base of the target root.

$$\underline{M} = \begin{bmatrix} 0\,0\,0 & \dots & 0\ 0\ 1 \\ 1\,0\,0 & & 0\ 0\ 0 \\ \vdots & \ddots & \vdots \\ 0\,0\,0 & \dots & 0\ 1\ 0 \end{bmatrix} \tag{6.70}$$

For the creation of the particular templates $\underline{t}_s(k)$ with $k = \{A, A\#, \dots, G\#\}$, the basic template of A major $\underline{t}_s(A)$ is used which is then shifted by the matrix $\underline{M}$ to the respective root:

$$\underline{t}_s(C) = \underline{M}^3\,\underline{t}_s(A). \tag{6.71}$$

To build the scale based feature vector $\underline{s}$ element-wise, the CHROMA vector $\underline{x}$ is next correlated with the particular key pattern $\underline{t}_s(k)$:

$$\underline{s} = \begin{bmatrix} \underline{x}^T \underline{t}_s(A) \\ \underline{x}^T \underline{t}_s(A\#) \\ \vdots \\ \underline{x}^T \underline{t}_s(G\#) \end{bmatrix} = \begin{bmatrix} \underline{x}^T \underline{t}_s(A) \\ \underline{x}^T \underline{M}\, \underline{t}_s(A) \\ \vdots \\ \underline{x}^T \underline{M}^{11} \underline{t}_s(A) \end{bmatrix}. \tag{6.72}$$

This operation is repeated for every root and the derived pattern by multiplying the template with the shifting matrix $\underline{M}$. This provides a 'confidence' for each of the 12 possible keys, which reflects the intensity and at the same time the probability of the respective key.

6.2.2.5 Chord-based

For the next type of CHROMA-based features, only a key's main four chords' notes are considered. These chords are: tonic, sub-dominant, dominant, and the relative minor chord of the tonic. Again, only notes in the respective key are allowed. Now, however, these are weighted differently: according to their number of appearance in the key's main four chords. Obviously, other definitions can be thought of, such as including the relative minor chords of the sub-dominant and dominant. Again, a 12 dimensional 'chord vector' $\underline{c}$ is created by repeated correlation of the CHROMA vector $\underline{x}$ with accordingly shifted chord templates $\underline{t}_c(k)$ for each root k (exemplified is the root C):

$$\underline{t}_c(C) = \left[2,\ 0,\ 1,\ \underline{3},\ 0,\ 1,\ 0,\ 2,\ 1,\ 0,\ 2,\ 0\right]^T. \tag{6.73}$$

In [52] a similar method is suggested. Their approach bases on templates related to the chord by use of the histogram of semi-tones and additional overlay of all triads belonging to a scale. These templates also tend to have values close to zero at semi-tones outside the scale.

6.2.2.6 PTR Major/Minor-Based

Alternatively to these music theory motivated templates, one can consider templates based on human perception, such as the Probe Tone Ratings (PTR) [53]. PTR were collected in listening experiments as follows: A chord-progression within a given key was played, then a note was presented to the participating subjects. These had to rate how well the note fits the progression. The observed validations show good correlation with hard templates consisting exclusively of semi-tones belonging to the scale and thus can be considered as histogram of the intensities of semi-tones within a key. As an advantage, PTR templates allow scale external semi-tones such as 'blue notes'. Fig. 6.5c, d depict the PTR templates for major and minor keys starting from the tonic. In this book 'minor' refers to the natural minor scale as is most prominent in popular music, in contrast to harmonic or melodic minor scales. These minor scales differ in their semitones used as compared to the respective relative major scale.

Given the difference of the major and minor PTR ratings, both are considered in the ongoing. As before, 12 dimensional vectors $\underline{p}_{maj}$ and $\underline{p}_{min}$ are obtained by repeated correlation of the CHROMA vector $\underline{x}$ with accordingly shifted PTR templates $\underline{t}_{Pmaj}(k)$ and $\underline{t}_{Pmin}(k)$ for each root k, leading to (exemplified is again the root C):

$$\underline{t}_{Pmaj}(C) = \left[3.8,\ 2.4,\ 3.3,\ \underline{6.6},\ 2.2,\ 3.5,\ 2.4,\ 4.8,\ 4.2,\ 2.6,\ 5.5,\ 2.0\right]^T \tag{6.74}$$

$$\underline{t}_{Pmin}(C) = \left[3.0,\ 3.5,\ 3.0,\ \underline{6.5},\ 2.7,\ 3.0,\ 5.2,\ 2.8,\ 3.3,\ 2.6,\ 4.8,\ 4.1\right]^T. \tag{6.75}$$

6.2.2.7 Derived Features

The root's neighbouring keys in the circle of fifths are now also considered. Starting for example with C major, the fifth above is G major, and the fifth below F major. The first such new type will be obtained by adding the dominant. Hence, every possible root is added by its corresponding fifth in order to reduce confusion with the dominant. A template is used also this time. However, this template is not applied to the CHROMA vectors directly. Instead, in a second step the previously computed feature vectors $\underline{s}$, $\underline{c}$, $\underline{p}_{maj}$, and $\underline{p}_{min}$ are repeatedly correlated with the accordingly shifted template $\underline{t}_{dom}(k)$ for each root k (exemplified is C):

$$\underline{t}_{dom}(C) = \left[0,\ 0,\ 0,\ \underline{1},\ 0,\ 0,\ 0,\ 0,\ 0,\ 0,\ 1,\ 0\right]^T. \tag{6.76}$$

E.g., the scale dominant feature vector $\underline{s}_{dom}$ is obtained as follows:

$$\underline{s}_{dom} = \begin{bmatrix} \underline{s}^T\, \underline{t}_{dom}(A) \\ \underline{s}^T\, \underline{t}_{dom}(A\#) \\ \vdots \\ \underline{s}^T\, \underline{t}_{dom}(G\#) \end{bmatrix} = \begin{bmatrix} \underline{s}^T\, \underline{t}_{dom}(A) \\ \underline{s}^T\, \underline{M}\, \underline{t}_{dom}(A) \\ \vdots \\ \underline{s}^T\, \underline{M}^{11}\, \underline{t}_{dom}(A) \end{bmatrix}. \tag{6.77}$$

Another option is to furthermore add the fifth below the root to the search mask and thus regard the whole cadence. As for the last feature type, a secondary repeated correlation with the template $\underline{t}_{cad}(k)$ for each root k is executed (root C in our example), where:

$$\underline{t}_{cad}(C) = \left[0,\ 0,\ 0,\ \underline{1},\ 0,\ 0,\ 0,\ 0,\ 1,\ 0,\ 1,\ 0\right]^T. \tag{6.78}$$

The presented set of CHROMA-derived features could be further extended. However, it has been shown that the already described features are sufficient and other features

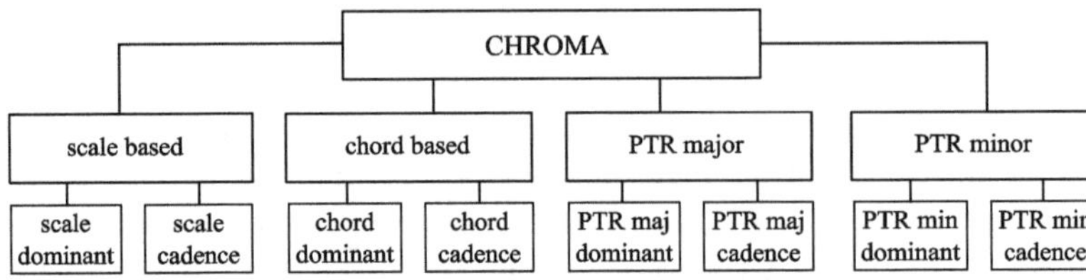

Fig. 6.6 Overview on the creation of music theoretically inspired and perception based features for tonal analysis [44]

decrease the performance in anaylsis systems. E.g., an exclusive combination of the tonic and the sub-dominant, showed the tendency of false decisions towards the dominant in chord and key determination tasks. The enlargement of the search mask to the sub-dominant on the other hand would increase the risk of favouring dominant keys [44].

Overall, 13 feature groups for tonal analysis were shown (cf. Fig. 6.6), each containing 12 features: CHROMA (or alternatively CENS), four basic, and eight further derived feature types.

6.2.3 Sound Descriptors

For the intelligent analysis of general sounds, most of the features as described for speech analysis are often used, including intensity, ZCR, linear prediction-based and cepstral coefficients as well as specialised spectral features. Apart from features designed for the specific task at hand, some examples of statistical spectral features are given in this section. These features are often used for general sound and sound quality analysis. On the other hand they can also be of used in speech or music analysis tasks.

Starting with the centre of gravity m_c, the i-th central moment is next introduced as

$$M_i := \frac{1}{E} \sum_{0}^{\infty} (m - m_c)^i |S(n, m)|^2. \tag{6.79}$$

Examples of statistical spectral features comprise:

- The **spectral standard deviation** which is a measure for how much the frequencies in a spectrum can deviate from the centre of gravity. It is equal to $\sqrt{M_2}$.
- The **spectral skewness** as a measure on how much the shape of the spectrum below the centre of gravity frequency is different from the shape above this frequency. It is calculated as $M_3/(M_2)^{1.5}$.
- The **kurtosis** as a measure of how much the shape of the spectrum around the centre of gravity differs from a Gaussian shape. It is equal to $M_4/\sqrt{M_2} - 3$.

- **Spectral band energies** and **energy densities** such as for the following seven octave based frequency intervals: 0–200 Hz, 200–400 Hz, 400–800 Hz, 800Hz–1.6 kHz, 1.6–3.2 kHz, 3.2–6.4 kHz, and 6.4–12.8 kHz.

Further, a set of LLD is standardised in the MPEG-7 standard for audio analysis.[1] This set is also often used for music analysis, but well suited for general sound analysis. The LLD are audio power, spectrum centroid and spread, fundamental frequency, harmonics, log attack time, harmonic spectral centroid with deviation, spread, and variation, temporal centroid, spectral centroid, and spectrum envelope with flatness, projection, and bias.

6.3 Textual Descriptors

For some tasks, especially recognition of emotion and speaker states and traits, the spoken content is of importance. The acoustic LLDs described in the previous sections only contain information on 'how' something is said and not on 'what' is being said. To obtain the chain of spoken words, automatic speech recognition (ASR) algorithms have to be used in real applications. For assessing the maximum gain in recognition performance that a system can reach when the textual content is considered, in most experiments often a manually transcribed ground truth is used.

Some of such studies have shown that methods of linguistic analysis of spoken (or sung) text can complement the acoustic information and thus enhance the combined recognition performance, e.g., in emotion recognition from speech [54–57] or music mood recognition [58].

This section presents different approaches for linguistic analysis. While they are mostly established for the processing of textual strings such as words or chord sequences in music, any other information that may be represented as string by symbolic entities can be modelled in a similar fashion [59]. In the ongoing—for the sake of simplification—we will speak of 'words' consisting of 'characters' representing the basic string units of analysis.

Often, only a fraction of these words convey relevant information about the target task of interest and many words are similar and related in their meaning. In order to reduce the information in a meaningful way, two methods are usually applied: stopping and stemming.

Stopping is the exclusion of words from the vocabulary for their low relevance in the context of the analysis. It is usually executed by expert rules such as exclusion of function words or a data-driven evaluation. A popular data-driven method is using a minimum word frequency f_{min} for a word in the database to become part of the vocabulary. Rare words are thus discarded. However, frequently appearing function words which may be irrelevant in many search tasks are left over. Therefore, an additional data-based feature selection by suitable criteria such as information gain can be used.

[1] ISO/IEC JTC 1/SC 29/WG 11 N7708.

Stemming on the other hand reduces different morphological forms of a word to its base form, i.e., the 'stem'. Thus, different flexions of the same word are clustered—e.g., *"loved"*, *"loving"*, *"loves"* is stemmed to *"love"*. There exist a number of popular stemming algorithms such as Porter's stemmer [60] that shall serve as an example of the underlying principle here: Each (English) word can be represented in the form $[C](VC)^m[V]$, where C (V) denotes a sequence of one or more consecutive consonants (vowels) and m is called the *measure* of the word.[2] Then, in five steps, replacement rules are applied to the word: First is the removal of plural and participle endings. Then, in steps two to five common word endings are replaced such as ATION $\rightarrow$ ATE or IVENESS $\rightarrow$ IVE. Usually, these rules contain conditions under which they may be applied. For example, the rule "($m > 0$) TIONAL $\rightarrow$ TION" is only applied if and only if the remaining stem has a measure greater than zero. This leaves, for example, the word *"rational"* unmodified while *"occupational"* is being replaced. Should more than one rule match in a step, only the rule with the longest matching suffix is applied.

A very compact approach to stemming is part of speech (POS) tagging. This technique is also known as grammatical tagging or word-category disambiguation. Examples of 'open' word classes are adjectives, adverbs, nouns, verbs without auxiliary verbs, and interjections [61]. In addition, 'closed' word classes contain auxiliary verbs, clitics, coverbs, conjunctions, determiners (articles, quantifiers, demonstrative adjectives, and possessive adjectives), particles, measure words, adpositions (prepositions, postpositions, and circumpositions), preverbs, pronouns, contractions, and cardinal numbers, of which sometimes only auxiliary verbs and particles are tagged [62]. As this task is ambiguous and depends on the context in which a word appears, techniques such as dynamic programming or HMMs are applied for automatic POS tagging. Also *sememes*, i.e., semantic units represented by lexemes, can be clustered into higher semantic concepts. Examples would be generally positive or negative terms [62]. As an advantage, the stemming of words to their base forms allows for out of vocabulary (OOV) words replacement to some extent. Furthermore, words may be clustered to the correct corresponding lexemes even if some small recognition errors from the speech or lyrics recognition appear.

When dealing with processing of spoken or sung language, the linguistic analysis is often based on the correct transcription by humans for higher level analysis. Therefore, it describes the performance under perfect speech or sung lyrics recognition conditions. This allows for direct comparability of results [62] for the higher level semantic analysis: A corpus comes with its transcription, while speech or singing recognition results usually differ. However, in the real world spoken or sung content has to be determined by an ASR engine first. Though recognition of speech or singing can be a rather difficult problem if it comes to spontaneous speech or even singing [63–65], this may lead to smaller differences for further linguistic analysis in special cases, as reported, e.g., in [66, 67] for paralinguistic analysis. The reason is that the perfect word chain is not always being needed as opposed to, e.g., ASR. Few minor mistakes may be caught by stemming and not all words are necessarily needed for

[2] $(VC)^m$ here means an m-fold repetition of the string VC

all further analysis if some should be deleted. As for insertions and substitutions, these are only critical if they change the 'tone' of the content.

For the alternative processing of written text, some text pre-processing will usually be needed. First, delimiters such as punctuation can be used for segmentation. Then, capital letters are often de-capitalised to avoid double entries for same words. Finally, it may be reasonable to allow for some word replacement rules or calculation of edit distance between written words and their counterparts in the vocabulary. This may cover misspelling of words or varieties such as in British English, American English, or Australian English (e.g., [68]).

We will next look at different methods for generating linguistic features.

6.3.1 Bag of Words

The basic idea behind Bag of Words (BoW) is the representation of symbolic information in a numeric feature space. Each feature thereby represents the occurrence of a specific 'word', i.e., symbolic entity, in the string of analysis. BoW, originally developed for document retrieval [69], was successfully applied to the fields of emotion [57] and interest (cf. [70, 71]) recognition from text and speech. BoW became a popular approach for these fields [62, 72]. The recognition is often based on speech turns or larger segments, such as paragraphs or the entire lyrics of a song. Every such sequence $\mathcal{S}$ can be described by the set of its contained word entities w_i, i.e., $\mathcal{S} = \{w_1, \ldots, w_S\}$, where $S = |\mathcal{S}|$ is the sequence length. The BoW method considers these words w_i as units of interest. For a given training set $\mathcal{L}$, all different words build the word inventory—the 'vocabulary' $\mathcal{V} = \{w_1, \ldots, w_V\}$, with $V = |\mathcal{V}|$ being the size of this vocabulary. Particularly in spoken or sung language analysis, also non-linguistic vocalisations like sighs and yawns [73], laughs [74, 75], cries [76], and coughs [77] can be integrated into such a vocabulary [62, 70] in speech [78] or singing decoding.

For each word w_i with $i \in \{1, \ldots, V\}$ in the vocabulary a corresponding feature x_i is created. This may easily lead to a high dimensional feature vector space. Each sequence $\mathcal{S}_j$ can then be mapped to a vector $\underline{x}_j$ in this feature space. Ways to determine the value of each feature x_i first include counting the number of occurrences of a word w_i in the sentence $\mathcal{S}_j$, resulting in the word frequency $f_{i,j}$. As a simplification, the binary general occurrence (or non-occurrence) can be used. The 'term frequency' can also be transformed in other ways (cf.[69]), for example by application of the logarithm—the term frequency transformation (TF):

$$\mathrm{TF}_{i,j} = log\left(c + f_{i,j}\right), \tag{6.80}$$

where the offset parameter c prevents definition problems in case of $f_{i,j} = 0$. It is often set to $c = 1$. Another measure is the inverse document frequency transformation (IDF). For $|\mathcal{L}|$ as the number of sequences in the training set $\mathcal{L}$, and L_i as the number of sentences where the word w_i appears, the IDF transformation is given by:

$$\mathrm{IDF}_{i,j} = f_{i,j} \cdot log \left[\frac{|\mathcal{L}|}{L_i} \right]. \tag{6.81}$$

The motivation for IDF is that words used in almost every sentence are often less informative. Combining the TF and the IDF transformations results in the TFIDF approach:

$$\mathrm{TFIDF}_{i,j} = log \left(1 + f_{i,j}\right) \cdot log \left[\frac{|\mathcal{L}|}{L_i} \right]. \tag{6.82}$$

After setting the components x_i to TF, IDF, TFIDF, or other term frequency representations, the final feature vector $\underline{x}_j$ for a sentence $\mathcal{S}_j$ can additionally be normalised, for example to the same Euclidean length:

$$\underline{x}_j^{norm} = \frac{\frac{1}{|\mathcal{L}|} \sum_{k=1}^{L} |\underline{x}_k|}{|\underline{x}_j|} \cdot \underline{x}_j. \tag{6.83}$$

This length is usually not chosen to be one in order to avoid very small numbers and potential arithmetic underflow. A good option is the average length of the $|\mathcal{L}|$ feature vectors.

A disadvantage of the BoW method is the modelling of isolated words without their 'left' and 'right' neighbouring context in a string. Thus, BoW ignores word positions or word dependencies. N-grams partly overcome this. In the next section, a simple extension combines these BoW and N-grams.

6.3.2 Bag of N-grams

The Bag of N-grams (BoNG) approach also represents text in a numeric feature space. The main difference when compared to BoW is the observation of a series of N consecutive words as semantic units of interest—i.e., 'N-grams' of words. The approach in general allows to combine N-grams of different number N of consecutive words similarly to 'backing-off'. By that, if a longer sequence of words is not observed, several shorter ones may replace the longer one. This leads to the parameters of the minimum N-gram length g_{min} and the maximum N-gram length g_{max}. For each N-gram, numeric features are computed as in BoW (cf. Sect. 6.3.1). Stopping and stemming need to be applied for single words, not for N-grams. Then, frequencies of N-grams are counted, and the features are optionally transformed and normalised as for BoW. Because of the combinatorial explosion of created N-grams—the number of combinations equals V^N—with increasing vocabulary size V or N-gram length N, the feature space dimension is in most cases considerably higher than the number of available training instances. This makes stopping and stemming particularly important to provide sufficient observations per N-gram. At the same time, larger amounts of irrelevant data have to be discarded by the feature selection, and it is more likely that a heuristic feature selection gets stuck in a local optimum given a larger feature space. Compared to BoW, the requirements on the automatic speech or singing recognition system are higher, as it has to recognise more consecutive words accurately.

6.3.3 Bag of Character N-grams

N-grams can also be created on the character level by observing N-grams of characters instead of words. This leads to Bag of Character N-grams (BoCNG). Like BoW and BoNG, these base on mapping from text to a numeric feature space. Successes of BoCNG was reported in the field of (spoken) document retrieval [79] and affect recognition [80]. As for BoNG, observation of N-grams with different lengths is possible in combination, determined by a minimum string length of c_{min} characters and a maximum string length of c_{max} characters. Word boundaries can optionally be ignored. For each character N-gram, a mapping to a numeric feature is realised as for words in BoW. Because observation of N-grams at character level naturally results in considerably more possible features than for BoW, more 'aggressive' stopping can be used to discard rare strings.

BoCNG has some interesting characteristics: Stemming on word level is implicitly modelled by using N-grams of characters: one or even more words can be mapped to a base form if they contain similar character substrings. The BoCNG approach—in contrast to BoW—has a finer resolution by observing the character level. Given successful feature selection, only strings of relevant lengths are kept in the feature space. Further, BoCNG can handle unseen compound words if these consist of substrings contained in the feature space. This may be relevant for 'open-vocabulary' languages such as German, which allow the formation of long compound words. Instead of characters in the sense of graphemes, phonemes from the ASR engine can be used, which may lead to an improvement [79].

In fact, other variants of features can be thought of and are used, such as N-grams of syllables. Compared to character N-grams the vocabulary size, and thus the number of combinations for higher N, are significantly reduced.

6.3.4 On-Line Knowledge

Apart from the data-driven approaches for linguistic analysis introduced so far, open-domain methods can be applied which base on knowledge sources (e.g., [81, 82]). On-line knowledge sources are publicly available on the Internet. In natural language processing such databases provide linguistic knowledge, such as information on words, concepts, or phrases, as well as on connections among them. Connections among such entities—again referred to as words or terms in the ongoing independently of their type—include common-sense knowledge, or lexical relations. Various representational schemes provide the information in a suitable and efficient way. In semantic networks, words or concepts are represented as nodes in a graph. Relations are represented by named links [83]. Another form of on-line knowledge sources in the linguistic domain are annotated dictionaries. There, properties of a term are stored as tags. However, dictionaries usually do not contain relations between terms. Some well-known examples of such linguistic open-domain information sources are now introduced and an approach for using these sources for content and sentiment analysis based on linguistic cues is described.

6.3.4.1 ConceptNet

ConceptNet is a semantic network of concepts, such as *"actor"* or *"to watch a movie"*. It is freely available for download[3] and provides commonsense knowledge in a machine-readable format. Knowledge is added by crowd-sourcing of non-specialised humans. The interface for edition by users[4] is capable to a certain extent to avoid false claims and other mistakes [84]. ConceptNet's storage format does not contain syntactic category information. Thus, it has no support for word sense disambiguation. This can, however, be overcome by formulating sufficiently specific concepts, since a concept can consist of an arbitrary amount of words. Concepts are stored in a normalised format. This format aims at ignoring minor syntactic variations that do not affect the meaning of the concept. A concept is normalised by [84]: removal of punctuation and stop words, running each word through Porter's stemmer, alphabetise the stems, such that the order of words does not matter. Figure 6.7 shows the histogram of concept size in ConceptNet. As can be seen, multi-word concepts form the largest part of the database.

Twenty one relations that encode the meaning of the connection between concepts interlink these. Relations names aim at intuitiveness, such as in *IsA* or *PartOf*. The unit of meaning representation is the *predicate*. Figure 6.8 shows an exemplary storage of predicates in ConceptNet.

Each predicate consists of two concepts and a relation, e.g., *"actor" PartOf "movie"* (*"An actor is part of a movie"*). Further, a concept can be part of many relations. In the example in Fig. 6.8, *"movie"* is also connected to *"fun"* by a *HasProperty* relation. Relations are always unidirectional, as in the majority of cases predicates are not invariant to order (cf. e.g., *"A movie is part of an actor"* for a non-sense inversion of order). Predicates may be negated, such as in *"A car cannot travel at the speed of light"*. Furthermore, each predicate has a confidence score on its reliability initialised at one. It can then be increased/decreased by users. Confidence values equal to or below zero indicate unreliable ones [84]. The current version ConceptNet 3 contains 250 556 concepts, and 390 885 predicates for the English language.

---

[3] http://conceptnet.media.mit.edu/

[4] http://commons.media.mit.edu/en/

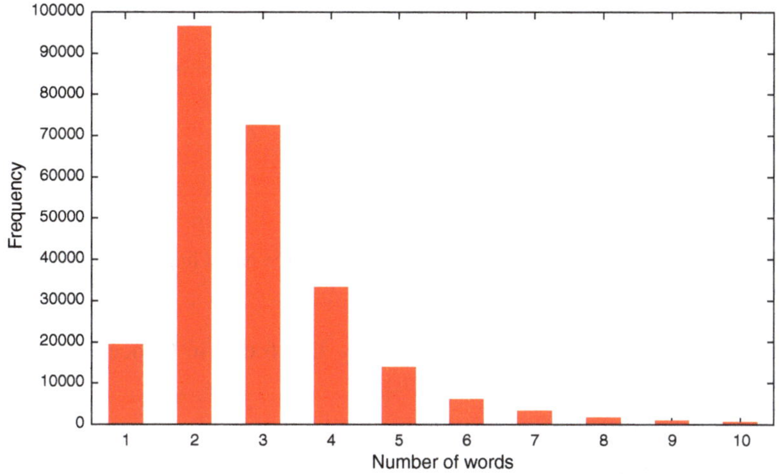

Fig. 6.7 Concept size in words and concept occurrence frequency in ConceptNet 3 [82]

Fig. 6.8 Exemplification of concepts and relations in ConceptNet 3

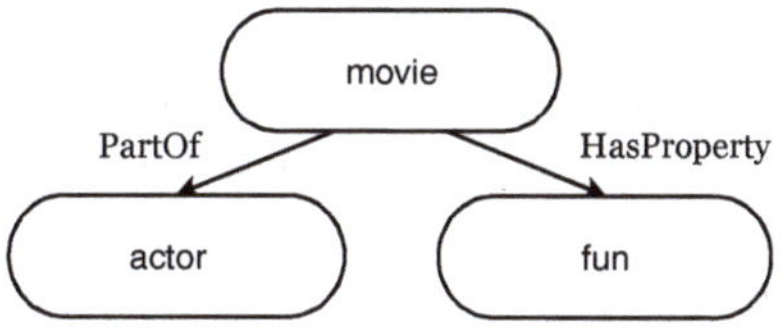

6.3.4.2 General Inquirer

General Inquirer [85] is a lexical database. Each entry consists of the term and a number of tags to characterise a specific property of the term. For example, there are 1 915 terms in the *Positiv* category—e.g., *"adore"*, *"master"*, or *"intriguing"*—, and 2 291 in the *Negativ* counterpart—e.g., *"accident"*, *"lack"*, or *"boring"*. There is partial support for POS information. General Inquirer also contains definitions and occurrence frequencies for rudimentary word sense disambiguation.

6.3.4.3 WordNet

WordNet is a database that organises lexical concepts in sets of synonymous words. These are called *synsets*. Its design is inspired by current psycholinguistic and computational theories of human lexical memory [86]. Entries are strictly separated by syntactic category membership. These categories include nouns, verbs, adjectives, and adverbs. Unlike ConceptNet, synsets are not linked by relations expressing commonsense knowledge. Synsets are rather connected by lexical or semantic relatedness, such as *hyponymy* (a word is a specific instance of a more general word), *meronymy* (a word is a constituent part of another word), or *antonymy* (a word is the opposite

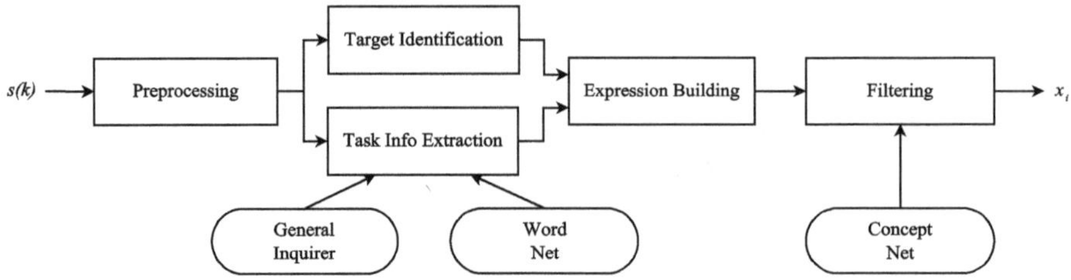

Fig. 6.9 Flowchart for open domain on-line knowledge source-based linguistic analysis.

of another word). These relations are partially also found in ConceptNet, e.g., the complement of meronymy is *PartOf*.

6.3.4.4 Methodology

Based on the on-line knowledge sources as described, this section now introduces an open domain approach towards linguistic analysis. Figure 6.9 visualises the principle of the algorithm and the incorporation of the on-line knowledge sources at two steps. The flow is as follows: First is preprocessing of the input sequence. Then, two parallel steps extract words that convey information on a task of interest, as well as theses task's targets—the words. This information is next combined into expressions. The expressions are filtered aiming at discarding irrelevant ones. Finally, a score value is obtained from the remaining expressions that can be used as linguistic feature for classification or regression.

First, the text is split into sequences $\mathcal{S}$ of words or similar entities. The sequences $\mathcal{S}$ are then analysed by a syntactic parser for POS tagging. The POS classes include adjective (JJ),[5] adverb (RB), determiner (DT), verb (VB), and noun (NN), and are attached to the words by "/" in examples in the ongoing. If it is not necessary to have comprehensive knowledge of the syntax, a chunker suffices for the chunking of longer sequences. The chunks equal phrases, such as a noun phrase (NP), verb phrase (VP), or prepositional phrase (PP). An additional benefit is the flat structure produced by a chunker, which is better suited for the processing steps that follow. As a unit of representation, *ternary expressions (T-expressions)* are extracted on a per-sentence basis. T-expressions were introduced for automatic question answering in [87] and adapted to product review classification [88]. Here, a T-expression is formatted as: *<target, verb, source>*. The 'target' thereby refers to a feature term of the subject of the sequence, e.g., a movie in the case of movie critic valence estimation. The verb is picked from the same phrase as the target. Should the verb not provide information of interest for the target, another according information source—mostly an adverb—is selected instead. By this logic, the T-expression of the sequence *"a/DT carefully/RB designed/VB plot/NN"* would be *<plot, designed, carefully>*. If no verbs exist in

[5] openNLP notation is followed for POS classes.

a sequence, T-expressions cannot be built, and a fall back strategy is applied: This second form is the *binary expression (B-expression)* [88] and is a co-occurrence of adjective plus target in the sequence. As an example, the B-expression for *"an/DT excellent/JJ setting/NN"* would be <*setting, excellent*>. The candidates for targets are identified from NPs. This stems from the observation of feature terms being nouns, as in [88] for sentiment analysis. NPs in the following form are considered for target identification: *NN*; *NN, NN*; *JJ, NN*; *NN, NN, NN*; *JJ, NN, NN*; *JJ, JJ, NN*. Words of other POS classes (DT, RB) not contributing to the target identification and punctuation are removed. The personal pronoun *"it"* is considered as reference to the subject of the sequence and accordingly used as target.

Next, target sources need to be identified for each target, i.e., words conveying the actual information of interest such as affect, gender, or personality, etc.To ensure that a given target source is being directed to the target in question, the search space needs to be restricted. This can be accomplished by finding border indications that appear between clauses or phrases within a sequence. These border indications are subordinating conjunctions, prepositional phrases, coordinating conjunctions, and punctuation such as commas or colons. The sequence is thus broken down into units of statements, and a target source is only associated to a given target if and only if both occur in the same section without a border indication separating them.

In the ongoing, an arbitrary target is exemplified by the concept of valence, such as in the review of movie critics shown later on. However, this target can be easily exchanged by other semantic concepts such as personality of a speaker or mood in music analysis, etc. All verbs and adjectives are selected from the target section, and General Inquirer is used to determine its value v. A word w_i is assigned a value $v(w_i) = 1$ if it has the General Inquirer tag *Positiv*, and a value $v(w_i) = -1$ if it is tagged *Negativ*. Should a word not exist in General Inquirer, WordNet synsets are help to lookup its synonyms, until a match is found. Words for which a valence was found then are the target words. If a target word is an adjective, a B-expression is built from it and stored in the result set. If it is a verb, its *siblings*—the direct neighbours—are first scanned for target adverbs. Given a match, a T-expression of the form <*target, verb, adverb*> is generated. If no match was found, the adverb part is left out. Non-target verbs are processed in the same fashion, and a T-expression is built if a target adverb is located within its siblings. Thus, e.g., *"a/DT carefully/RB designed/VB plot/NN"*, with *"designed"* not being a target verb, also results in an expression. According to the POS class of the target word, T-expressions and B-expressions are generated. It seems intuitive to use the distance between the target word and the target of an expression as measure for the strength of their relation. A maximum distance can be enforced to decide upon overall relation [89, 90], but [91] showed that this can degrade the performance. As an alternative, following [92], feature-opinion pairs similar to B-expressions are extracted. Then, the output value per pair is computed by the multiplicative inverse of the distance between the two words. Then, only the expression with the shortest distance between the target word and its target is kept for further processing. By this, one assumes a target word mostly being directed at the 'closer' target. This choice also reduces the probability of associating a target word with an unrelated target. Note that multiple expressions can exist per target,

to catch all target words in sentences containing more than one. Two target adverbs are, e.g., contained in *"a/DT carefully/RB designed/VB/, superbly/RB executed/VB plot/NN"*: *"carefully"* and *"superbly"*. Following the principle above, this result in two corresponding T-expressions:
<plot, designed, carefully> and *<plot, executed, superbly>*. The word distance further boosts or lowers the score of an expression employing a decay function. The weighted score s of an expression that contains the target t_i and the word source w_i is thereby calculated as:

$$s(w_i, t_i) = c \cdot v(w_i) \cdot \frac{1}{D(w_i, t_i)^e}, \tag{6.84}$$

where the value of w_i taken from General Inquirer is denoted by $v(w_i)$, and $D(w_i, t_i)$ is the distance of words between w_i and t_i. As in [92], the function is based on the multiplicative inverse of the distance, yet, with an additional constant factor c, and an exponent e needed for fine-tuning. With

$$\frac{1}{D(w_i, t_i)^e} = 1 \tag{6.85}$$

holding for any e if $D(w_i, t_i) = 1$, i.e., w_i and t_i are adjacent, $c > 1$ boosts the score. As opposed to this, c has little effect for $D(w_i, t_i) \gg 1$, i.e., words occur further apart. If choosing $e > 1$, the score decreases more rapidly for greater $D(w_i, t_i)$. This allows to weight the influence of the word distance. In the opposite case, $e < 1$ leads to a slower decrease of the score. A fallback mechanism takes place if no target, and no target words could be found in a sentence. In this case, the score is set to $s = 1$ for the class with a-priori higher count of instances as a last resort. Failure to extract target words can be caused by very short sequences, or colloquial language: Colloquial terms are sparsely contained in general purpose dictionaries.

The final step of the proposed algorithm—filtering and classification—determines if the expressions that were found earlier are actually directed at the the target of interest, such as a movie in movie valence estimation. It is assumed that an expression refers to the target in question, if its target word is a feature of the term that names the target of interest, such as *"movie"*. As building a manually assembled list of features to use limits domain-independence and is labour intensive, ConceptNet is used to identify features. As a drawback, however, ConceptNet does not contain named entities. An even larger scale encyclopedia such as Wikipedia could thus be additionally given that it contains such domain-specific knowledge. Feature terms are selected by the predicates: *<feature> PartOf <subject>*, *<feature> AtLocation <subject>*, and *<subject> HasProperty <feature>*. Expressions for a sequence $\mathcal{S}$ with no target in the feature list are filtered out, except if this leaves none at all. The final output of a sequence—the accumulated score S of the N expressions it contains—is:

$$S = \sum_{i=1}^{N} s(w_i, t_i). \tag{6.86}$$

A binary class can be chosen by the signum of S, or S serves as feature for classification or regression in combination with data-driven analysis.

6.4 Supra Segmental Features

After having discussed and introduced various types of acoustic and symbolic LLD, in this section we will have a look at the principle of supra segmental analysis by feature brute-forcing.

The basis is provided by statistical 'functionals', which are applied to an audio chunk and map each LLD's time series of varying length to a single value per functional. Examples of functionals are the mean, minimum, maximum, or standard deviation or the ones shown in Table 6.2. Such mappings are also referred to as *aggregate features* or *feature summaries*. Further, delta coefficients, moving average, or various filter types are commonly applied to low-level descriptors. Hierarchies of such post-processing steps have proven to lead to more robust features, e.g., in [9], hierarchical functionals, i.e., 'functionals of functionals' are used. This consequently leads to the novel principle of Analytic Feature (AF) generation [93]: A large number of LLD derivations and subsequent functional application in a systematic manner, i.e., applied to each LLD, results in brute-forcing of up to several thousands of audio features.

The principle of feature brute-forcing together with LLD extraction will be illustrated in the next section based on the open-source Speech and Music Interpretation by Large-space Extraction (openSMILE[6]) toolkit, a fast feature extractor and signal processing tool [94].

6.5 Audio Feature Extraction: The openSMILE Toolkit

openSMILE's aim is to unite features typically used fro the different types of audio signals—speech, music, and sound—as were introduced so far. This shall enable research in either domain to benefit from features from the other domains and to facilitate general Intelligent Audio Analysis.

A strong focus is put on fully supporting real-time, incremental processing. openSMILE provides a simple, scriptable console application where modular feature extraction components can be freely configured and connected via configuration files. Most of the individual feature extraction functions are usable as library functions and can be integrated into existing applications. Both incremental on-line processing for live applications and off-line batch processing is supported. Unit tests are provided for developers to ensure exact numeric compatibility with future versions.

[6] Available at: http://opensmile.sourceforge.net/.

Other related feature extraction tools used for speech research include, e.g., the *Hidden Markov Model Toolkit* (*HTK*) [20], the *PRAAT* Software [95], the *Speech Filing System*[7] (*SFS*), the Auditory Toolbox,[8] a Matlab^{TM} toolbox[9] by Raul Fernandez [96], the Tracter framework [97], and the *SNACK*[10] package for the Tcl scripting language. However, not all of these tools are distributed under a permissive open-source license, e.g., *HTK* and *SFS*. The *SNACK* package is without support since 2004.

For Music Information Retrieval many feature extraction programs under a permissive open-source license exist, e.g., the lightweight ANSI C library *libXtract*,[11] the Java based *jAudio* extractor [98], the Music Analysis, Retrieval and Synthesis Software *Marsyas*,[12] the *FEAPI* framework [99], the MIRtoolbox,[13] and the *CLAM* framework [100]. As for sound, there are hardly any dedicated extractors available. In general, very few feature extraction utilities exist that unite features from all audio domains, i.e., speech, music, and sound.

6.5.1 openSMILE's Architecture

This section introduces openSMILE's architecture as seen in Fig. 6.10.[14]

To provide comprehensive and standardised cross-domain feature sets, flexibility and extensibility, and incremental processing support, a number of requirements had to be met: First, incremental processing demands for the ability of sample-wise pushing of audio data from arbitrary input streams such as files or the sound card through the chain of processing (cf. Fig. 6.11). Then, a ring-buffer memory for features is needed and provides temporal context modelling and/or buffering. For an efficient design, re-usability of data is required to avoid duplicate computation by multiple feature extractors such as FFT spectra (cf. Fig. 6.11). Algorithms ideally are fast and 'lightweight' and were implemented in this respect in C and C++ without third-party dependencies for the core functions. A modular basis further enables arbitrary combination of features and invites the research community to add new feature extractor components, given an application programming interface (API) and a run-time plug-in interface. To handle asynchronous feature streams, universal timing information is available for processing of feature frames. Finally, to ensure high distribution and acceptance, platform independence seems mandatory. Apart

[7] http://www.phon.ucl.ac.uk/resource/sfs/

[8] http://cobweb.ecn.purdue.edu/malcolm/interval/1998-010/

[9] http://affect.media.mit.edu/publications.php

[10] http://www.speech.kth.se/snack/

[11] http://libxtract.sourceforge.net/

[12] http://marsyas.sness.net/

[13] https://www.jyu.fi/hum/laitokset/musiikki/en/research/coe/materials/mirtoolbox

[14] A more detailed description can be found in the openSMILE documentation available in the download package at http://sourceforge.net/projects/opensmile/.

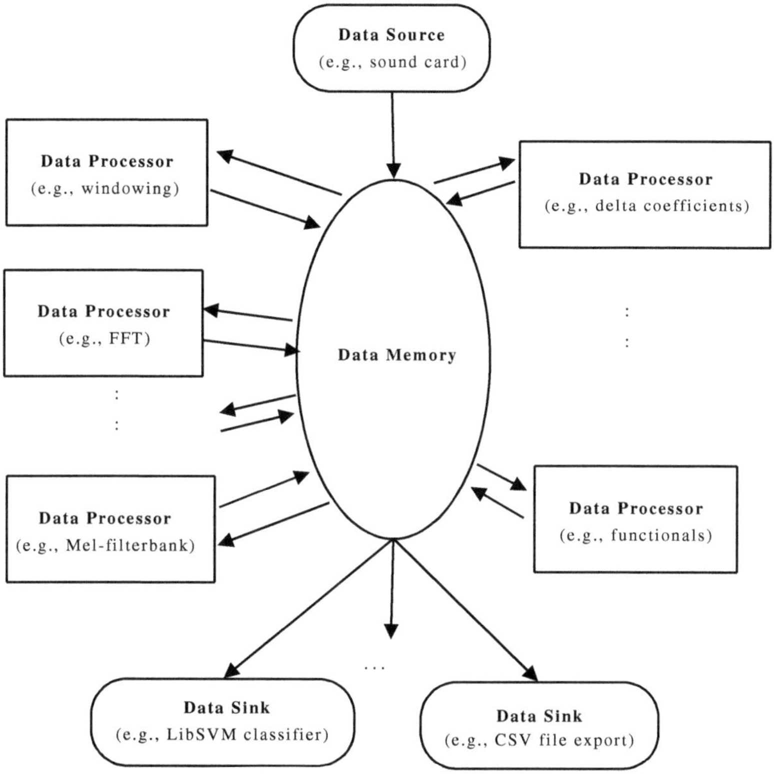

Fig. 6.10 openSMILE's architecture

from ensuring easy compilation on standard platforms, pre-compiled binaries are provided for standard Linux distributions, and newer Windows platforms.

Let us now look at openSMILE's modular architecture tailored towards incremental processing, and the features currently implemented. In Fig. 6.10 of the overall data-flow architecture of openSMILE, the *Data Memory* is the central link between all *Data Sources* (writing from external sources to the data memory), *Data Processors* (reading from the data memory, modifying it, and writing it back), and *Data Sinks* (reading from the data memory and writing to external devices).

The principle of the ring-buffer based incremental processing can be seen in the example in Fig. 6.11 by the three levels wave, frames, and pitch. The 'cWaveSource' component writes samples to the 'wave' level, with the write positions shown by vertical arrows. The 'cFramer' produces non-overlapping frames of size three from the wave samples. It then writes the produced frames to the 'frames' level. Finally, the 'cPitch' component (simplified in the example) calculates a pitch LLD from the frames. It then writes the LLD to the 'pitch' level. Since all boxes in the plot contain values (ie data), the buffers have been filled, and the write pointers have been warped.

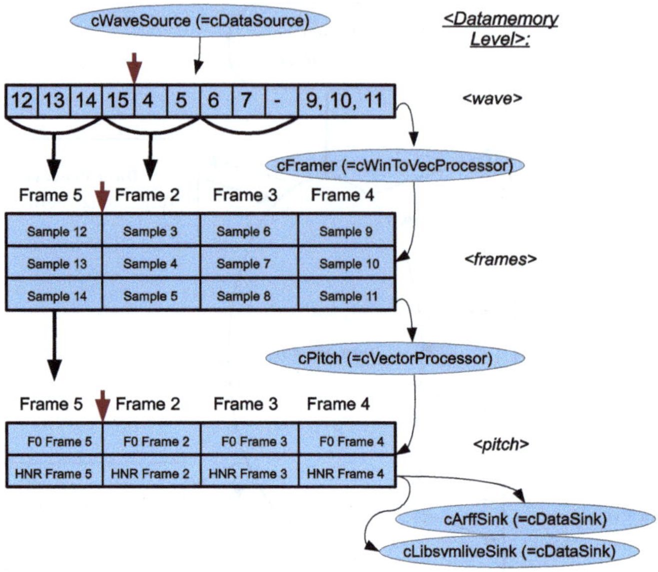

Fig. 6.11 Incremental data-flow in openSMILE's ring-buffer memories on LLD level. The (*light*) *arrows* pointing in between the columns depict the current write pointer [94]

Figure 6.12 next shows this incremental processing of higher order features such as functionals to project the time series to single feature values. Shown are two exemplary functionals, namely 'max' and 'min'. These are calculated over two overlapping frames from the pitch LLD. Then, they are saved to the level 'func'. The buffers-size is matched to the block-size of the reader or writer. In the pitch functionals example the read block-size of the functionals component thus would be two because two pitch frames are read at once. openSMILE supports multi-threading for fast computation. For utmost parallelisation on multi-core computers, each component can be run in a separate thread. Individual components can further be freely instantiated, configured, and connected to the *Data Memory* via a central configuration file. Further, on-line audio recording and live feature extraction is possible.

6.5.2 Available Feature Extractors

openSMILE provides a number of LLDs (cf. Table 6.1) for automatic extraction and the application of several filters, functionals, and transformations to these. Mel-spectra, MFCCs, and PLPs can be computed exactly in full compliance with the popular Hidden Markov Toolkit (HTK) [20], fostering compatibility and comparability. PLP computation can be carried out as in original works [22] or in modification (eg leaving out processing steps).

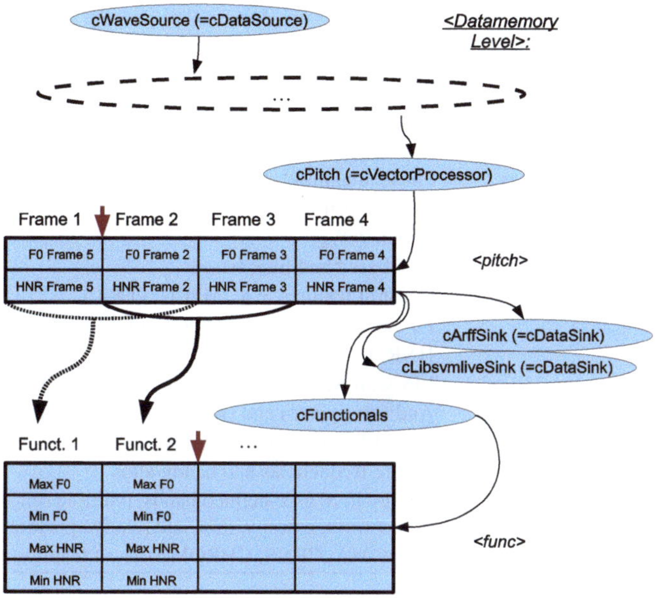

Fig. 6.12 Incremental data-flow in openSMILE's ring-buffer memories on functional level. The (*light*) *arrows* pointing in between the columns depict the current write pointer [94]

The spectral centroid $C(n)$ at time n for N FFT bins is obtained by

$$C(n) = \frac{\sum_{m=1}^{N} m \cdot S(m,n)}{\sum_{m=1}^{N} S(m,n)}, \tag{6.87}$$

where $S(m,n)$ is the spectral magnitude at time n in bin m. Spectral flux $F(n)$ is computed via

$$F(n) = \sqrt{\frac{1}{N}\sum_{m=1}^{N}\left(\frac{S(m,n)}{E(n)} - \frac{S(m,n-1)}{E(n-1)}\right)^2}, \tag{6.88}$$

where $E(n)$ is the energy of the frame at time n.

The p percent spectral roll-off is determined as the frequency or FFT bin below which p percent of the total signal energy are contained. Frequencies for centroid and roll-offs are normalised to 1 kHz.

LLDs can be processed frame by frame with the filters: weighted differential and raised-cosine lowpass as in [101], first order infinite impulse response (IIR) lowpass/highpass, comb-filter bank (cf. Sect. 11.3) with arbitrary number of filters, moving average smoothing filter, and regression (delta) coefficients (x^t) of arbitrary order t. These are computed from any feature contour $x(n)$ again in HTK-style [20]

Table 6.1 openSMILE's LLDs

Group	LLDs
Waveform	ZCR, extremes, DC
Signal energy	RMS and logarithmic
Loudness	Intensity and approximated loudness
FFT spectrum	Phase, magnitude (lin, dB, dBA)
ACF, Cepstrum	ACF, Cepstrum
Mel/Bark spectrum	Bands 0-N_{mel}
Semitone spectrum	FFT based and filter based
Cepstral	Cepstral features, e.g., MFCC, PLP-CC
Pitch	F_0 via ACF and SHS methods, probability of voicing
Voice quality	HNR, jitter, shimmer
LPC	LPCC, reflection coefficients, residual, LSP
Auditory	Auditory spectra and (RASTA-)PLP coefficients
	Model-based auditory loudness, Sharpness
Formants	Centre frequencies and bandwidths
Spectral	Energy in N user-defined bands, multiple roll-off points, centroid, entropy, flux, and relative positions of extrema
Tonal	CHROMA, CENS, CHROMA-based features

with the parameter W:

$$d(n) = \frac{\sum_{i=1}^{W} i \cdot (x(n+i) - x(n-i))}{2\sum_{i=1}^{W} i^2}. \tag{6.89}$$

Additional arithmetic operations include add, multiply, and power, for creation of custom features by combining existing operations.

Supported functionals comprising statistical, polynomial regression coefficients, and transformations are found in Table 6.2. They can be applied to LLDs or functionals in a hierarchical structure with unbounded depth as described in [9]. Their choice follows the CEICES standard of seven sites [62, 72]. This scheme is also employed for feature name assignment. The modular architecture allows to use any implemented processing functionality in arbitrary combination. For example, one may use a Mel-band filter-bank as functionals. This enables brute-forcing of unrestricted feature spaces of several thousands. The idea is not to compute more features than data points, but rather to provide a broad basis of new features for self-adaptation of feature spaces for new Intelligent Audio Analysis tasks where little is known on representative feature bases [93]. For exchange with other popular software modules, supported file formats include Weka's Attribute Relation File Format (ARFF) [102], the LibSVM format, Comma Separated Value (CSV), HTK [20] parameter files, and raw binary files.

Table 6.2 openSMILE's functionals

Group	Functionals
Extremes	Extreme values, positions, ranges
Means	Arithmetic, quadratic, geometric
Moments	Standard deviation, variance, kurtosis, skewness
Percentiles	Percentiles and percentile ranges
Regression	Linear and quadratic approximation coefficients, regression error, and centroid
Peaks	Number of peaks, mean peak distance, mean peak amplitude mean and standard deviation of rising/falling slopes
Segments	Number of segments by delta thresholding, mean segment length
Sample values	Values of the contour at configurable relative positions
Times/durations	Up- and down-level times, rise/fall times, duration
Onsets	Number of onsets, relative position of first/last on-/offset
DCT	DCT coefficients
LPC	Linear prediction (autoregressive) coefficients
Zero-crossings	Zero- and mean-crossing rate

A built-in audio activity detection can be used for audio stream chunking in real-time. For noise robustness, on-line mean and variance normalisation (MVN, cf. Chap. 9), can be used.

6.5.3 Performance

Given openSMILE's focus on real-time and on-line processing even when brute-forcing large feature spaces, algorithmic complexity and run-time benchmarks are of interest.

For the first, the extraction of LLD is always of linear asymptotic complexity ($O(n)$) when n is the number of frames. Since the number of frames is proportional to the length of the input, the asymptotic complexity wrt. the input length is also linear. This is independent of the complexity of the individual LLD extraction algorithms. E.g., a Fast Fourier Transform (FFT) is of $O(n \cdot log(n))$, however n in this case is the number of samples per frame, which is a constant throughout the processing.

The asymptotic algorithmic complexity of the functionals extraction (wrt. the input length) depends on the types of functionals. Descriptors which can be calculated in a single pass, or a constant number of passes, such as mean (single pass), standard deviation (two pass), higher moments, peaks, segments, etc. take time $O(n)$. Descriptors such as percentiles, however, require the inputs to be sorted by value. This is implemented using the Quick Sort algorithm [103], which takes an expected time of $O(nlog(n))$.

Run-time benchmarks were carried out under Ubuntu (11.10) Linux on an AMD FX-8120 at 3.1 GHz with 16 GB dual-channel DDR3-1866 RAM using only one of the eight available cores (i.e., running all openSMILE components in a single thread). Real-time factors (RTF) were computed by timing the CPU time required for extracting features from 30 min of monaural 16 kHz PCM (uncompressed) audio data similar to the benchmark in [94]. We used the latest SVN revision 822 (Dec/11/2012) for this benchmark. 12 MFCC coefficients with first and second order delta coefficients were extracted with an RTF of 0.008 12 MFCC 0.008. The INTERSPEECH 2011 Speaker State Challenge baseline feature set was extracted with an RTF of 0.037, and the INTERSPEECH 2012 Speaker Trait Challenge baseline feature set with an RTF of 0.041.

To conclude, openSMILE was introduced as an example of a feature extractor tailored to be an efficient, on-line as well as batch scriptable, open-source, cross platform, and extensible tool implemented in C++ with a well structured API. Despite being rather new, it is increasingly turning into a standard toolkit—in particular in the field of computational paralinguistics.[15] Moreover, the openEAR project [104] builds on openSMILE and extends it by integrated classification algorithms and data-trained models for various Intelligent Audio Analysis tasks [104]. Development of openSMILE is still active and even more features and signal processing components such as TEAGER energy, TOBI pitch descriptors, Gabor filterbanks, and modulation spectra are considered for integration.

Figure 6.13 gives a final overview on the principle of feature extraction.

6.6 Reduction and Selection of Features

Having discussed the principle of feature brute-forcing in the last section, it is next important to be able to reduce these to the most relevant ones. Otherwise, the ratio between parameters to be trained for a machine learning algorithm—which usually increases with increasing number of features—may become to large in comparison to the available amount of data.

Feature selection usually first requires a measure for the evaluation of a feature's merit. In terms of the quality of the resulting set of selected features, this is best solved by employing the target classifier or regressor in a 'wrapper' manner and its accuracy as evaluation measure [18, 102]. In order to save computation time as highly repeated training of and testing with a machine learning algorithm can easily become computationally expensive, one can chose an alternative learning algorithm that can be trained and evaluated faster. This comes, however, at the risk of introducing a bias as the feature set is not optimised for the exact learning algorithm that will be used later in a system. An alternative are 'filter' methods for the determination of

[15] openSMILE was awarded third place in the ACM Multimedia 2010 Open-Source Software Competition. It was further used as standard feature extractor for baseline computation and use by participants in six research challenges.

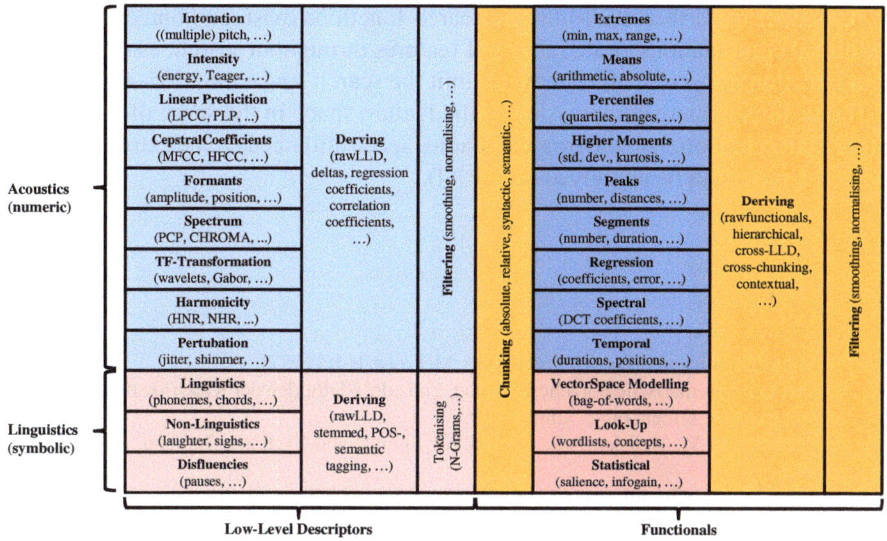

Fig. 6.13 Overview on the principle of audio feature brute forcing in several hierarchical layers.These are generally divided into LLD and the subsequent (optional) Functional level. Shown are further acoustic and linguistic features

the value or contribution of features or feature groups. Examples of filter functions are statistic and information theoretic measures such as CC or IGR.

Given the size of the data set and the feature space, a search algorithm and simple evaluation or ranking functions may additionally become mandatory, as exhaustive search of all possible feature combinations can become computationally prohibitive. A simple, yet highly efficient search method is 'conservative hill climbing', i.e., sequentially deciding for the best feature at the time starting from one and adding the 'next best', each. As this obviously is prone to nesting effects, one usually adds a back stepping option whether 'another previous candidate' would have better suited. This is known as 'floating search', and with the described forward addition as Sequential Forward Floating Search. A backward search starting from the full feature set as well as bi-directional searches are alternatives depending on the ratio of the feature inventory and the target space size. As a result of a typical search, one obtains a mixed view as for the brute-forced features, which is usually hard to interpret: Features in the 'optimal' set, are usually a mixture of all groups. Yet, it is not clear whether these are the best due to the suboptimal nature inherent in any search function and the fact that it de-correlates the space rather than ranks. By that, the value of a feature is unclear, as is whether a picked feature does not have a counter-part of similar characteristics that was not picked, as only one of a sort is needed. An alternative is a systematic 'scan' by feature groups, for examples per LLD type and per functional type.

A number of further measures and search functions exist, and one can also add additional combinations or alterations of features throughout search, usually by random injection or genetic algorithms to limit the search space [93, 105–107].

If one aims at mere compression of the feature space in the sense of a reduction rather than selection, i.e., the original feature space still needs to be extracted, PCA, LDA or similar can be employed (cf. [108]).

References

1. Parsons, T.: Voice and Speech Processing. McGraw-Hill (1987)
2. Ruske, G.: Automatische Spracherkennung, 2nd edn. Methoden der Klassifikation und Merkmalsextraktion. Oldenbourg, Munich (1993)
3. Oppenheim, A.V., Willsky, A.S., Hamid, S.: Signals and Systems, 2nd edn. Prentice Hall, (1996)
4. Wendemuth, A.: Grundlagen der digitalen Signalverarbeitung: Ein Mathematischer Zugang. Springer, Berlin (2005)
5. Wendemuth, A.: Grundlagen der stochastischen Sprachverarbeitung. Oldenbourg, München, Wien (2004)
6. Deller, J., Proakis, J., Hansen, J.: Discrete-Time Processing of Speech Signals. Macmillan Publishing Company, Yew York (1993)
7. O'Shaughnessy, D.: Speech Communication, 2nd edn. Adison-Wesley (1990)
8. Schuller, B., Rigoll, G.: Timing levels in segment-based speech emotion recognition. In: Proceedings of the 9th International Conference on Spoken Language Processing, INTERSPEECH 2006, ICSLP, ISCA, pp. 1818–1821, Pittsburgh, Sep 2006
9. Schuller, B., Wimmer, M., Mösenlechner, L., Kern, C., Arsić, D., Rigoll, G.: Brute-forcing hierarchical functionals for paralinguistics: a waste of feature space? In: Proceedings of the 33rd IEEE International Conference on Acoustics, Speech, and Signal Processing, ICASSP 2008, (IEEE) pp. 4501–4504, Las Vegas, NV, April 2008
10. Sohn, J., Kim, N.: A statistical model-based voice activity detection. IEEE Signal Process. Lett. **6**(1), 1–3 (1999)
11. Ramirez, J., Segura, J., Benitez, M., De La Torre, A., Rubio, A.: Efficient voice activity detection algorithms using long-term speech information. Speech Commun. **42**(3), 271–287 (2004)
12. Ramirez, J., Segura, J., Benitez, C., Garcia, L., Rubio, A.: Statistical voice activity detection using a multiple observation likelihood ratio test. IEEE Signal Process. Lett. **12**(10), 689–692 (2005)
13. R. Gemello, F. Mana, and R. D. Mori. Non-linear esimation of voice activity to improve automatic recognition of noisy speech. In: Proceedings of INTERSPEECH, 2005, ISCA pp. 2617–2620, Lisbon, Sept 2005
14. Mousazadeh, S., Cohen, I.: AR-GARCH in presence of noise: parameter estimation and its application to voice activity detection. IEEE Trans. Audio Speech Lang. Process. **19**(4), 916–926 (2011)
15. Zwicker, E., Fastl, H.: Psychoacoustics—Facts and Models, 2nd edn. Springer, Berlin (1999)
16. Kießling, A.: Extraktion und Klassifikation prosodischer Merkmale in der automatischen Sprachverarbeitung. Berichte aus der Informatik. Shaker, Aachen (1997)
17. Furui, S.: Digital Speech Processing: Synthesis, and Recognition. Signal Processing and Communications, 2nd edn. Marcel Denker Inc, New York (1996)
18. Schuller, B.: Automatische Emotionserkennung aus sprachlicher und manueller Interaktion. Doctoral thesis, Technische Universität München, Munich, Germany, June (2006)
19. Fant, G.: Speech Sounds and Features. MIT Press, Cambridge (1973)

20. Young, S., Evermann, G., Gales, M., Hain, T., Kershaw, D., Liu, X., Moore, G., Odell, J., Ollason, D., Povey, D., Valtchev, V., Woodland, P.: The HTK Book (v3.4). Cambridge University Press, Cambridge, (2006)
21. Kabal, P., Ramachandran, R.P.: The Computation of Line Spectral Frequencies Using Chebyshev Polynomials. IEEE Trans. Acoust. Speech Signal Process. **34**(6), 1419–1426 (December 1986)
22. Hermansky, H.: Perceptual linear predictive (PLP) analysis of speech. J. Acoust. Soc. Am. **87**, 1738–1752 (1990)
23. Hermansky, H., Morgan, N., Bayya, A., Kohn, P.: RASTA-PLP speech analysis technique. In: Proceedings of International Conference on Acoustics, Speech, and Signal Processing, vol. 1, pp. 121–124 (1992)
24. Rigoll, G.: A new algorithm for estimation of formant trajectories directly from the speech signal based on an extended Kalman-filter. In: Proceedings of IEEE International Conference on Acoustics, Speech, and Signal Processing (ICASSP), vol. 11, pp. 1229–1232. Tokyo (1986)
25. Broad, D.J., Clermont, F.: Formant estimation by linear transformation of the LPC cepstrum. J. Acoust. Soc. Am. **86**, 2013–2017 (1989)
26. McCandless, S.: An algorithm for automatic formant extraction using linear prediction spectra. IEEE Trans. Acoust. **22**, 134–141 (1974)
27. Gläser, C., Heckmann, M., Joublin, F., Goerick, C.: Combining auditory preprocessing and bayesian estimation for robust formant tracking. IEEE Trans. Audio Speech Lang. Process. **18**(2), 224–236 (2010)
28. Deng, L., Cui, X., Pruvenok, R., Huang, J., Momen, S., Chen, Y., Alwan A.: A database of vocal tract resonance trajectories for research in speech processing. In: Proceedings IEEE International Conference on Acoustics, Speech, and Signal Processing (ICASSP), p. 1. Toulouse May 2006.
29. Fulop, S.A.: Accuracy of formant measurement for synthesized vowels using the reassigned spectrogram and comparison with linear prediction. J. Acoust. Soc. Am. **127**, 2114–2117 (2010)
30. Miyanaga, Y., Miki, N., Nagai, N.: Adaptive identification of a time-varying ARMA speech model. IEEE Trans. Acoust. **34**, 423–433 (1986)
31. Steiglitz, K.: On the simultaneous estimation of poles and zeros in speech analysis. IEEE Trans. Acoust. **25**, 229–234 (1977)
32. Batliner, A., Steidl, S., Schuller, B., Seppi, D., Vogt, T., Devillers, L., Vidrascu, L., Amir, N., Kessous, L., Aharonson, V.: The impact of f0 extraction errors on the classification of prominence and emotion. In: Proceedings 16th International Congress of Phonetic Sciences, ICPhS 2007, pp. 2201–2204. Saarbrücken, Aug 2007
33. Hess, W.: Pitch Determination of Speech Signals. Springer, Berlin (1983)
34. Heckmann, M., Joublin, F., Nakadai, K.: Pitch extraction in human-robot interaction. In: Proceedings of IEEE/RSJ International Conference on Intelligent Robots and Systems (IROS), IEEE/RSJ, Taipei (2010)
35. Hermes, D.J.: Measurement of pitch by subharmonic summation. J. Acoust. Soc. Am. **83**(1), 257–264 (1988)
36. Ahmadi, S., Spanias, A.S.: Cepstrum-Based Pitch Detection Using a New Statistical V/UV Classification Algorithm. IEEE Trans. Speech Audio Process. **7**(3), 333–338 (May 1999)
37. Botros, N.: Speech-pitch detection using maximum likelihood algorithm. In: Proceedings of the First Joint BMES/EMBS Conference, vol. 2. (1999)
38. Markel, J.: The SIFT algorithm for fundamental frequency estimation. IEEE Trans. Audio Electroacoust. **20**, 367–377 (1972)
39. Boersma, P.: Praat, a system for doing phonetics by computer. Glot Int. **5**, 341–345 (2001)
40. Ross, M., Shaffer, H., Cohen, A., Freudberg, R., Manley, H.: Average magnitude difference function pitch extractor. IEEE Trans. Acoust. Speech Signal Process. **22**, 353–362 (1974)
41. Orlikoff, R.-F., Baken, R.: The effect of the heartbeat on vocal fundamental frequency perturbation. J. Sport Health Res. **32**(3), 576–582 (1989)

42. Haji, T., Horiguchi, S., Baer, T., Gould, W.: Frequency and amplitude perturbation analysis of electroglottograph during sustained phonation. J. Acoust. Soc. Am. **80**(1), 58–62 (1986)
43. Schuller, B.: Voice and speech analysis in search of states and traits. In: Salah, A.A., Gevers, T. (eds.) Computer Analysis of Human Behavior, Advances in Pattern Recognition, chapter 9, pp. 227–253. Springer, Heidelberg (2011)
44. Schuller, B., Gollan, B.: Music theoretic and perception-based features for audio key determination. J. New Music Res. **41**(2), 175–193 (2012)
45. Harte, C.A., Sandler, M.: Automatic chord identification using a quantised chromagram. In: Proceedings of the 118th Convention of the AES, May 2005
46. Schuller, B., Dorfner, J., Rigoll, G.: Determination of non-prototypical valence and arousal in popular music: Features and performances. EURASIP J. Audio Speech Music Process. (Special Issue Scalable Audio Content Anal.) **735854**, 19 (2010)
47. Schuller, B., Hörnler, B., Arsić, D., Rigoll, G.: Audio chord labeling by musiological modeling and beat-synchronization. In: Proceedings of the 10th IEEE International Conference on Multimedia and Expo, ICME 2009, IEEE, pp. 526–529. New York, July 2009
48. Müller, M.: Information Retrieval for Music and Motion. Springer, Berlin (2007)
49. Müller, M., Kurth, F., Clausen, M.: Chroma-based statistical audio features for audio matching. In: Proceedings of the IEEE Workshop on Applications of Signal Processing to Audio and Acoustics, pp. 275–278, Oct 2005
50. Müller, M., Kurth, F.: Towards structural analysis of audio recordings in the presence of mucical variations. EURASIP J. Adv. Signal Process. **89686** (2007)
51. Schuller, B., Dibiasi, F., Eyben, F., Rigoll, G.: Music thumbnailing incorporating harmony- and rhythm structure. In: Detyniecki, M., Leiner, U., Nürnberger, A. (eds.) Adaptive Multimedia Retrieval: 6th International Workshop, AMR 2008, Berlin, Germany, 26–27 June 2008. Revised Selected Papers. Lecture Notes in Computer Science, vol. 5811, pp. 78–88. (LNCS) Springer, Berlin (2010)
52. Gomez, E.: Estimating the tonality of polyphonic audio files: cognitive versus machine learning modelling strategies. In: Proceedings of the 5th International Conference on Music Information Retrieval, Barcelona (2004)
53. Krumhansl, C.L.: Cognitive Foundations of Musical Pitch. Oxford University Press, New York (1990)
54. Polzin, T.S., Waibel, A.: Emotion-sensitive human-computer interfaces. In: Proceedings of the ISCA Workshop on Speech and Emotion, pp. 201–206, Belfast (2000)
55. Devillers, L., Vasilescu, I., Lamel, L.: Emotion detection in task-oriented dialog corpus. In: Proceedings of the ICME 2003, IEEE, Multimedia Human-Machine Interface and Interaction, pp. 549–552, Baltimore (2003)
56. Schuller, B., Rigoll, G., Lang, M.: Speech emotion recognition combining acoustic features and linguistic information in a hybrid support vector machine-belief network architecture. In: Proceedings of the 29th IEEE International Conference on Acoustics, Speech, and Signal Processing, ICASSP 2004, IEEE, vol. I, pp. 577–580. Montreal, May 2004
57. Schuller, B., Müller, R., Lang, M., Rigoll, G.: Speaker independent emotion recognition by early fusion of acoustic and linguistic features within ensembles. In: Proceedings of Interspeech, Eurospeech, ISCA, pp. 805–809. Lisbo, Sept 2005
58. Schuller, B., Hage, C., Schuller, D., Rigoll, G.: "mister d.j., cheer me up!": musical and textual features for automatic mood classification. J. New Music Res. **39**(1), 13–34 (2010)
59. Eyben, F., Wöllmer, M., Valstar, M., Gunes, H., Schuller, B., Pantic, M.: String-based audiovisual fusion of behavioural events for the assessment of dimensional affect. In: Proceedings International Workshop on Emotion Synthesis, Representation, and Analysis in Continuous spacE, EmoSPACE 2011, held in conjunction with the 9th IEEE International Conference on Automatic Face & Gesture Recognition and Workshops, FG 2011, IEEE, IEEE, pp. 322–329. Santa Barbara, CA, March 2011
60. Porter, M.F.: An algorithm for suffix stripping. Program |textbf3(14), 130–137 (1980)
61. Batliner, A., Buckow, J., Huber, R., Warnke, V., Nöth, E., Niemann, H.: Prosodic feature evaluation: brute force or well designed? In: Proceedings of the 14th International Congress of Phonetic Sciences, vol. 3, pp. 2315–2318, San Francisco, (1999)

62. Batliner, A., Steidl, S., Schuller, B., Seppi, D., Laskowski, K., Vogt, T., Vidrascu, L., Amir, N., Kessous, L., Aharonson, V.: Combining efforts for improving automatic classification of emotional user states. In: Proceedings 5th Slovenian and 1st International Language Technologies Conference, ISLTC 2006, Slovenian Language Technologies Society, pp. 240–245. Ljubljana, Slovenia, Oct 2006
63. Athanaselis, T., Bakamidis, S., Dologlou, I., Cowie, R., Douglas-Cowie, E., Cox, C.: ASR for emotional speech: clarifying the issues and enhancing performance. Neural Netw. **18**, 437–444 (2005)
64. Wöllmer, M., Eyben, F., Keshet, J., Graves, A., Schuller, B., Rigoll, G.: Robust discriminative keyword spotting for emotionally colored spontaneous speech using bidirectional lstm networks. In: Proceedings 34th IEEE International Conference on Acoustics, Speech, and Signal Processing, ICASSP 2009, IEEE, IEEE, pp. 3949–3952. Taipei, Taiwan, April 2009
65. Steidl, S., Batliner, A., Seppi, D., Schuller, B.: On the impact of children's emotional speech on acoustic and language models. EURASIP J. Audio Speech Music Process. (Special Issue on Atyp. Speech 2010) **783954**, p. 14 (2010)
66. Seppi, D., Gerosa, M., Schuller, B., Batliner, A., Steidl, S.: Detecting problems in spoken child-computer interaction. In: Proceedings 1st Workshop on Child, Computer and Interaction, WOCCI 2008, ACM ICMI 2008 post-conference workshop, ISCA, p. 4. Chania, Greece, Oct 2008
67. Metze, F., Batliner, A., Eyben, F., Polzehl, T., Schuller, B., Steidl, S.: Emotion recognition using imperfect speech recognition. In: Proceedings of INTERSPEECH 2010, 11th Annual Conference of the International Speech Communication Association, ISCA, pp. 478–481. Makuhari, Sept 2010
68. Schuller, B., Müller, R., Rigoll, G., Lang, M.: Applying bayesian belief networks in approximate string matching for robust keyword-based retrieval. In: Proceedings 5th IEEE International Conference on Multimedia and Expo, ICME 2004, IEEE, vol. 3, pp. 1999–2002. Taipei, Taiwan, June 2004
69. Joachims, T.: Text categorization with support vector machines: learning with many relevant features. In: Nédellec, C., Rouveirol, C. (eds.) Proceedings of 10th European Conference on Machine Learning (ECML), Chemnitz, pp. 137–142. Springer, Heidelberg (1998)
70. Schuller, B., Köhler, N., Müller, R., Rigoll, G.: Recognition of interest in human conversational speech. In: Proceedings of INTERSPEECH 2006, 9th International Conference on Spoken Language Processing, ICSLP, ISCA, pp. 793–796. Pittsburgh, Sept 2006
71. Schuller, B., Müller, R., Eyben, F., Gast, J., Hörnler, B., Wöllmer, M., Rigoll, G., Höthker, A., Konosu, H.: Being bored? recognising natural interest by extensive audiovisual integration for real-life application. Image Vis. Compu. (Special Issue Visual Multimodal Anal. Hum. Spontaneous Behav. **27**(12), 1760–1774 (2009)
72. Batliner, A., Steidl, S., Schuller, B., Seppi, D., Vogt, T., Wagner, J., Devillers, L., Vidrascu, L., Aharonson, V., Kessous, L., Amir, N.: Whodunnit—searching for the most important feature types signalling emotion-related user states in speech. Comput. Speech Lang. (Special Issue Affect. Speech Real-Life Interact.) **25**(1), 4–28 (2011)
73. Russell, J., Bachorowski, J., Fernandez-Dols, J.: Facial and vocal expressions of emotion. Annu. Rev. Psychol. **54**, pp. 329–349 (2003)
74. Campbell, N., Kashioka, H., Ohara, R.: No laughing matter. In: Proceedings of Interspeech, pp. 465–468, Lisbon (2005)
75. Truong, K.P., van Leeuwen, D.A.: Automatic detection of laughter. In: Proceedings of Interspeech, pp. 485–488, Lisbon (2005)
76. Pal, P., Iyer, A., Yantorno, R.: Emotion detection from infant facial expressions and cries. Proc. ICASSP **2**, 809–812 (2006)
77. Matos, S., Birring, S., Pavord, I., Evans, D.: Detection of cough signals in continuous audio recordings using hmm. IEEE Trans. Biomed. Eng. **53**, pp. 1078–1083 (2006)
78. Schuller, B., Eyben, F., Rigoll, G.: Static and dynamic modelling for the recognition of non-verbal vocalisations in conversational speech. In: André, E., Dybkjaer, L., Neumann, H.,

Pieraccini, R., Weber, M. (eds.) Perception in Multimodal Dialogue Systems: 4th IEEE Tutorial and Research Workshop on Perception and Interactive Technologies for Speech-Based Systems, PIT 2008, Kloster Irsee, Germany, 16–18 June 2008. Lecture Notes on Computer Science (LNCS), vol. 5078, pp. 99–110. Springer, Berlin (2008)
79. Iurgel, U.: Automatic media monitoring using stochastic pattern recognition techniques. Ph.D thesis, Technische Universität München, Germany, (2007)
80. Schuller, B.: Recognizing affect from linguistic information in 3d continuous space. IEEE Trans. Affect. Comput. **2**(4), 192–205 (2012)
81. Schuller, B., Schenk, J., Rigoll, G., Knaup, T.: "the godfather" vs. "chaos": Comparing linguistic analysis based on online knowledge sources and bags-of-n-grams for movie review valence estimation. In: Proceedings of the 10th International Conference on Document Analysis and Recognition, ICDAR 2009, IAPR, IEEE, pp. 858–862. Barcelona July 2009
82. Schuller, B., Knaup, T.: Learning and knowledge-based sentiment analysis in movie review key excerpts. In: Esposito, A., Esposito, A.M., Martone, R., Müller, V., Scarpetta, G. (eds.) Toward Autonomous, Adaptive, and Context-Aware Multimodal Interfaces: Theoretical and Practical Issues: Third COST 2102 International Training School, Caserta, Italy, March 15–19, 2010, Revised Selected Papers. Lecture Notes on Computer Science, 1st edn, vol. 6456, pages 448–472. (LNCS) Springer, Heidelberg, (2011)
83. Jurafsky, D., Martin, J.H.: Speech and Language Processing. Prentice-Hall, Upper saddle river (2000)
84. Havasi, C., Speer, R., Alonso, J.: Conceptnet 3: a flexible, multilingual semantic network for common sense knowledge. In: Recent Advances in Natural Language Processing. Borovets, Sept 2007
85. Stone, P., Kirsh, J., Associates, C.C.: The General Inquirer: A Computer Approach to Content Analysis. MIT Press, Cambridge (1966)
86. Fellbaum, C. Wordnet: An Electronic Lexical Database. MIT Press, Cambridge (1998)
87. Katz, B.: From sentence processing to information access on the world wide web. In: Proceedings of the AAAI Spring Symposium on Natural Language Processing for the, World Wide Web, pp. 77–86 (1997)
88. Yi, J., Nasukawa, T., Bunescu, R., Niblack, W.: Sentiment analyzer: extracting sentiments about a given topic using natural language processing techniques. In: Proceedings of the Third IEEE International Conference on Data Mining, pp. 427–434, Nov 2003
89. Morinaga, S., Yamanishi, K., Tateishi, K., Fukushima, T.: Mining product reputations on the web. In: KDD '02: Proceedings of the Eighth ACM SIGKDD International Conference on Knowledge Discovery and Data Mining, ACM, pp. 341–349, New York (2002)
90. Turney, P.D., Littman, M.L.: Measuring praise and criticism: inference of semantic orientation from association. ACM Trans. Inf. Syst. **21**(4), 315–346 (2003)
91. Zhang, M., Ye, X.: A generation model to unify topic relevance and lexicon-based sentiment for opinion retrieval. In: SIGIR '08: Proceedings of the 31st Annual International ACM SIGIR Conference on Research and Development in Information Retrieval, ACM, New York, pp. 411–418 (2008)
92. Ding, X., Liu, B., Yu, P.S.: A holistic lexicon-based approach to opinion mining. In: Proceedings of the WSDM '08 International Conference on Web Search and Web Data Mining, ACM, New York, pp. 231–240 (2008)
93. Pachet, F., Roy, P.: Analytical features: a knowledge-based approach to audio feature generation. EURASIP J. Audio Speech Music Process. **153017**, 23 (2009)
94. Eyben, F., Wöllmer, M., Schuller, B.: Opensmile—the munich versatile and fast open-source audio feature extractor. In: Proceedings of the 9th ACM International Conference on Multimedia, MM 2010, ACM, Florence, pp. 1459–1462, Oct 2010
95. Boersma, P., Weenink, D.: Praat: doing phonetics by computer (v. 4.3.14). http://www.praat.org/ (2005)
96. Fernandez, R.: A Computational Model for the Automatic Recognition of Affect in Speech. Ph.D thesis, MIT Media Arts and Science (2004)

97. Garner, P.N., Dines, J., Hain, T., El Hannani, A., Karafiat, M., Korchagin, D., Lincoln, M., Wan, V., Zhang, L.: Real-time asr from meetings. In Proceedings of INTERSPEECH, ISCA, Brighton 2009
98. McEnnis, D., McKay, C., Fujinaga, I., Depalle, P.: Jaudio: a feature extraction library. In: Proceedings of ISMIR 2005, pp. 600–603 (2005)
99. Lerch, A., Eisenberg, G.: FEAPI: a low level feature extraction plug-in api. In: Proceedings of the 8th International Conference on Digital Audio Effects (DAFx), Madrid 2005
100. Amatriain, X., Arumi, P., Garcia,D.: A framework for efficient and rapid development of cross-platform audio applications. Multimedia Syst. **14**(1), 15–32 (2008)
101. Schuller, B., Eyben, F., Rigoll, G.: Fast and robust meter and tempo recognition for the automatic discrimination of ballroom dance styles. In: Proceedings 32nd IEEE International Conference on Acoustics, Speech, and Signal Processing, ICASSP 2007, IEEE, vol. I, pp. 217–220. Honolulu, April 2007
102. Witten, I.H., Frank, E.: Data Mining: Practical Machine Learning Tools and Techniques, 2nd edn. Morgan Kaufmann, San Francisco (2005)
103. Hoare, C.A.R.: Quicksort. Comput. J. **5**(1), 10–16 (1962)
104. Eyben, F., Wöllmer, M., Schuller, B.: Openear - introducing the munich open-source emotion and affect recognition toolkit. In: Proceedings of 3rd International Conference on Affective Computing and Intelligent Interaction and Workshops, ACII 2009, HUMAINE Association, IEEE, vol. I, pp. 576–581. Amsterdam, Sept 2009
105. Schuller, B., Arsić, D., Wallhoff, F., Lang, M., Rigoll, G.: Bioanalog acoustic emotion recognition by genetic feature generation based on low-level-descriptors. In: Proceedings International Conference on Computer as a Tool, EUROCON 2005, IEEE, vol. 2, pp. 1292–1295. Belgrade, Serbia and Montenegro, Nov 2005
106. Schuller, B., Reiter, S., Rigoll, G.: Evolutionary feature generation in speech emotion recognition. In: Proceedings of 7th IEEE International Conference on Multimedia and Expo, ICME 2006, IEEE, pp. 5–8. Toronto, July 2006
107. Schuller, B., Wallhoff, F., Arsić, D., Rigoll, G.: Musical signal type discrimination based on large open feature sets. In: Proceedings of 7th IEEE International Conference on Multimedia and Expo, ICME 2006, IEEE, pp. 1089–1092. Toronto, July 2006
108. Kroschel, K., Rigoll, G., Schuller, B.: Statistische Informationstechnik, 5th edn. Springer, Berlin (2011)

Chapter 7
Audio Recognition

Learning without thought is labor lost; thought without learning is perilous.

—Confucius

We will now deal with methods towards the actual classification or regression of audio data. A good overview on these is also found in [1].

7.1 Audio Recognition Requirements

A number of requirements speak for the consideration of diverse learning algorithms. In Table 7.1 typical such requirements are summarised.

According to these requirements, a number of learning algorithms were picked as examples in the next sections. These have proven to be reasonable choices throughout many applications as presented later in this book. They can be roughly divided into static and dynamic learners. This categorisation can best be understood by considering the chain of audio processing (cf. Chap. 4): Static learners operate on single feature vector basis (which means that multivariate time series of variable length have to be mapped to fixed size vectors), whereas their dynamic counterparts are able to handle such time series directly.

7.2 Static Learning Algorithms

7.2.1 Decision Trees

As a first learning algorithm, let us consider decision trees (DT). In principle, a DT produces a human-readable set of rules, which makes it very transparent and intuitive to understand. In case of numeric feature information, these are typically

B. W. Schuller, *Intelligent Audio Analysis*, Signals and Communication Technology, DOI: 10.1007/978-3-642-36806-6_7, © Springer-Verlag Berlin Heidelberg 2013

Table 7.1 Requirements for machine learning algorithms

Requirement	Example
Adequate modeling	Static /(async.)dynamic modeling
	Data-/Knowledge-driven
	Handling of missing features
	Handling of uncertainty
	Learning stability
	Model-/Instance-based
	Transparency
Optimal accuracy	Non-linear problem handling
	Discriminative learning
	Auto-weighting of features
	Tolerance wrt. dimension
	Adaptability
	Allowance for diverse spaces
Efficiency	Real-time recognition
	Short learning/adaptation time
Economic factors	Low computational cost
	Low memory requirement
	Low HW realization costs
	Space optimization w/o training
Optimal integration	N-best provision
	Confidence provision
	Class-wise confidences
	Distributable

comparisons with constants to decide to which next comparison to branch, until the class labels are reached as terminals. A DT is thus a specific directed acyclic graph (DAG). As such, it can be defined by a set of nodes V and a set $E \subseteq V \times V$ of edges, where each element $e = (v_1, v_2) \in E$ represents a connection from node v_1 to node v_2. A path of the length P through the tree is a sequence of $v_1, \ldots, v_P,\ v_k \in V$ with $(v_k, v_{k+1}) \in E,\ k = 1, \ldots, P-1$. Starting from an undirected graph, conditions of a tree are that the graph is acyclic and connected, i.e., each node needs to be reachable by a path from each other node. By that, each tree has exactly $|V| - 1$ edges. Further, there is exactly one 'root' w in the sense of a node that possesses no incoming edges, i.e., E contains no element of the form $(v, r), v \in V$. The 'leaves' are the nodes b without an outgoing edge, i.e., for which in E there exists no (b, v) with $v \in V$. All remaining nodes are referred to as 'inner' nodes [1, 2].

In the learning process, features are assigned to the inner nodes: Given a feature space of the dimension N a mapping

$$a : V \to \{1, \ldots, N\}$$

is defined. In this process, the edges are assigned the values on which the branch decisions are based upon. The values of the features as seen in the training are quantised to J_n values per feature n to reach a finite number of edges. Each inner

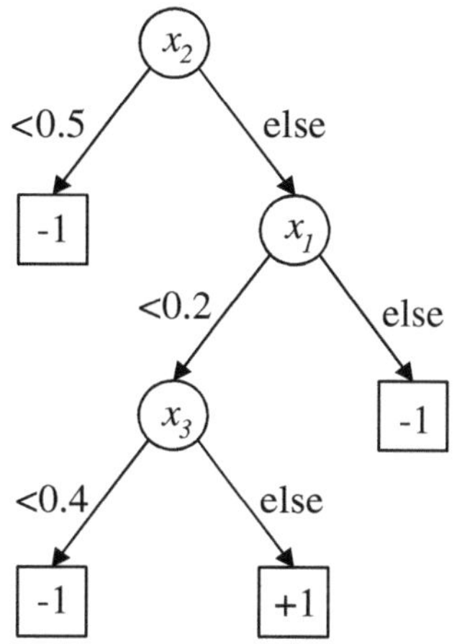

Fig. 7.1 Exemplary DT: A two-class problem is shown with three features. Circles represent the root and inner nodes, rectangles represent the leaves with the class labels

node v then has $J_{a(v)}$ outgoing edges. The leaves are assigned the according class labels.

In the recognition phase of an unknown pattern vector $\underline{x} = (x_1, \ldots, x_N)^T$, one starts at the root w and follows the path through the tree as follows: At each node v along the path choose the edge for which $x_{a(v)}$ is within this edge's interval until a leave is reached. The class to decide for is then this leave's class.

An example of a DT is shown in Fig. 7.1. In this example, quantisation of feature values was chosen as binary. This results in a simple threshold decision at each node.

An optimisation criterion is now to maximise the information gain in view of the correct classification and with the remaining features at each node. The Shannon entropy $H(Y)$ of the distribution of the class probabilities $(Y_1, \ldots, Y_M)$ can be used to this end:

$$H(Y_1, \ldots, Y_M) = -\sum_{i=1}^{M} Y_i \operatorname{ld}(Y_i). \tag{7.1}$$

For a training set $\mathcal{L}$ of pattern vectors $\underline{x}$ with known class attribution y, the needed average information $H(\mathcal{L})$ to assign an instance in $\mathcal{L}$ to a class $i \in \{1, \ldots, M\}$ is determined according to:

$$H(\mathcal{L}) = -\sum_{i=1}^{M} \hat{Y}_i \operatorname{ld}(\hat{Y}_i), \quad \hat{Y}_i = \frac{|\mathcal{L}_i|}{|\mathcal{L}|}, \tag{7.2}$$

where $\mathcal{L}_i$ is the set of elements in $\mathcal{L}$ with class attribution i.

In order to determine the contribution of each individual feature n to the aimed at class assignment, for each n the set $\mathcal{L}$ is divided into the subsets $\mathcal{L}_{n,j}$, $j = 1, \ldots, J_n$ based on the different values of n. The remaining average information $H(\mathcal{L}|n)$ needed after observation of the feature n for the class assignment results as the weighted average of the information $H(\mathcal{L}_{n,j})$, as required to classify an element of the subset $\mathcal{L}_{n,j}$:

$$H(\mathcal{L}|n) = \sum_{j=1}^{J_n} \frac{|\mathcal{L}_{n,j}|}{|\mathcal{L}|} H(\mathcal{L}_{n,j}). \tag{7.3}$$

By this equation the IG can be defined. It describes how the entropy, i.e., the information needed for the assignment, is reduced by addition of the feature n:

$$\mathrm{IG}(\mathcal{L}, n) = H(\mathcal{L}) - H(\mathcal{L}|n). \tag{7.4}$$

However, this definition tends to favour features with a high number of different values J_n: If all elements in $\mathcal{L}$, whose features n have the same value belong to the same class—this is in particular the case, if a feature has a different value for each element in $\mathcal{L}$—, then $H(\mathcal{L}|n)$ equals zero, and by that one obtains a maximal $\mathrm{IG}(\mathcal{L}, n)$. By introduction of the Information Gain Ratio (IGR) this can be compensated:

$$\mathrm{IGR}(\mathcal{L}, n) = \frac{\mathrm{IG}(\mathcal{L}, n)}{H\left(\frac{|\mathcal{L}_{n,1}|}{|\mathcal{L}|}, \ldots, \frac{|\mathcal{L}_{n,J_n}|}{|\mathcal{L}|}\right)}. \tag{7.5}$$

The term in the denominator is called split information and is computed according to Eq. (7.1). This is the information one obtains by the described split of the set $\mathcal{L}$ according to the values of the feature n.

A popular method for the training of a DT based on a training set $\mathcal{L}$ is the iterative dichotomiser 3 (ID3) algorithm [3]. ID3 constructs the DT recursively for the overall feature set by concatenation of sub-trees for each subset of the features. For a given set of features $\mathcal{M} \subseteq \{1, \ldots, N\}$ and training set $\mathcal{L}$, the steps are as follows:

1. If all elements in $\mathcal{L}$ belong to class i return a leaf labelled i.
2. If $\mathcal{M}$ is empty, return a leaf labelled by the most frequent class in $\mathcal{L}$.
3. Else search for the feature n' with the highest IG(R), i.e.,

$$n' = \arg\max_{n \in \mathcal{M}} \mathrm{IG}(\mathcal{L}, n).$$

4. For all $j = 1, \ldots, J_{n'}$ construct a DT by ID3 on the feature set $\mathcal{M} - \{n'\}$ and the training set $\mathcal{L}_{n',j}$. Return a tree with the root labelled by the feature n' whose edges lead to the constructed DTs (cf. Fig. 7.2)

ID3 is a greedy algorithm as in every step a feature is selected by a local optimisation criterion. A global optimum is not guaranteed. Further, ID3 always terminates, as with every recursive call the remaining set of features decreases and the case of an empty feature set is handled separately.

An extension of ID3 are the C4.5 or J48 variants that introduce pruning of sub-trees [2, 4] for increased efficiency. During pruning, a whole sub-tree can be replaced by a leaf, if the error probability is not significantly increased by this substitution. Note

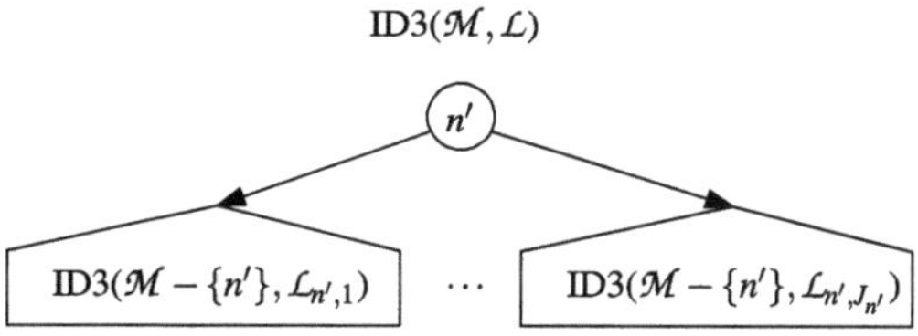

Fig. 7.2 Recursive call of the ID3 algorithm for a feature n' that maximises the IG(R) with respect to the classification of $\mathcal{L}$ within $\mathcal{M}$ [1]

that this reduces the number of features, i.e., an inherent feature selection by IG(R) is given. DTs are able to handle missing features both in training and recognition. Further, if both, the feature set and the training set are randomly sub-sampled for construction of an ensemble (cf. Sect. 7.4) of DTs, one speaks of Random Forests (RF), which are known as competitive classifier [5], e.g., to the further introduced classifiers.

7.2.2 Support Vectors

Support Vector Machines (SVM) and Support Vector Regression (SVR) were introduced in [6]. In principle, they base on statistical learning theory, and their theoretic foundation can be interpreted as analogon to electrostatics: Thereby, a training instance corresponds to a charged conductor at a given place in space, the decision function corresponds to an electrostatic potential function and the learning target function to Coulomb's energy [7].

The concept of SVM and SVR unites several theories of machine learning and optimisation: At first, a linear classifier or regressor—similar to a perceptron with linear activation function —is combined with a non-linear mapping into a higher dimensional decision space in order to be able to solve more complex decision tasks. The linear classifier is thereby built based on a subset of the learning instances—the so called 'support vectors'. By that, the danger of overfitting to the learning instances as a whole is limited. The choice of support vectors is achieved by a quadratic optimisation problem.

7.2.2.1 Support Vector Machines

In general, SVM are by that capable to discriminate between two classes, i.e., solve binary decision problems. We will at first focus on this task—the solving of multiple class problems can then be reached by diverse strategies such as one-versus-one pairwise decisions, one-versus-all decisions, or binary-tree-based grouping of decisions.

SVM are trained based on a set $\mathcal{L}$ of L learning instances, where each of the instances is accordingly labelled with its class. For $l = 1, \ldots, L$ the assignment of a pattern instance $\underline{x}_l$ to its class is denoted by $y_l \in \{-1, +1\}$. By definition the patterns $\underline{x}_l$ with $y_l = +1$ are the 'positive' instances, i.e., $\underline{x}_l \in \underline{X}_1$. If $y_l = -1$, $\underline{x}_l$ is a 'negative' instance, i.e., $\underline{x}_l \in \underline{X}_2$. By this we can denote $\mathcal{L}$ as:

$$\mathcal{L} = \{(\underline{x}_l, y_l) \mid l = 1, \ldots, L\} \text{ where } y_l \in \{+1, -1\}. \tag{7.6}$$

The assignment $y_l \in \{-1, +1\}$ simplifies the mathematical handling. In order to be able to strictly separate the according instances in the following, the normal vector $\underline{w}$ and the scalar bias b define the hyper plane $H(\underline{w}, b)$ given as

$$H(\underline{w}, b) = \{\underline{x} \mid \underline{w}^T \underline{x} + b = 0\}. \tag{7.7}$$

The task is now to find the hyper plane in such a way that the conditions

$$\begin{aligned} y_l = +1 &\Rightarrow \underline{w}^T \underline{x}_l + b \geq +1, \\ y_l = -1 &\Rightarrow \underline{w}^T \underline{x}_l + b \leq -1 \end{aligned} \tag{7.8}$$

are fulfilled. Under the condition that a hyper plane exists by which the separation of the (two) classes is possible without misclassification, a normalisation of the side conditions (7.8) can be realised by appropriate scaling of $\underline{w}$ and b [8]. Next, by application of the signed distance $D(\underline{x})$ of a point $\underline{x}$ to the hyper plane H

$$D(\underline{x}) = \frac{\underline{w}^T \underline{x} + b}{||\underline{w}||} \tag{7.9}$$

the margin of separation $\mu_{\mathcal{L}}$ is defined as the minimum of the magnitude of the distances of all points $\underline{x}_1 \ldots \underline{x}_l$ in $\mathcal{L}$ to H:

$$\mu_{\mathcal{L}}(\underline{w}, b) = \min_{l=1,\ldots,L} |D(\underline{x}_l)|. \tag{7.10}$$

In order to reach maximum discriminativity between the two classes, this margin needs to be maximised. To this end, we seek the hyper plane $H^* = H(\underline{w}^*, b^*)$ with maximal value $\mu^*_{\mathcal{L}}(\underline{w}^*, b^*)$ to separate the training instances set $\mathcal{L}$. The according instances $\underline{x}_l^{sv} \in \mathcal{L}$, which satisfy (7.10), are closest to the hyper plane H^* and are called support vectors of H^* with respect to $\mathcal{L}$. Their distance $D^*(\underline{x}_l^{sv})$ to the hyper plane H^* is—owing to the normalisation of the separation condition:

$$D^*(\underline{x}_l^{sv}) = \frac{\pm 1}{||\underline{w}||}. \tag{7.11}$$

As a consequence, a corridor between the positive and negative instances results of the width $2\,||\underline{w}||^{-1}$. Its border is given by the support vectors which are shown in Fig. 7.3.

Instead of the maximisation of the width of the corridor one can minimise the expression $\frac{1}{2}\underline{w}^T \underline{w}$. The resulting funtion to be minimised is strictly convex and

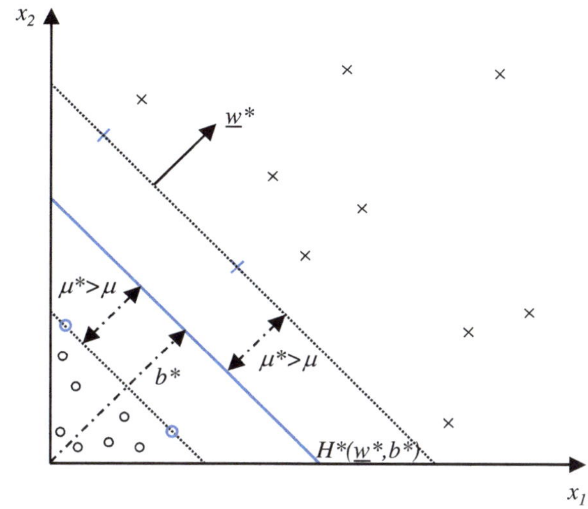

Fig. 7.3 Example of an optimal hyper plane $H^*(\underline{w}, b)$ (*lighter shaded*) in two dimensional space with maximum margin of separation μ^* (*dashed parallel lines*). "x" and "o" indicate exemplary instances of the two classes to be separated

posseses a unique minimum $\underline{w}^*$. From (7.8) result linear side conditions for the optimisation:

$$y_l\,(\underline{w}^T\underline{x}_l + b) - 1 \geq 0 \text{ with } l = 1, \ldots, L. \tag{7.12}$$

To solve this boundary value problem one can use Langrange multipliers. In [6] this is explained in detail.

In the general, non-trivial case, there does not exist—as opposed to the previously made assumption—a hyper plane to separate a training instances set $\mathcal{L}$ flawlessly. In this case the equations in (7.8) are extended by so called slack variables $\xi_l \geq 0, l = 1, \ldots, L$. This allows to stay with the approach, as vectors which cross the hyper plane may be placed on the 'wrong side':

$$\begin{aligned} y_l = +1 &\Rightarrow \underline{w}^T\underline{x}_l + b \geq +1 - \xi_l, \\ y_l = -1 &\Rightarrow \underline{w}^T\underline{x}_l + b \leq -1 + \xi_l. \end{aligned} \tag{7.13}$$

By that, the expression

$$\frac{1}{2}\underline{w}^T\underline{w} + G \cdot \sum_{l=1}^{L} \xi_l \tag{7.14}$$

needs to be minimised, where G is a free error weighting factor that needs to be determined. It can be shown that this optimisation—also called a 'primal problem'—is equivalent to a 'dual problem' of the maximisation of

$$\sum_{l=1}^{L} a_l - \frac{1}{2}\sum_{k=1}^{L}\sum_{l=1}^{L} a_k\, a_l\, y_k\, y_l(\underline{x}_k^T\underline{x}_l), \tag{7.15}$$

with the side conditions

$$0 \leq a_l \leq C, l = 1, \ldots, L, \tag{7.16}$$

$$\sum_{l=1}^{L} a_l\, y_l = 0. \tag{7.17}$$

The hyper plane is then defined by

$$\underline{w} = \sum_{l=1}^{L} a_l\, y_l\, \underline{x}_l, \tag{7.18}$$

$$b = y_{l^*}(1 - \xi_{l^*}) - \underline{x}_{l^*}^T \underline{w}_{l^*}. \tag{7.19}$$

Thereby l^* represents the index of the vector $\underline{x}_l$ with the largest coefficient a_l. The normal vector $\underline{w}$ is thus represented as weighted sum of training instances with the coefficients $a_l \leq C, l = 1, \ldots, L$, where C is another free parameter to be determined. By the introduction of the weighting coefficients the slack variables ξ_l disappear in the optimisation problem. The support vectors are then the training instances $\underline{x}_l$ that satisfy $a_l > 0$.

By this, L^2 terms of the form $\underline{x}_k^T \underline{x}_l$ result, which can be summarised as a matrix. One of the frequently used and highly efficient methods for the recursive computation of this matrix and by that solving of the dual problem is the Sequential Minimal Optimisation (SMO), which is introduced in detail in [9]. The classification by SVM is now given by the function $d_{\underline{w},b} : \underline{X} \rightarrow \{-1, +1\}$,

$$d_{\underline{w},b}(\underline{x}) = \text{sgn}(\underline{w}^T \underline{x} + b) \tag{7.20}$$

where

$$\text{sgn}(u) = \begin{cases} 1 & u \geq 0 \\ -1 & u < 0. \end{cases} \tag{7.21}$$

So far, we are only able to solve pattern recognition problems that assign the instances belonging to the (two) different classes with a certain acceptable error by a hyper plane in the space $\underline{X}$. This is referred to as linear seperation problem. Aiming at classes that can only be separated non-linearly, one applies the so called 'kernel trick' [10]. Figure 7.4 depicts an exemplary two-class problem in one-dimensional space, which can only be solved linearly by a mapping into a higher (two-)dimensional space—without error in the given example.

In general, such a transformation is given by the mapping

$$\Phi : \underline{X} \rightarrow \underline{X}^{'}, \quad \dim(\underline{X}^{'}) > \dim(\underline{X}). \tag{7.22}$$

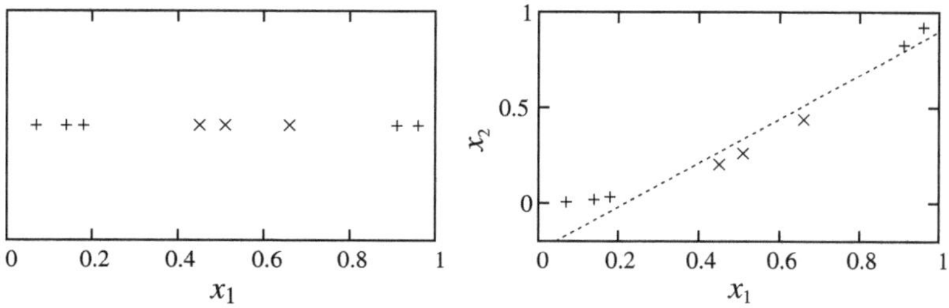

Fig. 7.4 Solving of an exemplary two-class problem by mapping into higher dimensional space: While in the one-dimensional (original) space the problem cannot be solved linearly, mapping by the function $\Phi : x_1 \mapsto (x_1, x_1^2)$ allows for error-free separation in the new two-dimensional space [1]

The normal vector $\underline{w}$ then results in

$$\underline{w} = \sum_{l:a_l>0} a_l\, y_l\, \Phi(\underline{x}_l). \tag{7.23}$$

The decision function $d_{\underline{w},b}(\underline{x})$ results—applying Φ—in:

$$d_{\underline{w},b}(\underline{x}) = \operatorname{sgn}(\underline{w}^T \Phi(\underline{x}) + b). \tag{7.24}$$

As

$$\underline{w}^T \Phi(\underline{x}) = \sum_{l:a_l>0} a_l\, y_l\, \Phi(\underline{x}_l)^T \Phi(\underline{x}), \tag{7.25}$$

the transformation Φ is explicitly neither needed for the estimation of the parameters of the classifier, nor for the classification. Instead a so called 'kernel function' $K^{\Phi}(\underline{x}, \underline{x}')$ is being defined, with the condition

$$K^{\Phi}(\underline{x}, \underline{x}') = \Phi(\underline{x})^T \Phi(\underline{x}'). \tag{7.26}$$

The kernel function additionally needs to be positively semi-definite, symmetric, and fulfil the Cauchy-Schwarz inequality. The optimal kernel function for a given classification or regression problem can only be found empirically. However, recently so called multi kernels try to overcome the search for optimal kernel functions [11]. Most frequently used kernel functions comprise:

- Polynomial kernel:

$$K_p^{\Phi}(\underline{x}, \underline{x}') = (\underline{x}^T \underline{x}' + 1)^p, \tag{7.27}$$

where p is the polynomial order,

- Gaussian kernel (radial basis function, RBF):

$$K_{\sigma}^{\Phi}(\underline{x}, \underline{x}') = e^{\frac{||\underline{x}-\underline{x}'||^2}{2\sigma^2}}, \tag{7.28}$$

  where σ is the standard deviation of the Gaussian, and
- Sigmoid kernel:

$$K_{k,\Theta}^{\Phi}(\underline{x}, \underline{x}') = \tanh(k(\underline{x}^T \underline{x}') + \Theta), \tag{7.29}$$

  where k is the amplification, and Θ the off-set.

The application of the kernel function K^{Φ} instead of the transformation Φ considerably reduces the required computation effort and allows for practical application of SVM and SVR when coping with high dimensional problems, as can be seen by the example of the polynomial kernel: to compute a polynomial of the order p in the space $\underline{X}$,

$$\binom{\dim(\underline{X}) + p}{p} \approx \frac{\dim(\underline{X})^p}{p!} \tag{7.30}$$

terms would need to be calculated, while the computation employing the polynomial kernel independently of p requires only approximately $\dim(\underline{X})$ operations. There exist manifold further kernels for special requirements, such as the KL divergence kernel frequently used in Gaussian Mixture Model (GMM)-SVM 'super vector' construction.

The last kernel that is introduced is a special solution for symbolic, i.e., non-numeric input: The recent string subsequence kernel (SSK) approach [12] makes use of a mapping from text information to a high dimensional feature space without explicit calculation of features. Based on the theory of Support Vector Machines, the idea of kernel mapping is extended for strings as input parameters. Thus, a special kernel for text information is provided. The idea behind is to observe small substrings in a given string. For a predefined substring length, all possible substrings form a feature space in which a string can be represented. The numeric value of each feature depends on the substring occurrence frequency in the string and on the degree of contiguity. For example, the substring *"ser"* exists in the word *"serene"* as well as in *"superb"*, but with a different degree of contiguity. This degree is weighted by a decay factor $\lambda \in [0, 1]$ which penalises non-contiguous substrings. Taking non-continuous substrings into account is a specific characteristic of the string kernel method.

The transformation of a string s into the feature space is done by a mapping $\Phi(s)$ which can be calculated numerically as described in [12]. Analog to the Support Vector Machines' theory, this mapping does not have to be done explicitly. An implicit calculation is done by using a kernel function:

$$K^{\Phi}(s, t) = \langle \Phi(s), \Phi(t) \rangle . \tag{7.31}$$

This kernel function is part of the decision function for SVM or SVR. The inner product calculated by the kernel can be seen as a numeric measure of similarity between two strings s and t. The calculation of this string subsequence kernel can further be simplified due to recursive computation [12], making the procedure practicable.

7.2.2.2 Support Vector Regression

Let us now have a very short introduction to SVR. Again, we first consider a set of training patterns $\mathcal{L}$, but now with numeric values $y_l \in \mathbb{R}$. The goal of SVR is to find a regression function $f(\underline{x})$ that has at the most a deviation of ϵ from the actually obtained targets and, at the same time, is as flat as possible. For a linear regression function,

$$f(\underline{x}) = \underline{w}^T \underline{x} + b \tag{7.32}$$

described by a vector $\underline{w}$ and a scalar b, this flatness can be achieved by minimising the dot product $\underline{w}^T \underline{w}$ under the conditions:

$$\begin{aligned} y_l - \underline{w}^T \underline{x}_l - b &\leq \epsilon, \\ \underline{w}^T \underline{x}_l + b - y_l &\leq \epsilon. \end{aligned} \tag{7.33}$$

Because there are only few cases where all y_l can be linearly estimated within a range between $\pm\epsilon$, non-negative slack variables ξ_l and ξ_l^* are introduced in analogy to SVM, allowing vectors to lie outside this range of $\pm\epsilon$:

$$\begin{aligned} y_l - \underline{w}^T \underline{x}_l - b &\leq \epsilon + \xi_l, \\ \underline{w}^T \underline{x}_l + b - y_l &\leq \epsilon + \xi_l^*. \end{aligned} \tag{7.34}$$

As in the case of SVM, the optimisation is done with Lagrangian multipliers, leading to:

$$\begin{aligned} &\frac{1}{2} \underline{w}^T \underline{w} + C \sum_{l=1}^{L} (\xi_l + \xi_l^*) - \sum_{l=1}^{L} (\eta_l \xi_l + \eta_l^* \xi_l^*) \\ &- \sum_{l=1}^{L} \alpha_l (\epsilon + \xi_l - y_l + \underline{w}^T \underline{x}_l + b) \\ &- \sum_{l=1}^{L} \alpha_l^* (\epsilon + \xi_l^* + y_l - \underline{w}^T \underline{x}_l + b). \end{aligned} \tag{7.35}$$

The optimisation problem is to minimise the Lagrangian multiplier with respect to $\underline{w}$, b and the Lagrangian multipliers $\alpha_l, \alpha_l^*, \eta_l, \eta_l^*$ ($l = 1, \ldots, L$). The complexity parameter $C > 0$ determines the penalty for regression errors larger than ϵ.

As a further analogy to SVM, the solution of the optimisation problem shows that the vector $\underline{w}_o$ for the regression function searched for can be written as a linear combination of vectors in the test set [13]:

$$\underline{w}_o = \sum_{l=1}^{L} (\alpha_l - \alpha_l^*) \underline{x}_l, \tag{7.36}$$

and thus, the linear regression function becomes

$$f(\underline{x}) = \sum_{l=1}^{L} (\alpha_l - \alpha_l^*) \underline{x}_l^T \underline{x} + b. \tag{7.37}$$

In consequently analogous manner to SVM theory, the algorithm for SVR is described by dot products between training vectors $\underline{x}_l$ and the new, unseen pattern vector $\underline{x}$, whereas only those training vectors are relevant for which the Lagrangian multipliers $(\alpha_l - \alpha_l^*) \neq 0$. These are the *support vectors* for SVR. Geometrically interpreted these are the training vectors which have an absolute estimation error of exactly ϵ.

As in the SVM case, the model is extended to solve non-linear regression tasks. This is done by applying the same kernel trick. The kernel function can be built into the regression function in Eq. (7.37), where it substitutes the dot product $\underline{x}_l^T \underline{x}$:

$$f(\underline{x}) = \sum_{l=1}^{L} (\alpha_l - \alpha_l^*) K^{\Phi} (\underline{x}_l, \underline{x}) + b. \tag{7.38}$$

The function in this form makes SVR an efficient algorithm for regression tasks.

7.2.3 Neural Networks

This section gives a short introduction to Artificial Neural Networks (ANN) with a focus on (bidirectional) Long-Short-Term Memory (BLSTM) networks.

ANNs are capable of learning practically arbitrary functions [14], and belong to the most popular learning algorithms, since McCulloch's and Pitts's first mathematical models in the year 1943 [15] that still provide the basis for today's ANN [16]. Their inspiration is given by neural networks in the central nervous system of vertebrates. The central information processing unit thereby is the neuron. Via its axon the neuron emits a certain activity by electrical pulses [17]. These impulses are propagated to the synaptic connection of other neurons via a branched network. The activity of a neuron is based on its cumulative input activation. In the nature, a higher activity results in a higher impulse frequency. Decisive is in general a threshold—usually approximated by a non-linear transfer function. Overall, a neural network consists of neurons and

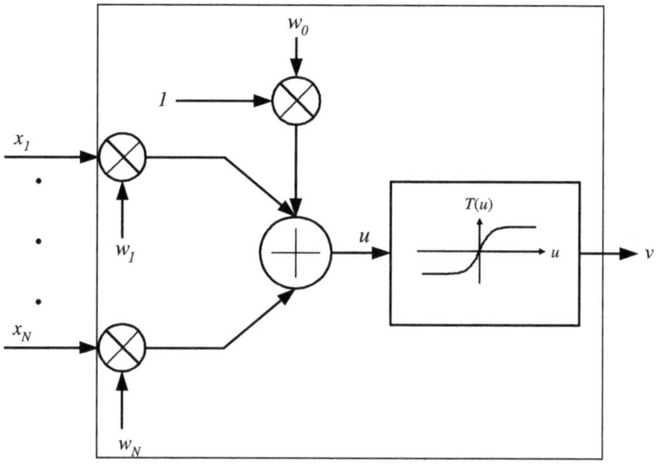

Fig. 7.5 Exemplary neuron

their directed connections. It is fully described by the network topology and weights, the computation type of its units and the encoding of the output. Figure 7.7 shows an example of an ANN. At the N input neurons the values of the feature vector $\underline{x} = \{x_i\}$ with $i = 1, \ldots, N$ are input. These values are weighted by the weights w_i with $i = 0, \ldots, N$ that can be written as $\underline{w} = \{w_i\}$. The weight w_0 is the 'bias'—a permanent additive offset. In the next part, a summation of the weighted inputs takes place. Its result u is then input into the—as stated usually non-linear—transfer function $T(u)$. The output of this function is v at the output of the neuron. In most cases, one aims at a steep decision function. An according visualisation of a neuron is given in Fig. 7.5.

Popular transfer functions are in particular the sigmoid function (cf. Fig. 7.6)

$$T(u) = \frac{1}{1 + e^{-\alpha u}}, \tag{7.39}$$

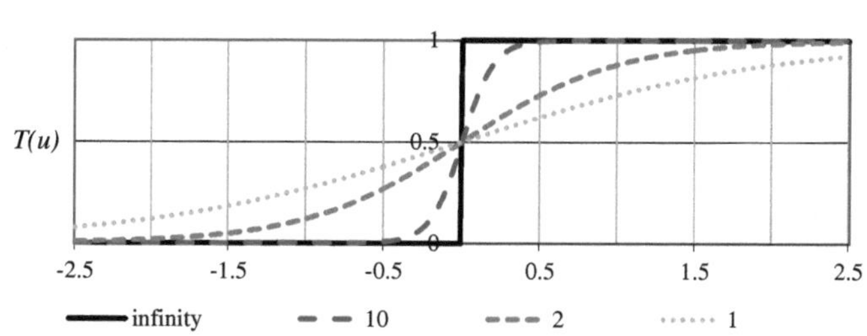

Fig. 7.6 Sigmoid function with different values for the steepness parameter α. In the case $\alpha \to \infty$ the function approximates a threshold decision

where α is the steepness parameter, the hyperbolic tangent function as special case of the sigmoid function with additive offset, and the unit step

$$T(u) = \begin{cases} 0 & \text{if } u < 0 \\ 1 & \text{if } u \geq 0 \end{cases}. \tag{7.40}$$

The sigmoid function is particularly popular owing to its approximation of an ideal threshold decision (cf. Fig. 7.6) while being differentiable. The latter will be needed throughout the training of the network.

A multiplicity of different network topologies exist, of which the most important will be introduced next.

7.2.3.1 Feed Forward Neural Networks

The most commonly used form of feed forward neural networks (FNN) is the multilayer perceptron (MLP) [18]: It consists of a minimum of three layers, one input layer—typically without processing—, one or more hidden layers, and an output layer. All connections feed forward from one layer to the next without any backward connections. MLPs classify all input feature vectors over time independently. In general, encoding of the outputs $\hat{y}_j$ with $j = 1, \ldots, M$ of the last layer that can be written as vector $\underline{\hat{y}}$ is required. A popular way is to provide one output neuron for regression and one per class in the case of classification. As an advantage, this provides a measure of confidence of the network: The 'softmax' function as a transfer function normalises the sum of all outputs to one in order to allow for interpretation as posterior probability $P(j|\underline{x})$ of the final output:

$$P(j|\underline{x}) = \hat{y}_j = \frac{e^u}{\sum_{j=1}^{M} e^u}. \tag{7.41}$$

In the recognition phase the computation is processed step-wisely from the input layer to the output layer. Per layer the weighted sum of the inputs from the previous layer is computed for each neuron and weighted by the non-linearity. Using the softmax function at the outputs, and the named encoding, the recognised class is assigned by maximum search. As an alternative, one can choose, e.g., a binary encoding of the classes with the network's outputs.

7.2.3.2 Back Propagation

Among the multiplicity of learning algorithms for ANNs, the gradient descent-based back propagation algorithm [19] is among the most popular ones and allowed for the break-through of ANN. Let $\underline{W} = \{\underline{w}_j\}$ summarise the weight vectors $\underline{w}_j$ of a layer with $j = 1, \ldots, J$ and J being the number of neurons in this layer. As target function to measure the progress of (supervised) learning, the MSE $E(\underline{x}, \underline{W})$ between the gold standard y and the network output $\hat{y} = f(\underline{x}, \underline{W})$ is used—for simplification we consider the case of a single output as in regression—an extension to multiple outputs is straight forward:

$$E(\underline{x}, \underline{W}) = |y - \hat{y}|^2 \tag{7.42}$$

Other target functions are frequently used, such as McClelland error or cross-entropy. After an initialisation of weights, e.g., by random, three steps follow for the back propagation:

1. Forward pass as 'normal' pass as in the recognition phase.
2. Computation of the MSE according to Eq. (7.42).
3. Backward pass with weight adaptation by the corrective term:

$$w_i \rightarrow w_i + \Delta w_i = w_i - \beta \cdot \frac{\delta E(\underline{x}, \underline{W})}{\delta w_i}, \tag{7.43}$$

   where β is the step size, which is to be determined empirically, and w_i is an individual weight within a neuron.

As a stopping criterion of the iterative updating of the weights one can either use a maximum number of iterations or a minimal change of the error [20]. A 'good' parameter set can only be determined empirically and based on experience. However, approaches exist to learn these. To avoid over fitting, a sufficient number of training instances is required as compared to the number of parameters in the network and the dimensionality of the feature vector. An alternative is resilient propagation that incorporates the last change of weights into the current change of weights [21]. By learning the weights, ANNs are able to cope with redundant feature information. The learning process is further discriminative as the information over all classes is learnt at a time [17]. Their highly parallel processing is one of the main advantages for efficient implementation. If the temporal context of a feature vector is relevant, this context must be explicitly fed to the network, e.g., by using a fixed width sliding window that combines several feature vectors to a 'super vector', as in [22].

7.2.3.3 Recurrent Neural Networks

Another technique for introducing past context to neural networks is to add backward (cyclic) connections to FNNs. The resulting network is called a recurrent neural network (RNN). RNNs can theoretically map from the entire history of previous inputs to each output. The recurrent connections implicitly form a kind of memory, which allows input values to persist in the hidden layer(s) and influence the network output in the future. RNNs can be trained by back propagation through time (BPTT) [23]. In BPTT, the network is first unfolded over time. The training then is similar as if training a FNN with back propagation. However, each epoch must run through the output observations in sequential order. Details are found in [23]. If in a RNN future

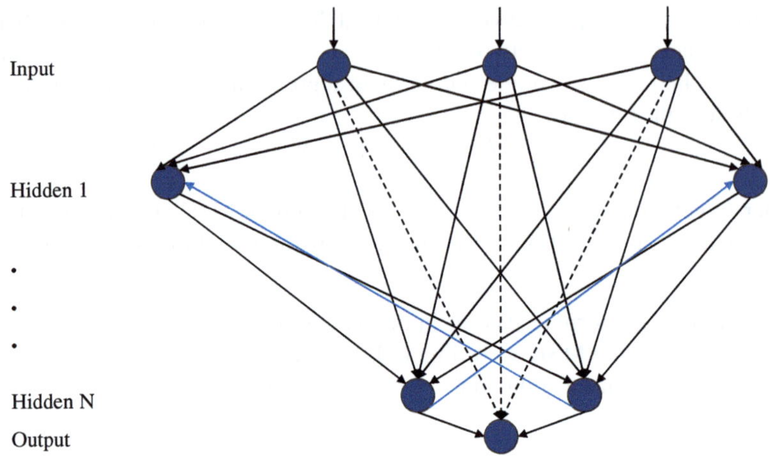

Fig. 7.7 A RNN with two hidden layers and a single output neuron for regression or binary classification. Dashed connection are an example of an architectural variation. The blue connections are examples of recurrent connections. Other popular ways of recurrent connections include such from the output nodes of a layer to its own input nodes

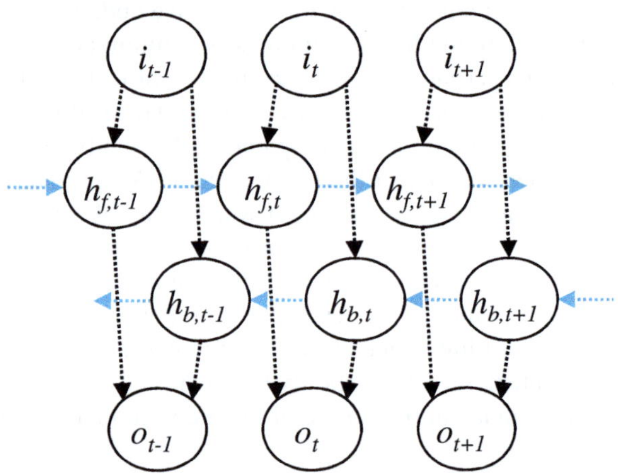

Fig. 7.8 Structure of a bidirectional network with input i, output o, as well as two hidden layers that processes the input sequence forwards (h_f) and backwards (h_b) over time t

context is also required, a delay between the input values and the output targets can be introduced. An example is shown in Fig. 7.7.

A more elegant incorporation of future temporal context is provided by a bidirectional recurrent neural network (BRNN). Two (sets of) separate hidden layers are used instead of one, both connected to the same input and output layers. The first processes the input sequence forwards and the second backwards. The network therefore has always access to the complete past and the future temporal context in a symmetrical way, without bloating the input layer size or displacing the input values from the corresponding output targets. Figure 7.8 visualises this principle.

However, they must have the complete input sequence at hand before it can be processed.

7.2.3.4 Long Short-Term Memory

Although BRNNs have access to both past and future information, the range of temporal context is limited to a few frames due to the 'vanishing gradient' problem [24]. The influence of an input value decays or blows up exponentially over time, as it cycles through the network with its recurrent connections and gets dominated by new input values. To overcome this deficiency, a method called Long Short-Term Memory (LSTM) was introduced in [25]. In a LSTM hidden layer, the non-linear units are replaced by LSTM memory blocks (cf. Fig. 7.10). Each block contains one or more self connected linear memory cells. By that, they are able to overcome the vanishing gradient problem and can learn the optimal amount of contextual information relevant for the learning task. Figure 7.9 depicts this vanishing gradient problem for RNN and how it is overcome by LSTM (right).

A LSTM layer is composed of recurrently connected memory blocks, each of which contains one or more memory cells, along with three multiplicative 'gate' units: the input, output, and forget gates. The gates perform functions analogous to read, write, and reset operations. More specifically, the cell input is multiplied by the activation of the input gate, the cell output by that of the output gate, and the previous cell values by the forget gate (cf. Fig. 7.10). Usually, one can employ the same non-linear transfer function for these gates, denoted as T_g in the ongoing. A popular choice is a hyperbolic tangent function. The transfer function of the 'original' neuron (top neuron Fig. 7.10) is often chosen as a sigmoid function and referred to by T_i in the ongoing, as it functions as the actual input neuron of a LSTM cell. The output transfer function of the LSTM cell after the 'error carousel' (EC) is denoted as T_o from now on. Sigmoid or softmax functions are popular choices for this function. The outgoing weight of the EC is chosen as 1 to realise the storage effect by an auto-transition of one.

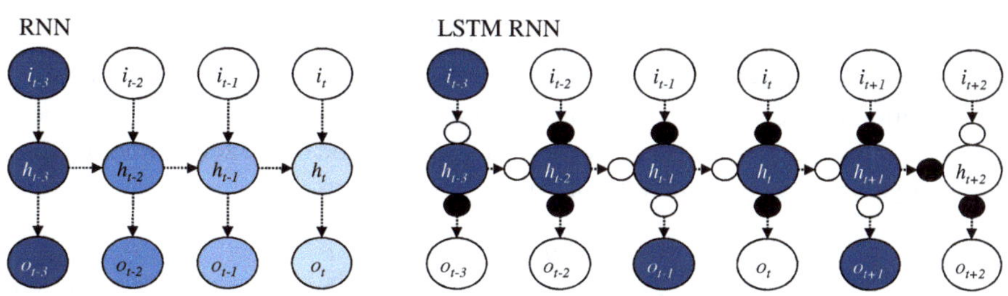

Fig. 7.9 Vanishing gradient problem of a RNN *(left)* and overcoming it by use of LSTM *(right)*. Lighter shading indicates decreased memory of past events. i_t, h_t, o_t represent the input, hidden, and output layers at time t, respectively

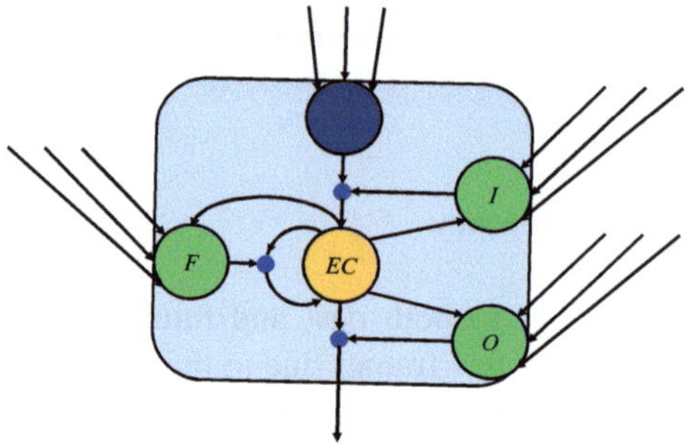

Fig. 7.10 LSTM memory block consisting of one memory cell: input, output, and forget gate collect activations from inside and outside the block which control the cell through multiplicative units (depicted as small circles); input, output, and forget gate scale input, output, and internal state respectively; a recurrent connection of fixed weight 1.0 maintains the internal state

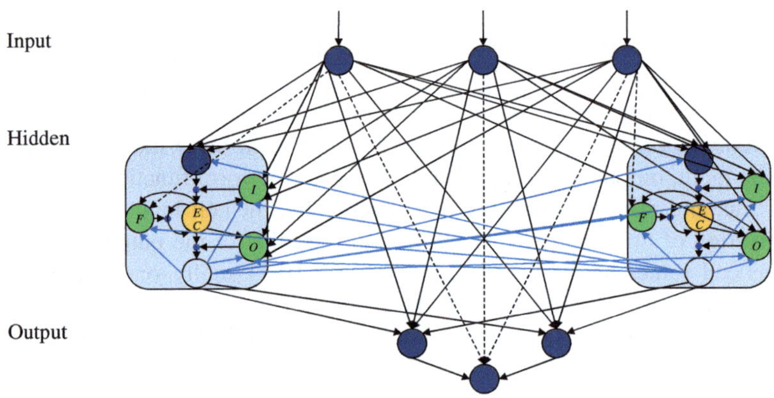

Fig. 7.11 An exemplary layout of a RNN with LSTM cells

The overall effect is to allow the network to store and retrieve information over long periods of time. For example, as long as the input gate remains closed, the activation of the cell will not be overwritten by new inputs and can therefore be made available to the net much later in the sequence by opening the output gate.

Figure 7.11 depicts LSTM cells' exemplary integration in a RNN. If $\alpha_{\text{in},t}$ denotes the activation of the input gate at time t *before* the activation function T_g has been applied and $\beta_{\text{in},t}$ represents the activation *after* application of the activation function, the input gate activations (forward pass) can be written as

$$\alpha_{\text{in},t} = \sum_{i=1}^{I} w_{i,\text{in}} x_{i,t} + \sum_{h=1}^{H} w_{h,\text{in}} \beta_{h,t-1} + \sum_{c=1}^{C} w_{c,\text{in}} s_{c,t-1} \tag{7.44}$$

and

$$\beta_{\text{in},t} = T_g(\alpha_{\text{in},t}), \tag{7.45}$$

respectively. The variable $w_{i,j}$ corresponds to the weight of the connection from unit i to unit j while 'in', 'for', and 'out' refer to input gate, forget gate, and output gate, respectively (cf. Eqs. 7.46 and 7.50). Indices i, h, and c count the inputs $x_{i,t}$, the cell outputs from other blocks in the hidden layer, and the memory cells, while I, H, and C are the number of inputs, the number of cells in the hidden layer, and the number of memory cells in one block. Finally, $s_{c,t}$ corresponds to the *state* of a cell c at time t, meaning the activation of the linear cell unit.

Similarly, the activation of the forget gates before and after applying T_g can be calculated as follows:

$$\alpha_{\text{for},t} = \sum_{i=1}^{I} w_{i,\text{for}} x_{i,t} + \sum_{h=1}^{H} w_{h,\text{for}} \beta_{h,t-1} + \sum_{c=1}^{C} w_{c,\text{for}} s_{c,t-1} \tag{7.46}$$

$$\beta_{\text{for},t} = T_g(\alpha_{\text{for},t}). \tag{7.47}$$

The memory cell value $\alpha_{\text{c},t}$ is a weighted sum of inputs at time t and hidden unit activations at time $t-1$:

$$\alpha_{\text{c},t} = \sum_{i=1}^{I} w_{i,\text{c}} x_{i,t} + \sum_{h=1}^{H} w_{h,\text{c}} \beta_{h,t-1}. \tag{7.48}$$

To determine the current state of a cell c, the previous state is scaled by the activation of the forget gate and the input $T_i(\alpha_{\text{c},t})$ by the activation of the input gate:

$$s_{\text{c},t} = \beta_{\text{for},t} s_{\text{c},t-1} + \beta_{\text{in},t} T_i(\alpha_{\text{c},t}). \tag{7.49}$$

The computation of the output gate activations follows the same principle as the calculation of the input and forget gate activations, however, this time the *current* state $s_{c,t}$ is considered, rather than the state from the previous time step:

$$\alpha_{\text{out},t} = \sum_{i=1}^{I} w_{i,\text{out}} x_{i,t} + \sum_{h=1}^{H} w_{h,\text{out}} \beta_{h,t-1} + \sum_{c=1}^{C} w_{c,\text{out}} s_{c,t} \tag{7.50}$$

$$\beta_{\text{out},t} = T_g(\alpha_{\text{out},t}). \tag{7.51}$$

Finally, the memory cell output is determined as

$$\beta_{\text{c},t} = \beta_{\text{out},t} T_o(s_{\text{c},t}). \tag{7.52}$$

Note that the initial version of the LSTM architecture contained only input and output gates. Forget gates were added later [26] in order to allow the memory cells to reset themselves whenever the network needs to *forget* past inputs.

LSTM networks can be trained by BPTT. They have shown remarkable performance in a variety of pattern recognition tasks, including phoneme classification [27], handwriting recognition [28], keyword spotting [29], affective computing [30], and driver distraction detection [31]. Combining bidirectional networks with LSTM leads to bidirectional LSTM (BLSTM). Further details on the LSTM technique can be found in [28].

7.3 Dynamic Learning Algorithms

Audio is sequential, and an endpointed audio stream $\underline{X} = \{\underline{x}_1, \underline{x}_2, \ldots, \underline{x}_T\}$ accordingly yields a series of T feature vectors. So far, however, we mostly dealt with classification of single feature vectors without use of temporal context. One exception were the different types of RNN that modelled such context, as discussed above. But even these are not able to 'warp' in time, i.e., to handle different tempo deviations between, e.g., two musical pieces or stretching or shortening, e.g., of vowels while speaking. The most frequently encountered algorithm for audio sequence classification are HMMs [32] as a simple form of DBNs. This property is owed to their ability of dynamic modelling throughout different hierarchy levels and a well-formulated stochastic framework. In ASR, for example, the extracted feature stream is first modelled on the phoneme level. On a higher level, these phonemes are then used to form words. Each class i is modelled by a HMM that represents the probability $P(\underline{X}|i)$, where $\underline{X}$ is called the 'observation', which is generated by the HMM.

A Markov model can be seen as finite state automaton that may change its state at any step in time. In a HMM, at each step in time t a feature vector $\underline{x}_t$ is being generated depending on the current state s and the emission probability $b_s(\underline{x})$. The probability of a transition from state j to state k is expressed by the state transition probability $a_{j,k}$ [33]. The probabilities $a_{0,j}$ are needed to enter the model in a state j with a certain probability. In order to simplify calculation, a non-emitting initial state s_0 and a non-emitting final state s_F can be defined [1]. In Fig. 7.12 the structure of such a model is depicted. In the example, the most frequently used type of HMM for audio processing is depicted—the so-called left-right model. In this model type, the state number cannot decrease over time. In the 'linear' model, no state can be skipped. Other topologies allow for a state skip, such as the Bakis model in which one state may be skipped. If any state can be reached from any other state with a probability above zero, the topology is referred to as 'ergodic'.

One speaks of a 'hidden' Markov model, as the sequence of states remains unknown—only the observations sequence is known [32]. Note the 'Markov property' that the conditional probability distribution of the hidden variable $s(t)$ at time step t, given the values of the hidden variable s at all times, depends only on the hidden variable $s(t-1)$, i.e., values at earlier steps in time have no influence [34].

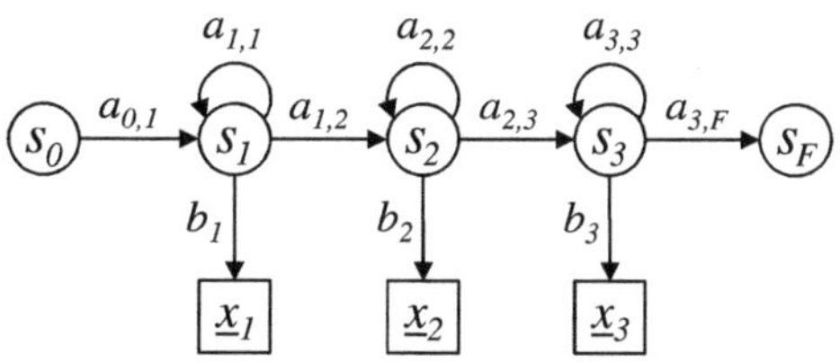

Fig. 7.12 Example of an instantiated linear left-right HMM with three emitting states. Squares indicate observations, circles represent switching states, arrows denote conditional dependencies

Further, the observation $\underline{x}(t)$ only depends on the value of the current state's hidden variable $s(t)$.[1]

The needed probability $P(\underline{X}|i)$ can be computed by summation over all possible state sequences:

$$P(\underline{X}|i) = \sum_{Seq} a_{s_0,s_1} \prod_{t=1}^{T} b_{s_t}(\underline{x}_t) a_{s_t,s_{t+1}}, \tag{7.53}$$

where Seq stands for the set of all possible state sequences. For the efficient computation of this probability, the forward algorithm is typically applied as is introduced in Sect. 7.3.1. Instead of a summation over all state sequences, the Viterbi algorithm considers only the most probable state sequence, which results in a speed-up at the cost of the global optimum [34]:

$$\hat{P}(\underline{X}|i) = \max_{Seq} \left\{ a_{s_0,s_1} \prod_{t=1}^{T} b_{s_t}(\underline{x}_t) a_{s_t,s_{t+1}} \right\}. \tag{7.54}$$

In the recognition phase the class i is decided for according to the model that is assigned the highest probability $P(\underline{X}|i)$. This requires the parameters $a_{j,k}$ and $b_s(\underline{x}_t)$ to be known for each model. Just as for the previous static classifiers, these are determined in a training phase given a large set of training instances. The popular method to this end is the forward-backward algorithm which is also described in Sect. 7.3.1.

In most Intelligent Audio Analysis application scenarios the emission probabilities $b_s(\underline{x}_t)$ are modelled by Gaussian mixtures. Such mixtures are linear superpositions of Gaussian functions. With the number of mixture components M and the 'mixture weight' of the m-th component $c_{s,m}$ the emission probability density function (PDF) can be determined as [34]:

$$b_s(\underline{x}_t) = \sum_{m=1}^{M} c_{s,m} \mathcal{N}(\underline{x}_t; \underline{\mu}_{s,m}, \underline{\Sigma}_{s,m}), \tag{7.55}$$

[1] Note that for better readability, the time t is in this section used in the subscript or argument following [32].

where $\mathcal{N}(\cdot; \underline{\mu}, \underline{\Sigma})$ is a multivariate Gaussian density with mean vector $\underline{\mu}$ and the covariance matrix $\underline{\Sigma}$. Apart from such 'continuous' HMMs, also 'discrete' HMMs are used. These use conditional probability tables for discrete observations $b_s(\underline{x}_t)$.

7.3.1 Estimation

The parameters of HMMs can be determined by the Baum-Welch estimation [35]—a case of generalised Expectation Maximisation (EM). If the ML estimates of the means and covariances per state s are to be computed, one has to take into account that each observation vector $\underline{x}$ contributes to the parameters of a state. This comes, as the overall probability of an observation bases on the summation of all possible state sequences. Thus, the Baum-Welch estimation assigns each observation to each state in proportion to the state probability at the observation of the respective feature vectors. With $L_{s,t}$ as the probability to be in state s at time step t, the Baum-Welch estimation for the means and covariances of a single Gaussian PDF is obtained as (the hat symbol marks estimated parameters in the following equations):

$$\hat{\underline{\mu}}_s = \frac{\sum_{t=1}^{T} L_{s,t}\underline{x}_t}{\sum_{t=1}^{T} L_{s,t}} \tag{7.56}$$

$$\hat{\underline{\Sigma}}_s = \frac{\sum_{t=1}^{T} L_{s,t}(\underline{x}_t - \underline{\mu}_s)(\underline{x}_t - \underline{\mu}_s)^T}{\sum_{t=1}^{T} L_{s,t}}. \tag{7.57}$$

The 'up-mixing' to several mixture components is reached in a simple way by considering the mixture components as sub-states. In these sub-states, the state transition probabilities correspond to the mixture weights. The state transition probabilities are estimated by the relative frequencies

$$\hat{a}_{j,k} = \frac{A_{j,k}}{\sum_{s=1}^{S} A_{j,s}}, \tag{7.58}$$

where $A_{j,k}$ denotes the number of transitions from state j to state k, and S denotes the number of states of the HMM.

For the computation of the probability $L_{s,t}$ the forward-backward algorithm is applied. The 'partial' forward probability $\alpha_s(t)$ for a HMM that represents the class i is defined as:

$$\alpha_s(t) = P(\underline{x}_1, \ldots, \underline{x}_t, s_t = s|i). \tag{7.59}$$

This can be interpreted as the joint probability of the observation of the first t feature vectors and being in state s at time step t. The following recursion allows for an efficient computation of the forward probability, where S is the number of emitting states:

$$\alpha_s(t) = \left[\sum_{j=1}^{S} \alpha_j(t-1) a_{j,s} \right] b_s(\underline{x}_t) \tag{7.60}$$

The according backward probability represents the joint probability of the observation from time step $t+1$ to T:

$$\beta_s(t) = P(\underline{x}_{t+1}, \ldots, \underline{x}_T | s_t = s, i). \tag{7.61}$$

It can be determined by the recursion:

$$\beta_j(t) = \sum_{s=1}^{S} a_{j,s} b_s(\underline{x}_{t+1}) \beta_s(t+1). \tag{7.62}$$

To compute the probability to be in a state at a given time step, one has to multiply the forward and backward probabilities:

$$P(\underline{X}, s_t = s|i) = \alpha_s(t) \cdot \beta_s(t). \tag{7.63}$$

By that, L_{st} can be determined by:

$$L_{st} = P(s_t = s|\underline{X}, i) = \frac{P(\underline{X}, s_t = s|i)}{p(\underline{X}|i)} = \frac{1}{p(\underline{X}|i)} \cdot \alpha_s(t) \cdot \beta_s(t). \tag{7.64}$$

Assuming the last state S at the moment in time of the last observation $\underline{x}_T$ needs to be taken, the probability $P(\underline{X}|M_t)$ equals $\alpha_S(T)$. By that, the Baum-Welch estimation can be executed as described.

The Viterbi algorithm is usually applied in the recognition phase. It is similar to the forward probability. However, the summation is replaced by a maximum search to allow for the following forward recursion:

$$\phi_s(t) = \max_j \{\phi_j(t-1) a_{j,s}\} b_s(\underline{x}_t), \tag{7.65}$$

where $\phi_s(t)$ is the ML probability of the observation of the vectors $\underline{x}_1$ to $\underline{x}_t$ and being in state s at time step t for a given HMM representing class i. Thus, the estimated ML probability $\hat{P}(\underline{X}|i)$ equals $\phi_S(T)$.

7.3.2 Hierarchical Decoding

HMM are in particular suited for decoding, i.e., segmenting and recognising continuous audio streams. In addition, their probabilistic formulation allows for elegant hierarchical analysis in order to unite knowledge at different levels as stated. Typical tasks include continuous speech recognition or chord labelling in music. Let $\mathcal{S}$ be a 'sequence' such as a spoken sentence or musical phrase. Then, the sequence $\underline{X}$ of T feature vectors stems from the phrase $\mathcal{S}$ [36]. The classifier now provides an estimate $\hat{\mathcal{S}}$ for the sequence aiming at the best match with the actual sequence $\mathcal{S}$. According to Bayes' decision rule a decision is optimal if the classifier picks the class which—based on the current observation—has the highest probability. For the optimal decision it thus needs to hold:

$$p(\hat{\mathcal{S}}|\underline{X}) = \max_{\mathcal{S}_j} p(\mathcal{S}_j|\underline{X}), \tag{7.66}$$

where $\mathcal{S}_j$ are the possible observed sequences. It is thus required to determine the probability for all possible sequences $\mathcal{S}_j$. As in practice it is hardly possible to determine these, Bayes' law is applied for re-formulating as follows:

$$p(\mathcal{S}_j|\underline{X}) = p(\underline{X}|\mathcal{S}_j)\frac{p(\mathcal{S}_j)}{p(\underline{X})} \tag{7.67}$$

As the probability $p(\underline{X})$ depends only on the feature vector series $\underline{X}$ and thus is independent of $\mathcal{S}_j$, it can be neglected within the maximum search over all sequences $\mathcal{S}_j$:

$$\underbrace{p(\underline{X}|\mathcal{S}_j)}_{\text{AM}} \cdot \underbrace{p(\mathcal{S}_j)}_{\text{LM}} \stackrel{!}{=} \max, \tag{7.68}$$

where the AM and LM represent the acoustics and semantics or syntax, and can be modelled by the sequence of audio events—in the example of continuous speech recognition these would be words, in the case of chord recognition, these would be the chords. In order to weight the influence of the LM, an exponential factor Λ—the so-called LM scaling factor—can additionally be introduced leading to:

$$p(\hat{\mathcal{S}}|\underline{X}) = \max_{\mathcal{S}_j} p(\underline{X}|\mathcal{S}_j) \cdot p(\mathcal{S}_j)^{\Lambda}. \tag{7.69}$$

The LM scaling factor is usually determined empirically or can be learnt in semi-supervised manner [37] and is often in the range of 10 ± 5.

The sequence that maximises the expression is output as best estimation $\hat{\mathcal{S}}$:

$$\hat{\mathcal{S}} = \arg\max_{\mathcal{S} \in \mathcal{U}} p(\underline{X}|\mathcal{S}) \cdot p(\mathcal{S})^{\Lambda}, \tag{7.70}$$

where $\mathcal{U}$ represents all allowed sequences. Let us now assume that every sequence $\mathcal{S}_j$ is a sequence of audio events $a_1, a_2, a_3, \ldots, a_A$. In the following a single sequence $\mathcal{S}_j$ is highlighted. For this sequence then holds:

$$p(\mathcal{S}_j) = p(a_1, a_2, \ldots, a_A) \tag{7.71}$$

If we further assume that the acoustic realisations of the audio events are independent of each other, the audio events can be modelled individually:

$$p(\underline{X}|\mathcal{S}_j) = p(\underline{x}_1, \ldots, \underline{x}_i)p(\underline{x}_{i+1}, \ldots, \underline{x}_j) \ldots p(\underline{x}_{k+1}, \ldots, \underline{x}_A) \tag{7.72}$$

It is assumed that these audio events occur without pauses and pauses are treated as audio events. Note that the audio event boundaries $i, j, \ldots, k$ and audio event number A are unknown and need to be determined by the classifier.

In the same way each audio event can be constructed by a sequence of audio sub-events (ASE) one level lower in hierarchy again assuming independence. In the case of speech these could be phonemes, triphones, or syllables, etc. In the case of chord arpeggios, these could be note events. If the ASE are modelled by HMM, the Viterbi algorithm can be applied on all three layers [36]: for the search of the state sequence within the HMMs, for the sequence of the individual ASE HMMs in each audio event, and for the sequence of the audio events, i.e., $\hat{\mathcal{S}}$.

At the audio event transitions the LM can be applied to model higher level information by transition probabilities [38]. These can for example be N-grams that model the conditional probability of a sequence of consecutive audio events, e.g., two or three. The Viterbi path determines the optimal path through all layers and by that the optimal sequence recognition with the optimal sequence of audio events—for an illustrative example see 7.13 where an according 'Trellis' is shown [32].

If the number of audio events—the 'vocabulary' size—is very large, the Viterbi search can become very computationally demanding and thus slow. Though at time step t only a single column needs to be analysed in the Trellis diagram (cf. Fig. 7.13), all emission probabilities in all states for all ASE in all audio events need to be computed. In the case of large vocabulary continuous speech recognition (LVCSR) this may easily require computation of more than 100 000 normal distributions in 10 ms [36]. One can thus make use of the fact that usually many paths in the Trellis are not promising in the sense that they lead to the overall best path, which is searched for. The 'beam search' thus prunes these candidates accepting a sub-optimal solution (usually less then one percentage point increase in error probability) at considerable speed-up and reduced memory consumption. This is reached by a smart list management in five consecutive steps [39]:

First, at time step t a list of all active states is set-up. This contains all the points in the Trellis diagram whose validation exceeds a given threshold. Each element in the list is stored by the audio event number, the ASE number, the state number, and its validation.

Then, from this list all possible subsequent states are computed that can be reached by the Viterbi path-diagram. To this end, the path diagram is applied in forward direction by overwriting the place-holders in the transition from $(t-1)$ to (t) each according to higher validation. The algorithm works in a recursive manner as usual and the effect is the same as when applying the path diagrams as in the usual case in backward direction.

Next, the list of subsequent states is reduced by deleting those states below the threshold—this is the actual pruning. This threshold is best constantly adapted to the

Fig. 7.13 Viterbi search of the optimal audio event sequence, Trellis diagram for the hierarchical recognition of audio events that consist of audio sub-events (ASE). The backtracking path is shown over time, and squares represent feature vector observations. HMMs (one per ASE) are shown schematically in Bakis topology. After backtracking the sequence of audio events 2, 1, 3 is recognised

current step in time. By that, the 'beam width' is broadened or narrowed according to the validation of the concurring paths' ascent or decline. This width is decisive for the trade-off between higher accuracy (broadened width) and higher speed (narrowed width).

Subsequently, at audio event transitions the value of the LM is added in the computation and it is jumped to the first state of the first model of the new audio event. In addition the required back-tracking information is stored.

Finally, the best audio event sequence is obtained at its end by the usual back-tracking, and the recognised audio events and their boundaries are output.

In practical applications, this particularly efficient search algorithm can reach reductions of the number of states to be computed of 1:1 000 [36]. The overall approach integrates knowledge of information on different levels in hierarchy to avoid early wrong decisions.

7.4 Ensemble Learning

Up to now, a number of learning algorithms was presented. In order to benefit from diverse advantages of these, one can aim at a synergistic heterogeneous combination of these. Alternatively, or in addition, homogeneous combination of the same learning algorithms, but instantiated differently, can help overcome training instability [40]. Examples of training unstable classifiers include ANNs, DTs or rule-based approaches. Overall, such combinations are known as 'ensembles' or 'committees' of learning algorithms. Owing to the increased computation effort, often so-called 'weak' classifiers are preferred in the construction of ensembles.

The aim is to reach a minimum mean square error (MSE) E of the algorithm. If the MSE is interpreted as expectation value E over all instances' feature vectors $\underline{x}$, one obtains:

$$\begin{aligned} E &= \mathrm{E}\{(\hat{y} - y)^2\} \\ &= \mathrm{Var}\{\hat{y} - y\} + \mathrm{E}\{\hat{y} - y\}^2 \\ &= \mathrm{Var}\{\hat{y}\} + \mathrm{E}\{\hat{y} - y\}^2, \end{aligned} \tag{7.73}$$

where $\hat{y}$ is the output of the learning algorithm and y is the target output.

The term $\mathrm{E}\{\hat{y} - y\}^2$ is known as square bias. It resembles the systematic deviation of the learning algorithm from the target. $\mathrm{Var}(\hat{y})$ is the variance of the output of the learning algorithm. For the minimisation of E one thus has to ideally reduce bias and variance. However, in practice, mostly only one of these two is significantly reduced in the majority of ensemble methods.

The task is thus now to construct ensembles and find a mechanism for the final decision. A simple solution for this decision is majority voting—the example in the ongoing for this type will be Bootstrap-Aggregating or Bagging for short that mainly reduces the variance. In addition, one can introduce weights for individual instances or results. This will be exemplified by Boosting, which in principle reduces both—bias and variance—however, variance to a significantly lower extent [41]. More elaborately, but requiring additional training partitions and more computational effort, one can also use a learning algorithm to train this weighting. To this end, Stacking will be introduced and an efficient example of a Tandem architecture will be shown.

7.4.1 Bootstrapping

Bagging [40] constructs ensembles of the same learning algorithm that is trained on different sub-sets of the training set $\mathcal{L}$. These sub-sets are sampled by sampling with replacement. This is the actual bootstrapping process. The cardinality of the samples is usually chosen as $|\mathcal{L}|$. Following a sampling with replacement strategy, on average 63.2 % of the training instances are covered in each sub-set, whereas the remaining percentage consists of duplicates. A variant that ensures that all samples are contained in each sub-set is called Wagging. The final decision is made by unweighted majority vote over the 'class votes' per classifier. As for regression, the mean over the results of the re-instantiated instances of the regressor is computed as final decision.

Boosting or Arcing [42] introduces a weighting for the voting (or averaging) process. Weights are chosen indirectly proportional to the error probability in order to emphasise the 'difficult cases' [43]. An option of realising weighting is to sample these instances repeatedly according to the weight. By that, the construction of ensembles follows an iterative procedure wherein the observed error probabilities are chosen by individual learning algorithms. Usually, one obtains better results as in Bagging, however, downgrades may also occur [5]. In any case, the computational effort is higher owing to the iterative procedure. One of the most popular Boosting algorithms is Adaptive Boosting, or AdaBoost for short. Adaptive refers to the iterative focus on the cases producing errors. Let $\underline{x}_l, l = 1, \ldots, L$ be the feature vectors in the training set $\mathcal{L}$, $L = |\mathcal{L}|$. As for SVMs or DTs, the original AdaBoost algorithm is suited only for two classes. The variant AdaBoost.M1, however, is an extension suited multiple classes M. By that, the class assignment for the instance $\underline{x}_l$ is given by $y_l \in \{1, \ldots, M\}$. To each instance $\underline{x}_l \in \mathcal{L}$ weights w_l are assigned. These are all initialised as $w_l = 1/L$ and are—as indicated—considered during computation of a weighted error measure and the training of the classifier. The core of the algorithm is now the determination of the weights β_t for the classifier with index t, where β_t depends on the error probability ε_t of the classifier.

Given the training set $\mathcal{L}$ and a number T of time steps $t = 1, \ldots, T$ the following steps are carried out:

1. A classifier with the decision $\hat{y}_t : \mathcal{X} \rightarrow \{1, \ldots, M\}$ is trained on $\mathcal{L}$ considering the weights w_l. As indicated, this can be realised by sampling a sub-set according to the weights as probability distribution.
2. The weighted classification error ε_t is computed:

$$\varepsilon_t = \sum_{l:\hat{y}_{t,l} \neq y_l} w_l. \tag{7.74}$$

3. If $\varepsilon_t > 1/2$, then repeat steps 1–3; terminate after N repetitions.
4. Else compute classifier β_t as

$$\beta_t = \begin{cases} 10^{-10} & \text{if } \varepsilon_t = 0 \\ \frac{\varepsilon_t}{1-\varepsilon_t} & \text{else} \end{cases}, \tag{7.75}$$

where the constant 10^{-10} is arbitrarily chosen to avoid division by zero in Eq. (7.77) below.

5. If $\varepsilon_t \neq 0$, then the new weights $w_l^{'}$ as used in the following iterations result in:

$$w_l^{'} = \begin{cases} w_l \beta_t & \text{if } \hat{y}_l = y_l \\ w_l & \text{else.} \end{cases} \tag{7.76}$$

6. The weights $w_l^{'}$ are normalised for their sum to be one.

The decision $\hat{y}_{\text{Ada}}$ of the ensemble classifier is then

$$\hat{y}_{\mathrm{Ada}}(\underline{x}) = \arg\max_y \sum_{t:\hat{y}_t(\underline{x})=y} \log \frac{1}{\beta_t}. \tag{7.77}$$

Looking at Eq. (7.77), the decisions of the classifiers considered as 'strong'—i.e., those with a small β_t—is weighted higher than those of the classifiers accordingly considered as 'weak'. In particular, classifiers with a recognition rate merely above chance level will benefit most form boosting. If $\varepsilon_t \leq 1/2$ for $t = 1, \ldots, T$, one can show that for the average error $\varepsilon_{\mathrm{Ada}}$ of $\hat{y}_{\mathrm{Ada}}$ holds [43]:

$$\varepsilon_{\mathrm{Ada}} \leq \exp\left(-2\sum_{t=1}^{T}\gamma_t^2\right), \quad \gamma_t = 1/2 - \varepsilon_t. \tag{7.78}$$

The condition $\varepsilon_t \leq 1/2$ can always be met for a two-class problem, however, for multi-class problems this is a strong limitation for weak classifiers. This can be overcome by reducing multi-class problems to multiple binary decisions such as one-versus-all, one-versus-one, half-versus-half or other groupings. An alternative is provided by the AdaBoost.M2 algorithm which integrates this formulation of multi-class problems by binary decisions—for details refer to [43].

A downside of Boosting is its susceptibility to noisy data, as mis-classified instances owing to noise may be classified correctly by chance, but are still assigned a high weight. This is for example given for problems with uncertain ground truth. Further, a high number of learning instances is usually required.

To benefit from the better minimisation of variance as in Bagging and the reduction of bias as in Boosting, these two can be combined sequentially: Usually, sub-ensembles built by AdaBoost are extended by Bagging to turn sub-ensembles into ensembles. This is often done with Wagging instead of Bagging and known as Multi-Boosting [41]—often the most efficient approach. The parameters of choice are the number and size of sub-ensembles. Usually K sub-ensembles of size K are built, resulting in K^2 instantiations of the classifier. A parallel combination is, however, not possible owing to the diverse weighting strategies of these two algorithms.

7.4.2 Meta-Learning

The principle of meta-learning is to unite strengths of several heterogeneous learning algorithms—now usually on the same training set. In Stacking [44], a higher level learning algorithm—the meta learner—learns literally speaking 'whom to trust when': After seeing the decisions of each lower level learning algorithm's—the base learner's— result, it comes to the final decision [45]. The meta-level is also known as level-1 and the base-level as level-0— this holds also for the type of data on these levels. On level-1, only pre-decisions are seen as input data. On level-0, the original data is seen. In order to train the level-1 learning algorithm, a J-fold cross-validation is needed (cf. Sect. 7.5.1) to assure disjoint data from training of the level-0 learning algorithms. The choice of learning algorithms for these two levels is often based on experience and exploration, as a full comprehension is still missing in the literature. However, statistical classifiers, DTs, and SVMs as introduced previously can be reasonably combined on level-0 [46]. In contrast, these seem to be less suited on level-1, where mostly Multiple Linear Regression (MLR) is chosen. MLR is different from simple linear regression only by use of multiple input variables. In the case of regression, confidences $P_{k,i}(\underline{x}) \in [0; 1]$ are assumed per base learner $k = 1, \ldots, K$, and each class $i = 1, \ldots, M$. If the level-0 classifier k only decides for exactly one class i without provision of its confidence, i.e., $\hat{y}_k = i$, the level-1 decision by MLR is as follows:

$$P_{k,i}(\underline{x}) = \begin{cases} 0 & \text{if } \hat{y}_k(\underline{x}) \neq i \\ 1 & \text{else.} \end{cases} \tag{7.79}$$

Applying non-negative weighting coefficients $\alpha_{k,i}$ per class and learner, the computation of the MLR per class i is obtained by:

$$\text{MLR}_i(\underline{x}) = \sum_{k=1}^{K} \alpha_{k,i} P_{k,i}(\underline{x}). \tag{7.80}$$

During the recognition phase the class i with the highest $\text{MLR}_i(\underline{x})$ is chosen for an observed unknown feature vector $\underline{x}$, i.e., the decision $\hat{y}$ is:

$$\hat{y} = \arg\max_i \text{MLR}_i(\underline{x}). \tag{7.81}$$

A high value of $\alpha_{k,i}$ thus shows a high confidence in the performance of learner k for the determination of class i [40]. For the determination of the coefficients $\alpha_{k,i}$ the Lawson- and Hanson method of the least squares can be used, which will not be described here. The optimisation problem to be solved results per each learner $k = 1, \ldots, K$ in the minimisation of the following expression, in which j represents the index of the training sub-set of the J-fold cross-validation:

$$\sum_{j=1}^{J} \sum_{l=1}^{L} (y_l - \sum_{i=1}^{M} \alpha_{k,i} P_{k,i,j}(\underline{x}))^2. \tag{7.82}$$

In [45] it is shown that the meta-classification on the basis of the actual confidences of the level-0 learners results in an improvement in the majority of cases as opposed to Eq. (7.79). This is known as StackingC—short for Stacking with Confidences [46]. In [45] a description on obtaining confidence values for diverse learners is given.

Simpler alternatives use either an unweighted majority vote or one based on the mean confidences. This can also be applied in the case of regression.

Overall, ensemble learning linearly increases the computation effort. Whereas Bagging and Stacking methods can be distributed on several CPUs for parallelisation, this is not possible in the iterative Boosting process. The lowest error rate is usually obtained by StackingC, which, however, requires an extra training set for the meta-learner. It is further also suited for 'strong' learners. Finally, one can integrate Bagging and Boosting in Stacking.

7.4.3 Tandem Learning

The strengths of diverse learning algorithms can also be combined in sequential manner. An example is Tandem learning, here exemplified by a static learner that incorporates LSTM and discriminative learning abilities—namely a BLSTM RNN—, with a dynamic learner —a multi-stream HMM—that has warping abilities and 'sees' the BLSTM predictions and the original feature vectors. The structure of this multi-stream decoder can be seen in Fig. 7.14: s_t and $\underline{x}_t$ represent the HMM state and the audio feature vector, respectively, while b_t corresponds to the discrete frame-level prediction of the BLSTM network (shaded nodes). Squares denote observed nodes and white circles represent hidden nodes. In every time frame t the HMM uses two (not statistically) 'independent' observations: The audio features $\underline{x}_t$ and the BLSTM prediction feature b_t. The vector $\underline{x}_t$ also serves as input for the BLSTM, whereas the size of the BLSTM input layer i_t corresponds to the dimensionality of the audio feature vector. The vector $\underline{o}_t$ contains one probability score for each of the P different audio target classes at each time step. b_t is the index of the most likely class:

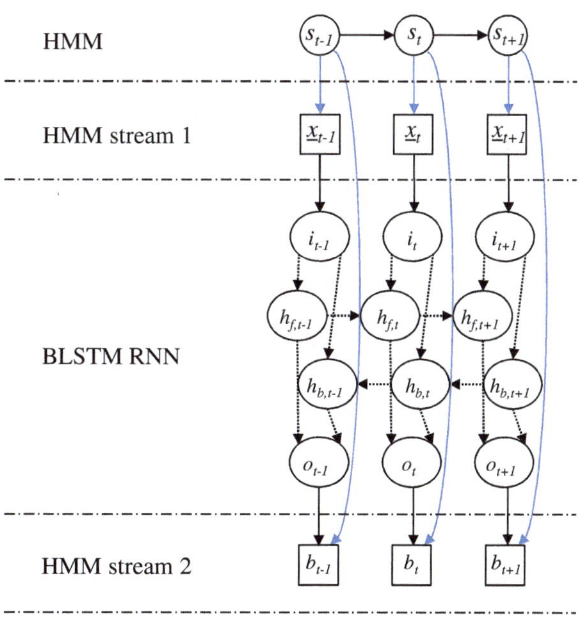

Fig. 7.14 Architecture of the multi-stream BLSTM-HMM decoder: s_t: HMM state, $\underline{x}_t$: acoustic feature vector, b_t: BLSTM class prediction feature, i_t, o_t, $h_{f,t}/h_{b,t}$: input, output, and hidden nodes of the BLSTM network; squares correspond to observed nodes, white circles correspond to hidden nodes, *shaded circles* represent the BLSTM network

$$b_t = \arg\max_j(\underline{o}_{t,1}, \ldots, \underline{o}_{t,j}, \ldots, \underline{o}_{t,P}) \tag{7.83}$$

In every time step the BLSTM generates a class prediction according to Eq. (7.83) and the HMM models $\underline{x}_{1:T}$ and $b_{1:T}$ as two independent data streams. With $\underline{y}_t = [\underline{x}_t; b_t]$ being the joint feature vector consisting of continuous audio features and discrete BLSTM observations and the variable a denoting the stream weight of the first stream (i.e., the audio feature stream), the multi-stream HMM emission probability while being in a certain state s_t can be written as

$$p(\underline{y}_t|s_t) = \left[\sum_{m=1}^{M} c_{s_t m}\mathcal{N}(\underline{x}_t; \underline{\mu}_{s_t m}, \underline{\Sigma}_{s_t m})\right]^a \times p(b_t|s_t)^{2-a}. \tag{7.84}$$

Thus, the continuous audio feature observations are modelled via a mixture of M Gaussians per state while the BLSTM prediction is modelled using a discrete probability distribution $p(b_t|s_t)$. The index m denotes the mixture component, $c_{s_t m}$ is the weight of the m'th Gaussian associated with state s_t, and $\mathcal{N}(\cdot; \underline{\mu}, \underline{\Sigma})$ represents a multivariate Gaussian distribution with mean vector $\underline{\mu}$ and covariance matrix $\underline{\Sigma}$. The distribution $p(b_t|s_t)$ is trained to model typical class confusions that occur in the BLSTM network.

7.5 Evaluation

7.5.1 Partitioning and Balancing

We now deal with typical ways of evaluating audio recognition systems' performance. We thereby focus on measurements that judge the reliability of the recognition result as these are of major interest in the extensive body of literature on intelligent speech, music, and sound analysis. However, as shown in the requirements section, a number of further aspects could be considered, such as real-time ability.

Evaluation should ideally be based on test partition(s) of suited audio databases that have not been 'seen' during system optimisation. Such optimisation includes data-based tuning of any steps in the chain of audio analysis including enhancement, feature extraction and normalisation, feature selection, parameter selection for the learning algorithm, etc. Thus, besides a training partition, a 'development' partition is needed for the above named optimisation steps. During the final system training, however, training and development partitions may be united in order to provide more learning material to the system. In general, one wishes all partitions to be somewhat large. For test, this is needed in order to provide significant results. Popular 'percentage splits' are thus 40 %:30 %:30 % for training, development, and test. In case of very large databases, as often given in ASR, the test partition is often chosen smaller, as around 10 %.

A solution to use as much data as possible for all partitions is the cross-validation. The overall corpus is thereby partitioned into J sets of equal size. These should be stratified, i.e., each set should show the same distribution of instances among classes or the continuum in case of numeric labels. If this is given, one speaks of J-fold stratified cross-validation (SCV). The evaluation is repeated J times with changing 'role' of the partitions. In each cycle $i = 1, \dots, J$ partition i can for example be used as test set and the remaining ones are united for training. After J cycles, each partition has then been used for testing once, and at the same time the maximum amount of training data was provided in each cycle. The final result is then usually provided as mean of the cycles. In addition, one can provide the standard deviation or similar measures to provide an impression on the 'stability' of the alteration of the learning material and the dependence on the test partition. If one needs an additional development partition, this could, e.g., be partition $(i + 1) \bmod J$. Popular values for J are three—this allows for transparent swapping of train, develop, and test sets without too high computational effort—or ten, which is reasonable if the database is very small, and too little training data would be provided by a third of the data. In general, one usually obtains better results with increasing J, as increasingly more training material is provided. This is, however, non-linear. In the extreme case, a single instance is left out at each cycle. This is known as leave-one-out (LOO).

A number of further criteria need to be respected for partitioning of a database: For example, independence of speakers, interprets, or sound sources, i.e., in the test partition the audio should be as independent as possible depending on the task of interest. In the case of cyclic iteration, this leads to a variant of LOO, where all instances of one aspect are clustered and left out at a time. An example is Leave One Speaker Out (LOSO) in intelligent speech analysis. Next, one wishes to keep good balance of all factors throughout the partitions. In particular the development partition should be similar in its characteristics to the test one in order to optimise the system in the right way. Next, partitioning should ideally be transparent and easy to reproduce. Thus, random partitioning can be a sub-optimal choice, as one has to provide the instance list or random seed and random function in order to allow for others to reproduce the partitions. As evaluation results depend on the (optimal) partitioning, one should make the choice also straight forward, such as by partitioning by sub-sequent speaker or song ID or similar.

In many cases, instances will be highly imbalanced across classes or the number scale. This can lead to preference of the 'majority class' which is reasonable if one wants to recognise as many instances correctly as possible. However, this comes at the cost of the under-represented class, and in extreme cases, such classes are completely ignored. If it is thus of higher importance to have a good balance in the recognition, balancing of the training set instances is advisable. Note that this is not required for all learning algorithms, as many can explicitly or implicitly model the class priors in the decision process. An example is the maximum a-posteriori (MAP) strategy for statistic learners such as HMMs, where the class priors are by intention multiplied with the model's generation probability to favour the majority class. As opposed to this, the maximum likelihood estimation (MLE) principle does not use priors. For other learning algorithms shown so far, such as SVMs or DTs, this is not directly possible and balancing of instances can be the preferred option.

Three different strategies are usually employed to balance the instances in the training set [47–49]: The first is down-sampling, in which instances from the over-represented classes are randomly removed until each class contains the same number of instances. This procedure usually withdraws a lot of instances and with them valuable information, especially in highly unbalanced situations: It always outputs a training dataset size equal to the number of classes multiplied with number of instances in the class with least instances. In highly unbalanced experiments this procedure thus leads to a pathologically small training set. The second method used is up-sampling, in which instances from the classes with proportionally low numbers of instances are duplicated to reach a more balanced class distribution. This way no instance is removed from the training set and all information can contribute to the trained classifier. To not falsify the classification results, it is important that only the training instances are upsampled. Naturally, one never balances test set instances. Likewise, replacement of instances is allowed so that equal class distribution is also achievable in highly unbalanced experiments. At the same time, it is ensured that each original instance is preserved in the training material. A mixed up-, and down-sampling strategy can be also be followed where instances from the majority class are deleted and from the minority class(es) are multiplied. This compromise keeps the overall number of instances at reasonable size, as with sheer up-sampling the problem of learning may become computationally too expensive. A third variant is assignment of different weighting of instances for the computation of the classifier objective function. In practice, this is often actually often solved by classifier internal up-sampling, and may lead to less stable results, while not providing any advantage in our respect, as obtainable performances are not higher, which is why this variant is not further pursued in this book. However, this may be well of interest in an on-line system which needs to be adapted, e.g., when a user labels a new song to adapt his audio-playing device. The latter are known as 'cost-sensitive' approaches where one 'punishes' confusions that should not occur in the case of discrete classes.

The question remains, how to pick the instances that are multiplied in the training set or deleted from it. While one can inject random into the selection process, this contradicts the above requested transparency and reproducibility of experiments by others. An easy strategy that does, however, not provide perfect balance, is thus the use of integer up-sampling factors for the minority classes. In addition, there are specialised algorithms that attempt to balance instances in an intelligent way. The idea is to up- or downsample those instances which are of particular relevance, as they are the 'hard' and 'interesting' cases and should not be emphasised on or at least not lost. An example of such an approach is the Synthetic Minority Over-sampling Technique [50].

7.5.2 Evaluation Measures

In the following, evaluation criteria for classifiers are considered at first. These will be followed by such for regressors where a continuous relation between the output of the learning algorithm and the target needs to be evaluated. In the case of classification, however, we need to compare discrete predicted class labels and compare these with the ground truth target. For simplification— without limitation of the general case— let the rejection class be assumed to be inherently modelled, i.e., rejection is one of the target classes. By that, we can consider the classification task as a mapping $\mathcal{X} \to \{1, \ldots, C\}, \underline{x} \mapsto \hat{y}$.

Evaluation criteria are defined as related to the test set's $\mathcal{T}$ instances, and the individual instances are each assigned to exactly one target class $i \in \{1, \ldots, C\}$:

$$\mathcal{T} = \bigcup_{i=1}^{C} \mathcal{T}_i = \bigcup_{i=1}^{C} \{\underline{x}_{i,n} \mid n = 1, \ldots, T_i\}, \tag{7.85}$$

where T_i is the number of instances in the test set that belong to class i. By that, the test set has the size $|\mathcal{T}| = \sum_{i=1}^{C} T_i$. Note, however, that attempts exist to find evaluation criteria where several classes may be assigned to one instance. This requirement is for example given in the case of the classification of a speaker's emotion, where one is not only 'surprised', but e.g., 'happily surprised' or 'angrily surprised' which led to the introduction of soft emotion profiles [51]. Similarly, music genre or ballroom dance style are often ambiguous in music analysis, cf. musical pieces that allow for either Rhumba or Foxtrott as choice of dance.

We will first consider evaluation measures for classification in the general case of two or more classes (i.e., $M \geq 2$) [1]. The most common measure is the probability that an instance of the test set is classified correctly. This is usually referred to as (weighted) accuracy *WA*, or weighted average recall or recognition rate.

$$\begin{aligned} \mathrm{WA} &= \frac{\#\,\text{correctly classified test instances}}{\#\,\text{test instances}} \\ &= \frac{\sum_{i=1}^{M} \left|\{\underline{x} \in \mathcal{T}_i \mid \hat{y} = i\}\right|}{|\mathcal{T}|}. \end{aligned} \tag{7.86}$$

If this rate is given per class i, one speaks of the class-specific recall RE_i:

$$\mathrm{RE}_i = \frac{\left|\{\underline{x} \in \mathcal{T}_i \mid \hat{y} = i\}\right|}{T_i}. \tag{7.87}$$

With $p_i = T_i/|\mathcal{T}|$ as the prior probability of class i in the test set further holds:

$$\mathrm{WA} = \sum_{i=1}^{M} p_i \,\mathrm{RE}_i. \tag{7.88}$$

The weighting by p_i in Eq. (7.88) leads to the name weighted accuracy. A special case of the calculation of accuracy is the word accuracy as encountered in the recognition of continuous speech. This accuracy is calculated by consideration of three types of errors: deletion, insertion, and substitution of words. With D, I, and S being the numbers for each type of these errors and N being the number of words in the test set, the word accuracy $\mathrm{WA}_{\mathrm{words}}$ is obtained by:

$$\mathrm{WA}_{\mathrm{words}} = \frac{N - D - I - S}{N}. \tag{7.89}$$

Note that, $N - (S + D)$ would be the number of correctly recognised words. Further, dynamic alignment of the recogniser output string and the reference transcription is needed to decide on the minimal number of errors, as a substitution could be counted as a deletion plus an insertion. This is also known as shortest Levenshtein distance. As word accuracy is also a type of WA—the accuracy depends on the frequency of occurrence of a specific word in the test set—it is also referred to as WA in the ongoing. However, from the context it will be clear that it is computed as word accuracy.

If balance of instances among classes is (highly) unbalanced, one can prefer to exchange the priors p_i for all classes by the constant weight $\frac{1}{M}$. This is known as unweighted accuracy *UA* or unweighted average recall:

$$\mathrm{UA} = \frac{\sum_{i=1}^{M} \mathrm{RE}_i}{M}. \tag{7.90}$$

The numerator in Eq. (7.87) equals the number of instances in $\mathcal{T}$, for which the decision was correctly made for class i. This is the number of 'true positives' TP_i as opposed to the false positives *FP* for class i:

$$\mathrm{FP}_i = \left|\left\{\underline{x} \in \mathcal{T} - \mathcal{T}_i \mid \hat{y} = i\right\}\right|. \tag{7.91}$$

With *TP* and *FP* we can define the precision *PR*:

$$\mathrm{PR}_i = \frac{\mathrm{TP}_i}{\mathrm{TP}_i + \mathrm{FP}_i}. \tag{7.92}$$

As increasing RE_i may come at the cost of decreasing PR_i, as many instances are assigned by mistake to class i, the wish for a measure that unites these two arises. This is given by their harmonic mean, known as F_1-measure (the subscript '1' is used for equal weighting of recall and precision—other common weights are doubling one up, i.e., F_2- or $F_{\frac{1}{2}}$-measure:

$$F_{1,i} = 2\frac{\mathrm{RE}_i \mathrm{PR}_i}{\mathrm{RE}_i + \mathrm{PR}_i}. \tag{7.93}$$

Considering decisions against class i, one can further introduce 'true negatives' TN_i and 'false negatives' FN_i:

$$\mathrm{TN}_i = \left|\left\{\underline{x} \in \mathcal{T} - \mathcal{T}_i \mid \hat{y} \neq i\right\}\right|, \tag{7.94}$$
$$\mathrm{FN}_i = \left|\left\{\underline{x} \in \mathcal{T}_i \mid \hat{y} \neq i\right\}\right|. \tag{7.95}$$

It is further of interest to investigate which classes are 'confused' with which. The according 'confusion matrix' $\underline{C} = (c_{i,j})$ thus has the entries:

$$c_{i,j} = \left|\left\{\underline{x} \in \mathcal{T}_i \mid \hat{y} = j\right\}\right|. \tag{7.96}$$

This matrix $\underline{C}$ contains all named measures as follows:

$$\mathrm{WA} = \frac{\mathrm{tr}(\underline{C})}{|\mathcal{T}|}, \tag{7.97}$$
$$\mathrm{RE}_i = \frac{c_{i,i}}{\mathcal{T}_i} = \frac{c_{i,i}}{\sum_{j=1}^{M} c_{i,j}}, \tag{7.98}$$
$$\mathrm{PR}_i = \frac{c_{i,i}}{\sum_{j=1}^{M} c_{j,i}}, \tag{7.99}$$
$$\mathrm{TP}_i = c_{i,i}, \tag{7.100}$$
$$\mathrm{FP}_i = \sum_{i \neq j} c_{j,i}. \tag{7.101}$$

In the case of binary decisions, the term $\mathrm{TP}_1/\mathcal{T}_1$ corresponds to the detection probability or 'true positive rate' (TPR), and $\mathrm{FP}_1/\mathcal{T}_2$ to false alarm probability or 'false positive rate' (FPR). Graphical evaluation often makes use of the Receiver Operating Characteristic (ROC, TPR vs. FPR) or its alternative, the Detection Error Trade-off (DET, false negative rate vs. FPR) curve. Such a plot demands for multiple evaluations of the learning algorithm's model or knowledge of confidences per instance in order to adjust a threshold for curve plotting. Popular measures to represent the plots in a single number are the 'area under the curve' (AUC) or the 'equal error rate' (EER). In case of more than two classes, i.e., $M > 2$, these measures are usually given as per one-versus-all.

We now shift to evaluation criteria for continuous value estimation, i.e., regression. Again, these are defined as related to the test set's $\mathcal{T}$ instances, and the individual instances are now each assigned to a continuous value $\hat{y} \in \mathbb{R}$. The test set has the size $|\mathcal{T}|$. In the case of regression, the common evaluation measure is the Pearson's correlation coefficient CC:

$$\mathrm{CC} = \frac{\sum_{n=1}^{|\mathcal{T}|}\left(\hat{y}_n - \overline{\hat{y}}\right)(y_n - \overline{y})}{\sqrt{\sum_{n=1}^{|\mathcal{T}|}\left(\hat{y}_n - \overline{\hat{y}}\right)^2 \cdot \sum_{n=1}^{|\mathcal{T}|}(y_n - \overline{y})^2}}, \tag{7.102}$$

with the averages

$$\overline{\hat{y}} = \frac{1}{|\mathcal{T}|}\sum_{n=1}^{|\mathcal{T}|}\hat{y}_n, \tag{7.103}$$

and

$$\overline{y} = \frac{1}{|\mathcal{T}|}\sum_{n=1}^{|\mathcal{T}|}y_n. \tag{7.104}$$

In addition, the the Mean Linear Error (MLE)—often referred to as Mean Absolute Error *MAE*—can be given:

$$\text{MAE} = \frac{1}{|\mathcal{T}|}\sum_{n=1}^{|\mathcal{T}|}|\hat{y}_n - y_n|, \tag{7.105}$$

MAE can be very intuitive, such as in the case of age determination in years of a speaker. Then, the MAE would be the absolute error in years, by which the regressor is mistaken 'on average'. However, in case of tasks where relative difference is more important than absolute numbers and the gold standard is less certain, such as for likability of a speaker or interest of a speaker on a continuous scale, CC is usually more representative and has a minimum and maximum independent of the task. CC is thus written without a leading zero in this book for better readability. This is different for MLE and MAE, as the number range varies.

As a general remark, it is important to note that all these evaluation measures naturally depend on the choice of the test instances. Apart from that, meaningful significance analyses should be considered as the difference between two results also depends on the quantity of test instances [52, 53]. Frequently employed tests contain, e.g., the one-sided z-test [54], which is the preferred choice in this book, and the common level of 0.05 is the minimum requirement for the claim of significance. Note that usually significance tests base on the independence assumption of tests [55]. As a consequence, this would require different data-sets for testing. However, as the test set is typically kept fixed in this field of research, the premise to reject the null hypothesis is comparably strict [55].

References

1. Kroschel, K., Rigoll, G., Schuller, B.: Statistische informationstechnik, 5th edn. Springer, Berlin (2011)
2. Quinlan, JR., C4.5: Programs for machine learning. Morgan Kaufmann, Burlington (1993)
3. Quinlan, J.: Learning efficient classification procedures and their application to chess end games. In machine learning: an artificial intelligence approach, pp .106–121. Tioga Publishing, Palo Alto (1983)
4. Quinlan, J.: Simplifying decision trees. Int. J. Man Mach. Stud. **27**, 221–234 (1987)

5. Quinlan, J., Bagging, Boosting and C4.5. In Proceedings 14th National Conference on AI, vol. 5, pp. 725–730, AAAI Press, Menlo Park (1996)
6. Cortes, C., Vapnik, V.: Support-vector networks. Mach. Learn. **20**(3), 273–297 (1995)
7. Hochreiter, S., Mozer, M., Obermayer, K.: Coulomb classifiers: generalizing support vector machines via an analogy to electrostatic systems. Adv. Neural Inf. Process. Sys. **15**, (2002)
8. Cristianini, N., Shawe-Taylor, J .: An introduction to support vector machines and other kernel-based learning methods. Cambridge University Press, Cambridge (2000)
9. Platt, J.: Sequential minimal optimization: a fast algorithm for training support vector machines. Technical report MSR-98–14, Microsoft Research, New York (1998)
10. Schölkopf, B., Smola, A.: Learning with kernels: support vector machines, regularization, optimization, and beyond (Adaptive computation and machine learning). MIT Press, Cambridge (2002)
11. Yang, H., Xu, Z., Ye, J., King, I., Lyu, M.: Efficient sparse generalized multiple kernel learning. IEEE Trans. Neural Netw. **22**(3), 433–446 (2011)
12. Lodhi, H., Saunders, C., Shawe-Taylor, J., Cristianini, N., Watkins, C.: Text classification using string kernels. J. Mach. Learn. Res. **2**, 419–444 (2002)
13. Smola, A., Schölkopf, B.: A tutorial on support vector regression. Stat. Comput. **14**(3), 199–222 (2004)
14. Niemann, H.: Klassifikation von mustern. published online, 2nd, revised and extended edition (2003)
15. McCulloch, W., Pitts, W.: A logical calculus of the ideas immanent in nervous activity. Bulletin Math. Biophy. **5**, 115–133 (1943)
16. Schuller, B.: Automatische emotionserkennung aus sprachlicher und manueller interaktion. Doctoral thesis, Technische Universität München, Munich (2006)
17. Rigoll, G.: Neuronale netze. Expert-Verlag (1994)
18. Deller, J., Proakis, J., J. Hansen.: Discrete-time processing of speech signals. Macmillan Publishing Company, New York (1993)
19. Rumelhart, D., Hinton, G., Williams, R.: Learning internal representations by error propagation. In parallel distributed processing: explorations in the microstructure of cognition, vol. 1, pp. 318–362. MIT Press, Boston (1987)
20. Schalkoff, R.:Artificial neural networks. McGraw-Hill, New York (1994)
21. Riedmiller, M., Braun, H.: Rprop—A fast adaptive learning algorithm. In Proceedings of the International Symposium on Computer and Information Science, vol. 7, (1992)
22. Lacoste, A., Eck, D.: Onset detection with artificial neural networks. In MIREX (2005)
23. Werbos, P.: Backpropagation through time: what it does and how to do it. Proc. IEEE **78**(10), 1550–1560 (1990)
24. Hochreiter, S., Bengio, Y., Frasconi, P., Schmidhuber, J.: Gradient flow in recurrent nets: the difficulty of learning long-term dependencies. In: Kremer, S.C., Kolen, J.F. (eds.) A field guide to dynamical recurrent neural networks, pp. 1–15. IEEE Press, New York (2001)
25. Hochreiter, S., Schmidhuber, J.: Long short-term memory. Neural Comput. **9**(8), 1735–1780 (1997)
26. Gers, F., Schmidhuber, J., Cummins, F.: Learning to forget: continual prediction with LSTM. Neural Comput. **12**(10), 2451–2471 (2000)
27. Graves, A., Schmidhuber, J.: Framewise phoneme classification with bidirectional LSTM and other neural network architectures. Neural Netw. **18**(5–6), 602–610 (2005)
28. Graves, A.: Supervised sequence labelling with recurrent neural networks. Ph.D thesis, Technische Universität München, Munich (2008)
29. Wöllmer, M., Schuller, B., Batliner, A., Steidl, S., Seppi, D.: Tandem decoding of children's speech for keyword detection in a child-robot interaction scenario. ACM Transactions on Speech and Language Processing, (Special Issue Speech Lang. Process. Children's Speech Child-mach. Interact. Appl.). vol. 7(4), August 2011, 22 pages
30. Wöllmer, M., Schuller, B., Eyben, F., Rigoll, G.: Combining long short-term memory and dynamic bayesian networks for incremental emotion-sensitive artificial listening. IEEE J. Sel. Top. Sig. Process. (Special Issue Speech Process. Nat. Interact. Intell. Environ). **4**(5), 867–881 (October 2010)

31. Wöllmer, M., Blaschke, C., Schindl, T., Schuller, B., Färber, B., Mayer, S., Trefflich, B.: Online driver distraction detection using long short-term memory. IEEE Trans. Intell. Transp. Syst. **12**(2), 574–582 (2011)
32. Rabiner, L.R.: A tutorial on hidden markov models and selected applications in speech recognition. Proc. IEEE **77**, 257–286 (1989)
33. O'Shaughnessy, D.: Speech communication. Adison-Wesley, 2nd edn, Boston (1990)
34. Jelinek, F.: Statistical methods for speech recognition. MIT Press, Cambridge (1997)
35. Baum, L.E., Petrie, T., Soules, G., Weiss, N.: A maximization technique occurring in the statistical analysis of probabilistic functions of markov chains. Ann. Math. Stat. **41**(1), 164–171 (1970)
36. Ruske, G.: Automatische spracherkennung, 2nd edn. Methoden der Klassifikation und Merkmalsextraktion, Oldenbourg (1993)
37. White, C.M., Rastrow, A., Khudanpur, S., Jelinek, F.: Unsupervised Estimation of the Language Model Scaling Factor. In Proceedings of the Interspeech, pp. 1195–1198, Brighton (2009)
38. Furui, S.: Digital speech processing: synthesis, and recognition, 2nd edn. Signal Processing and Communications. Marcel Denker Inc., New York (1996)
39. Lowerre, B.: The harpy speech recognition system. Ph.D thesis, Carnegie Mellon University, Pittsburgh (1976)
40. Breiman, L.: Bagging predictors. Mach. Learn. **24**(2), 123–140 (1996)
41. Webb, G.: Multiboosting: A technique for combining boosting and wagging. Mach. Learn. **40**, 159–198 (2000)
42. Valinat, L.: A theory of the learnable. Commun. ACM **27**(11), 1134–1142 (1984)
43. Freund, Y., Schapire, R.: Experiments with a New Boosting Algorithm, pp. 148–156. In Proceedings of the International Conference on, Machine Learning (1996)
44. Wolpert, D.: Stacked generalization. Neural Netw. **5**, 241–259 (1992)
45. Ting, K., Witten, I.: Issues in Stacked Generalization. J. Artif. Intell. Res. **10**(1), 271–289 (Jan. 1999)
46. Seewald, A.: Towards understanding stacking—Studies of a general ensemble learning scheme. Ph.D thesis, Technische Universität Wien, Vienna (2003)
47. Schuller, B.: Steidl, S., Batliner, A.: The interspeech 2009 emotion challenge. In Proceedings INTERSPEECH 2009, 10th Annual Conference of the International Speech Communication Association, pp. 312–315, Brighton, September ISCA, ISCA (2009)
48. Schuller, B., Müller, R., Eyben, F., Gast, J., Hörnler, B., Wöllmer, M., Rigoll, G., Höthker, A., Konosu, H.: Being bored? recognising natural interest by extensive audiovisual integration for real-life application. Image Vision Comput. (Special Issue Visual Multimodal Anal. Human Spontaneous Behav). **27**(12), 1760–1774 (2009)
49. Schuller, B., Schenk, J., Rigoll, G., Knaup, T.: "the godfather" versuss. "chaos": Comparing linguistic analysis based on online knowledge sources and bags-of-n-grams for movie review valence estimation. In Proceedings 10th International Conference on Document Analysis and Recognition, ICDAR 2009, pp.858–862, IAPR, IEEE, Barcelona (2009)
50. Chawla, N.V., Bowyer, K.W., Hall, L.O., Kegelmeyer, W.P.: Synthetic minority over-sampling technique. J. Artif. Intell. Res. **16**, 321–357 (2002)
51. Mower, E., Mataric, M., Narayanan, S.: A framework for automatic human emotion classification using emotional profiles. IEEE Trans. Audio Speech Lang. Process. **19**(5), 1057–1070 (2011)
52. Rozeboom, W.: The fallacy of the null-aypothesis significance test. Psychol. Bull. **57**, 416–428 (1960)
53. Nickerson, R.S.: Null hypothesis significance testing: a review of an old and continuing controversy. Psychol. Bull. **5**, 241–301 (2000)
54. Eysenck, H.: The concept of statistcal significance and the controversy about one-tailed tests. Psychol. Bull. **67**, 269–271 (1960)
55. Gillick, L., Cox, S.J.: Some statistical issues in the comparison of speech recognition algorithms. In Proceedings International Conference on Audio Speech and Signal Processing (ICASSP), vol. 1, pp. 23–26, IEEE, Glasgow (1989)

Chapter 8
Audio Source Separation

I just wondered how things were put together.
—Claude Elwood Shannon

In order to enhance the (audio) signal of interest in the case of added audio sources, one can aim at their separation. Albeit being very demanding, Audio Source Separation of audio signals has many interesting applications: In Music Information Retrieval (MIR), it allows for polyphonic transcription or recognition of lyrics in singing after decomposing the original recording into voices and/or instruments such as drums or guitars, or vocals, e.g., for 'query by humming' [1]. In ASR, the separation of the target speaker from others, background noises or music [2] may help to improve the accuracy. Given multiple microphone tracks, ICA [3] is usually among the first choices. Traditional ICA, however, limits the number of sources that can be separated to the number of available input channels, which makes basic ICA unsuitable for many audio recognition and retrieval applications where only mono- or stereophonic audio is available. To improve performance of ICA in challenging scenarios, source localisation information can be integrated as a constraint, which is promising for ASR in hands-free human-machine interaction [4]. However, this also requires knowledge about the localisation of the microphones used for recording, which is again not given in typical audio mining and retrieval tasks.

On the other hand, fully blind separation of multiple sources from mono- or stereophonic signals is considered infeasible as of today. To summarise, in most Intelligent Audio Analysis applications, prior knowledge has to be exploited in audio source separation, as will be detailed in the following section. A general framework for such 'informed' source separation has recently been presented in [5]. In the light of Intelligent Audio Analysis, such informed methods are particularly interesting, as they leverage machine intelligence for the highly challenging problem of underdetermined source separation. Among the most promising approaches towards separation of monophonic sources are those centred around NMF [6–11], which will be the focus of this chapter. NMF can also be applied in different places along the Intelligent Audio Analysis processing chain, e.g., for audio feature extraction and

B. W. Schuller, *Intelligent Audio Analysis*, Signals and Communication Technology, DOI: 10.1007/978-3-642-36806-6_8, © Springer-Verlag Berlin Heidelberg 2013

classification such as in noisy conditions [13–15]. This will be introduced towards the end of this section.

Let us now introduce the theoretical foundations of NMF. For clarity, the following notation will be used: For a matrix $\underline{A}$, the notation $\underline{A}_{i,:}$ denotes the i-th row of $\underline{A}$ (as a row vector), and let us analogously define $\underline{A}_{:,j}$ for the j-th column of $\underline{A}$ (as a column vector). Further, let $\underline{A} \otimes \underline{B}$ denote the elementwise product of matrices $\underline{A}$ and $\underline{B}$. The division of matrices is always to be understood as elementwise.

8.1 Methodology

Let us now discuss in detail how NMF can be used for source separation and NMF-based feature provision. The basic procedure is the extraction of an arbitrary number of sources—the 'components'— from audio by non-negative factorisation of a spectrogram in matrix representation $\underline{V} \in \mathbb{R}_+^{M \times N}$ into a spectral basis $\underline{W} \in \mathbb{R}_+^{M \times R}$ and activation matrix $\underline{H} \in \mathbb{R}_+^{R \times N}$:

$$\underline{V} = \underline{W}\,\underline{H}. \tag{8.1}$$

This yields R component spectrograms $\underline{V}^{(j)}$, $j = 1, \ldots, R$ either by multiplication of each basis vector $\underline{w}^{(j)} := \underline{W}_{:,j}$ with its activations $\underline{h}^{(j)} := \underline{H}_{j,:}$, as in [7], or by a more advanced 'Wiener filter' approach, as described in [6, 10]:

$$\underline{V}^{(j)} = \underline{V} \otimes \frac{\underline{w}^{(j)}\underline{h}^{(j)}}{\underline{W}\,\underline{H}}. \tag{8.2}$$

The spectrograms can be obtained from short-time Fourier transformation (STFT) and subsequent transformation to magnitude, power or Mel-scale spectrograms. Each $\underline{V}^{(j)}$ is then transformed back to the time domain by inverse STFT, using the original phase.

Several NMF algorithms can be used for the factorisation according to (8.1). These minimise a distance function $d(\underline{V}|\underline{W}\,\underline{H})$ by multiplicative updates of the matrices. The starting point can be a random initialisation. $d(\underline{V}|\underline{W}\,\underline{H})$ can be chosen as the β-divergence or one of its special instances, the Itakura-Saito (IS) [10] divergence, Kullback-Leibler (KL) divergence, or squared Euclidean distance (ED) [16]. Further, to support overcomplete decomposition, i.e., choosing R such that $R(M+N) > MN$, sparse NMF variants [7] exist for either of the named distance functions, as well as the sparse Euclidean NMF variant used in [17]. In addition, non-negative matrix deconvolution (NMD) [6, 8] has been proposed as a context-sensitive NMF extension. In NMD, each component is characterised by a sequence of spectra, rather than by an instantaneous observation. Alternatively, sequences of spectral feature vectors can be modelled as 'supervectors' in a sliding window approach to use standard NMF for context-sensitive factorisation [13]. More precisely, the original spectrogram $\underline{V}$ is transformed to a matrix $\underline{V}'$ such that every column of $\underline{V}'$ is the row-wise concatenation of a sequence of short-time spectra (in the form of row vectors):

$$\underline{V}' := \begin{bmatrix} \underline{V}_{:,1} & \underline{V}_{:,2} & \cdots & \underline{V}_{:,N-T+1} \\ \vdots & \vdots & \ldots & \vdots \\ \underline{V}_{:,T} & \underline{V}_{:,T+1} & \cdots & \underline{V}_{:,N} \end{bmatrix}, \tag{8.3}$$

where T is the desired context length. That is, the columns of $\underline{V}'$ correspond to overlapping sequences of spectra in $\underline{V}$. If signal reconstruction in the time domain is desired, the above named spectrogram transformations, including Mel filtering and transformation according to (8.3), can be reversed.

The basic NMF method as explained above is entirely unsupervised. In many practical applications, such as speech or music separation, prior knowledge about the problem structure can be exploited. A simple yet very effective method to integrate a-priori knowledge into NMF-based source separation is to perform supervised or semi-supervised NMF. This means that parts of the first NMF factor are predefined as a set of spectra characteristic for the sources to be separated rather than choosing random initialisations of both factors. This can be useful in audio enhancement, e.g., in a 'cocktail party' situation with several simultaneous speakers [6, 17], or noise versus a speaker of interest [18]. The initialisation spectra may themselves stem from NMF decomposition of training material or can be based on simpler methods such as median filtering or simply random sampling of training spectrograms. This procedure is outlined in Fig. 8.1 as a flowchart. An alternative supervised NMF method, depicted in Fig. 8.2, is to assign components computed by unsupervised NMF to classes such as 'drums' and 'non-drums' by means of a supervisedly trained classifier as in [19]. This allows dealing with observations that cannot be described as a linear combination of pre-defined spectra, but assumes that unsupervised NMF by itself can extract meaningful units, such as notes of different instruments. Given an assignment of NMF components to sources as described above, it is straightforward to synthesise the audio signals of interest by overlaying component spectrograms.

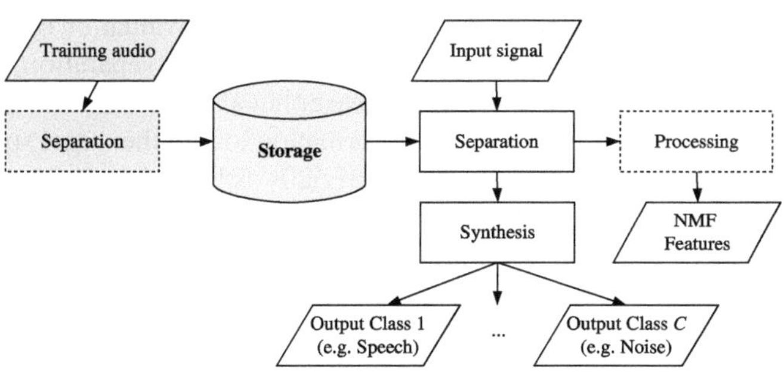

Fig. 8.1 Supervised NMF: A set of spectral components (which can themselves be computed by NMF from training audio) serve as constant basis for NMF; the activations can be exported as features or be used to synthesise audio signals for the sources [12]

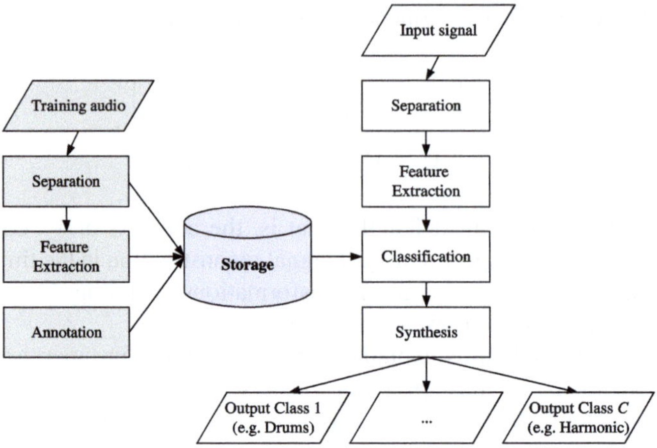

Fig. 8.2 Unsupervised NMF followed by supervised component classification, as in musical instrument separation: A classifier is built from labelled separated components. Steps required to train the classifier are *gray shaded* [12]

Besides using source separation as pre-processing for Intelligent Audio Analysis, the activations computed by NMF can be used directly for classification, as indicated by the flowchart in Fig. 8.1. This approach will be presented in more detail in Sect. 8.3.

8.2 Performance

To get an idea of the separation performance by basic NMF in a challenging task, let us consider the separation of two simultaneously speaking speakers from a monaural signal in the ongoing. Fig. 8.3 visualises the separation quality in terms of source-distortion ratio (SDR) depending on the targeted RTF. SDR, as introduced by [20], can be considered as the most popular objective metric for the evaluation of audio source separation as of today. In the considered scenario of speaker separation, it takes into account the suppression of the interfering speaker but also penalizes the introduction of artifacts due to signal separation, i.e., information loss in the target speech—note that perfect interference reduction can be trivially achieved by outputting a zero signal. These experiments are based on the procedure proposed in [6] and the results correspond to those reported in [12]. NMF is used over NMD based on the finding in [6] of no significant difference in separation quality by either of these two bases. The effect of using different numbers of iterations, DFT window sizes and the NMF cost function is assessed; the importance of these parameters on separation quality and computational complexity has been pointed out in [6, 12]. 12 pairs of male and female speakers—ensuring that the speech spectra do not fully overlap—were selected randomly from the TIMIT database (cf. also Sect. 10.4.3). Per pair, two

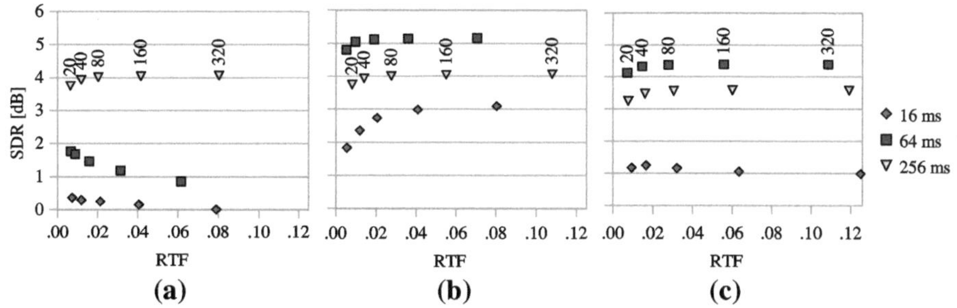

Fig. 8.3 Benchmark results for monaural speaker separation by supervised NMF, in terms of RTF and signal-to-distortion ratio (SDR) [20]. Mixed signals from pairs of male/female speakers (24 speakers total) from the TIMIT database. The open-source openBliSSART toolkit is used, and computation is performed on a consumer grade GPU (NVIDIA GeForce GTX 560). The number of NMF iterations (20–320), the DFT window size (16, 64, 256 ms) and the NMF cost function are adjusted. **a** Euclidean distance. **b** KL divergence. **c** Itakura-saito divergence

randomly selected sentences of roughly equal length were mixed, and a NMF basis $\underline{W}$ was computed from the other sentences spoken by each speaker. To this end, unsupervised NMF (250 iterations) was applied to the concatenated spectrograms of these sentences and only the first factor was kept. Separated signals for both speakers were obtained by supervised NMF with $\underline{W}$, by summing up component spectra corresponding to either speaker, and applying inverse STFT as discussed above. Computations base on a 2.4 GHz desktop PC with 4 GB of RAM, using a consumer grade GPU (NVIDIA GeForce GTX 560) with 336 CUDA cores. The NMF implementation from the open-source toolkit openBliSSART [12] is used. RTFs are computed by the elapsed GPU time over the length of the mixed signals. The number of separation iterations was chosen from {20, 40, 80, 160, 320} due to the quick saturation of the convergence of multiplicative update NMF algorithms in audio source separation [9]. The different DFT window sizes considered are powers of two, ranging from 2^6 to 2^{12}, or 8–256 ms assuming 16 kHz sample rate. From Fig. 8.3, it can be seen that the best average results are obtained by using the KL divergence as cost function. The Euclidean distance allows faster separation at the expense of quality, but here, reasonable results are only achieved for long window sizes (256 ms), which limits the practical applicability in contexts where real-time operation is required. Finally, the IS divergence enables robust separation, but is inferior to KL divergence both in terms of separation quality and RTF. Generally, it can be observed that in case of inadequate modeling of the sources (indicated by overall low SDR), more iterations do not necessarily improve separation quality, despite the fact that they linearly increase computational complexity; in fact, more iterations sometimes degrade quality, e.g., for the Euclidean cost function and 16 or 64 ms window size.

8.3 NMF Activation Features

Let us now move on to describe how NMF can be used directly for audio recognition, instead of performing signal pre-processing by audio source separation. The core idea is to use supervised or semi-supervised NMF (cf. above), and then directly exploit the matrix $\underline{H}$ for classification. In this case, NMF seeks a minimal-error representation of the signal (in terms of the cost-function) with only a set of given spectra. As outlined in Sect. 8.1, the $\underline{H}$ matrix measures the contribution of spectra to the original signal. Thus, by using a matrix $\underline{W}$ that contains spectra of different target classes, the rows of $\underline{H}$ provide information whether the original signal consists of components of these target classes. Furthermore, in this framework, additive noise can be modelled by simply introducing additional NMF components corresponding to noise.

For discrimination of C different audio signal classes $c \in \{1, \ldots, C\}$, the matrix $\underline{W}$ is built by column-wise concatenation:

$$\underline{W} := \underline{W}_1 | \underline{W}_2 | \cdots | \underline{W}_C | \underline{W}_N.$$

where each $\underline{W}_c$ contains 'characteristic' spectra of class c and the optional matrix $\underline{W}_N$ contains noise spectra. Similarly to the source separation application, there are a variety of methods for computing $\underline{W}_c$ and $\underline{W}_N$, such as base learning by NMF as in the supervised speaker separation example above, or simply by randomly sampling training spectrograms.

Based on this, NMF activation features can be derived from $\underline{H}$. In the example shown in Fig. 8.4, an exemplary scheme for static audio classification based on NMF activations is shown that delivered remarkable performance in discrimination of linguistic and non-linguistic vocalisations [15]. In this scheme, it is supposed that base learning by NMF is used. An activation feature vector $\underline{a} \in \mathbb{R}^R$ is calculated such that $\underline{a}_i$ is the Euclidean length of the i-th row of $\underline{H}$. For independence of the length and power of the signal, a_i is normalised such that $|\underline{a}|_1 = 1$. The 'NMF activation features' then are the components of the vector $\underline{a}$. This vector can be passed on to a suited classifier, or the activations per class can be summed up to derive class posteriors. In dynamic classification, e.g., the index of the most likely class per frame can be used as in [14, 21].

Let us now conclude the discussion of audio source separation and feature extraction by NMF by showing an exemplary application to keyword recognition in highly non-stationary noise [21]. This example is based on the CHiME (Computational Hearing in Multisource Environments) challenge task of recognising command words in a reverberated indoor domestic environment with multiple noise sources and interfering speakers [22].

NMD bases are learnt for each of the 51 words in the vocabulary, and an additional NMD noise base is computed from a set of noise samples in the training data. Speech separation is performed in a procedure similar to the speaker separation example above. Additionally, NMF activation features are computed using a base matrix $\underline{W}$ assembled from spectrogram 'patches' in the training data, in a 'sliding window NMF' framework (cf. above) with $T = 20$. As each speech spectrogram patch is

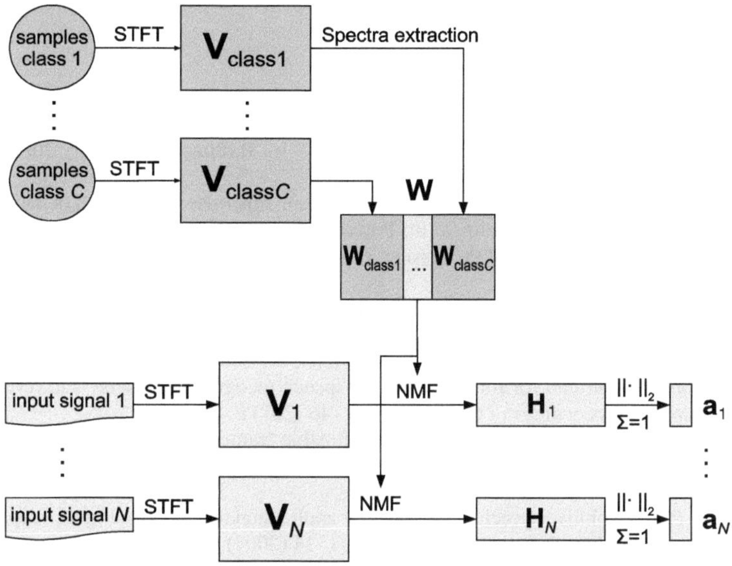

Fig. 8.4 Exemplary block diagram for extraction of NMF activation features for discrimination of C classes in N input signals [15]. Matrices denoted by $\underline{V}$ are spectrograms. The matrix $\underline{W}$ consists of spectra computed from training data for supervised NMF. Activation features are the resulting $\underline{H}$ (activation) matrices. $||\cdot||_2$ indicates that the Euclidean norm of each matrix row is computed, and $\sum = 1$ is a normalisation for the components of each vector $\underline{a}_i$ sum up to 1

associated with word likelihoods, the index of the most likely word per frame can be computed from the frame-wise activations of each spectrogram patch and used as a discrete feature. In this calculation of NMF activation features, $\underline{W}_N$ is pre-defined by training noise samples. Table 8.1 shows the WAs on the 35 keywords by SNR and on average, obtained by a baseline HMM recogniser adapted to noisy speech features, the results achieved by considering NMD speech separation as pre-processing, the results by usage of NMF activation features in HMM decoding, and combination of both. From the results, it is evident that both methods are complementary—the interested reader is referred to [21] for a more in-depth discussion.

Table 8.1 Effect of NMD speech separation and NMF activation features on speech recognition results (WA) reported in [21] on the Computational Hearing in Multisource Environments (CHiME) task [22]

WA [%]	SNR [dB]						Average
	−6	−3	0	3	6	9	
Baseline	54.5	61.1	72.8	81.7	86.8	91.3	74.7
NMD speech separation	75.6	79.2	84.1	87.7	88.3	90.6	84.2
NMF activation features	67.2	75.1	85.0	89.8	92.0	93.4	83.7
Combination	79.1	82.8	88.7	91.2	92.7	93.5	88.0

References

1. Schuller, B., Rigoll, G., Lang, M: Hmm-based music retrieval using stereophonic feature information and framelength adaptation. In: Proceedings 4th IEEE International Conference on Multimedia and Expo, ICME 2003, vol. II, pp. 713–716. Baltimore, MD, July 2003 (IEEE, IEEE)
2. Weninger, F., Feliu, J., Schuller, B.: Supervised and semi-supervised supression of background music in monaural speech recordings. In: Proceedings 37th IEEE International Conference on Acoustics, Speech, and Signal Processing, ICASSP 2012, pp. 61–64, Kyoto, Japan, March 2012 (IEEE, IEEE)
3. Hyvärinen, A., Karhunen, J., Oja, E.: Independent Component Analysis. John Wiley & Sons Inc., New York (2001)
4. Maas, R., Schwarz, A., Zheng, Y., Reindl, K., Meier, S., Sehr, A., Kellermann, W.: A two-channel acoustic front-end for robust automatic speech recognition in noisy and reverberant environments. In: Proceedings of CHiME, pp. 41–46 (2011)
5. Ozerov, A., Vincent, E., Bimbot, F.: A general flexible framework for the handling of prior information in audio source separation. IEEE Trans. Audio Speech Lang. Process. **20**(4), 1118–1133 (2012)
6. Smaragdis, P.: Convolutive speech bases and their application to supervised speech separation. IEEE Trans. Audio Speech Lang. Process. **15**(1), 1–14 (2007)
7. Virtanen, T.: Monaural sound source separation by nonnegative matrix factorization with temporal continuity and sparseness criteria. IEEE Trans. Audio Speech Lang. Process. **15**(3) (2007)
8. Wang, W., Cichocki, A., Chambers, J.A.: A multiplicative algorithm for convolutive non-negative matrix factorization based on squared Euclidean distance. IEEE Trans. Signal Process. **57**(7), 2858–2864 (2009)
9. Schuller, B., Lehmann, A., Weninger, F., Eyben, F., Rigoll, G.: Blind enhancement of the rhythmic and harmonic sections by nmf: Does it help? In: Proceedings International Conference on Acoustics including the 35th German Annual Conference on Acoustics, NAG/DAGA 2009, pp. 361–364, Rotterdam, The Netherlands: Acoustical Society of the Netherlands. DEGA, DEGA (2009)
10. Févotte, C., Bertin, N., Durrieu, J.-L.: Nonnegative matrix factorization with the Itakura-Saito divergence: with application to music analysis. Neural Comput. **21**(3), 793–830 (2009)
11. Duan, Z., Mysore, G.J., Smaragdis, P.: Speech enhancement by online non-negative spectrogram decomposition in non-stationary noise environments. In: Proceedings of Interspeech, Portland, OR, USA (2012)
12. Weninger, F., Schuller, B.: Optimization and parallelization of monaural source separation algorithms in the openblissart toolkit. J. Signal Process. Syst. **69**(3), 267–277 (2012)
13. Gemmeke, J.F., Virtanen, T.: Noise robust exemplar-based connected digit recognition. In: Proceedings of ICASSP, pp. 4546–4549, Dallas, TX, March 2010
14. Schuller, B., Weninger, F., Wöllmer, M., Sun, Y., Rigoll, G.: Non-negative matrix factorization as noise-robust feature extractor for speech recognition. In: Proceedings of 35th IEEE International Conference on Acoustics, Speech, and Signal Processing, ICASSP 2010, pp. 4562–4565, Dallas, TX, March 2010 (IEEE, IEEE)
15. Schuller, B., Weninger, F.: Discrimination of speech and non-linguistic vocalizations by non-negative matrix factorization. In: Proceedings of the 35th IEEE International Conference on Acoustics, Speech, and Signal Processing, ICASSP 2010, pp. 5054–5057, Dallas, TX, March 2010 (IEEE, IEEE)
16. Lee, D., Seung, H.: Algorithms for non-negative matrix factorization. In: Proceedings of NIPS, pp. 556–562, Vancouver, Canada (2001)
17. Schmidt, M.N., Olsson, R.K.: Single-channel speech separation using sparse non-negative matrix factorization. In: Proceedings of Interspeech, pp. 2–5, Pittsburgh, Pennsylvania (2006)
18. Ozerov, A., Févotte, C., Charbit M.: Factorial scaled hidden markov model for polyphonic audio representation and source separation. In: Proceedings of WASPAA, pp. 121–124, Mohonk, NY, United States (2009)

19. Helén, M., Virtanen, T.: Separation of drums from polyphonic music using non-negative matrix factorization and support vector machine. In Proceedings of EUSIPCO, Antalya, Turkey (2005)
20. Vincent, E., Gribonval, R., Févotte, C.: Performance measurement in blind audio source separation. IEEE Trans. Audio Speech Lang. Process. **14**(4), 1462–1469 (2006)
21. Weninger, F., Wöllmer, M., Geiger, J., Schuller, B., Gemmeke, J., Hurmalainen, A., Virtanen, T., Rigoll, G.: Non-negative matrix factorization for highly noise-robust asr: to enhance or to recognize? In: Proceedings 37th IEEE International Conference on Acoustics, Speech, and Signal Processing, ICASSP 2012, pp. 4681–4684, Kyoto, Japan, March 2012 (IEEE, IEEE)
22. Christensen, H., Barker, J., Ma, N., Green, P.: The CHiME corpus: a resource and a challenge for Computational Hearing in Multisource Environments. In: Proceedings of Interspeech, pp. 1918–1921, Makuhari, Japan (2010)

Chapter 9
Audio Enhancement and Robustness

Our view (...) is that it is an essential characteristic of experimentation that it is carried out with limited resources, and an essential part of the subject of experimental design to ascertain how these should be best applied; or, in particular, to which causes of disturbance care should be given, and which ought to be deliberately ignored.

—Sir Ronald A. Fisher

Once an audio recognition system that functions under idealistic conditions is established, the primary concern shifts towards making it robust in a real-world. The previous chapter touched this issue by illustrating how audio source separation can be exploited to recover a clean speech signal from a mixture. Extraction of the desired signal, however, is not a necessary pre-condition for robust audio recognition. Rather, several options exist for system improvement along the chain of processing, and have proved to be promising especially in the monaural case. Thus, we will next have a look at this issue following the overview given in [1].

First, filtering or spectral subtraction of the signal before can be applied directly after the audio capture. This is realised, for example, in the advanced front-end feature extraction (AFE) or Unsupervised Spectral Subtraction (USS). Then, auditory modelling can be introduced in the feature extraction process. The main influence of noise on audio is irreversible loss of information caused by its random behaviour and a distortion in the feature space that can be compensated by a suited audio representation in the noise condition [2, 3]. Examples of features in this direction include the MFCCs, PLP coefficients [4, 5] or RASTA-PLP features [6, 7] (cf. Sect. 6.2.1). Next along the chain of processing is the option to enhance the extracted features aiming at removal of effects as introduced by noise [8–10]. Exemplary techniques are normalisation methods such as (Cepstral) Mean Subtraction (CMS) [11], MVN [12], or HEQ [9]. Such feature enhancement can also be realised in a model based way, such as by jointly using a Switching Linear Dynamic Model (SLDM) for the dynamic behaviour of audio plus a Linear Dynamic Model (LDM) for additive noise [13]. Later in the chain, one could tailor the learning algorithm to

B. W. Schuller, *Intelligent Audio Analysis*, Signals and Communication Technology, 149
DOI: 10.1007/978-3-642-36806-6_9, © Springer-Verlag Berlin Heidelberg 2013

be able to cope with noisy signal input. Alternatives besides HMMs [14], such as Hidden Conditional Random Fields (HCRF) [15], Switching Autoregressive Hidden Markov Models (SAR-HMMs) [16], or other more general DBN structures provide higher flexibility in modelling. For example, the extension of an SAR-HMM to an Autoregressive Switching Linear Dynamical System (AR-SLDS) [17] allows for an explicit noise model leading to higher noise robustness. Another solution is to match the AM (or even LM) or feature space to noisy conditions. This requires a recogniser trained on noisy audio [18]. However, the performance highly dependends on how similar the noise conditions for training and testing are [19]. One can thus distinguish between low to highly matched conditions training. Further, it can be difficult to provide knowledge on the type of noisy condition. This can be eased by so-called multi-condition training, where clean and noisy material with different types of noise is mixed. This is usually not as good as perfectly matched condition between the current test setting and the one learnt previously. However, it provides a good compromise by improving on average over different noise conditions. Besides using noisy material for training, model adaptation can be used to quickly adapt the recogniser to a specific noise condition encountered in the test scenario. This covers widely used techniques such as maximum a posteriori (MAP) estimation [20], maximum likelihood linear regression (MLLR) [21], and minimum classification error linear regression (MCELR) [22].

Given the multiplicity of developed techniques for noise robustness in Intelligent Audio Analysis, a selection of relevant techniques and a good coverage of the different stages along the chain of processing is aimed at in this section. As these techniques are often also tailored to the specific type of noise at hand, relevant special cases such as white noise or babble noise are covered, which are very challenging for speech processing. In the ongoing, let us take a detailed look at the above mentioned options in particular for audio signal preprocessing, feature enhancement, and audio modelling. For the sake of better readability, 'audio of interest' such as speech, music, or specific sounds of interest as opposed to noise will partly simply be written as 'audio' in this chapter.

9.1 Audio Signal Preprocessing

The preprocessing of the audio signal for its enhancement shall compensate noise influence prior to the feature extraction [23–25]. Apart from explicit BASS as was shown in the last chapter, one of the frequently used audio and particular speech signal preprocessing [26] standards is the advanced front-end feature extraction introduced in [27] based on two-step Wiener filtering in the time domain. Spectral subtraction such as USS [10] can lead to similar effects at lower computational requirements in comparison to Wiener filtering [28, 29]. These techniques can also be subsumed under broader audio signal preprocessing despite being carried out in the (magnitude) spectogram domain. These two techniques will now be introduced in more detail.

9.1.1 Advanced Front-End Feature Extraction

The processing in the AFE [27] is shown in Fig. 9.1: Subsequent to noise reduction the denoised waveforms are processed and cepstral features are computed and blindly equalised.

Preprocessing in the AFE is based on two-stage Wiener filtering. After denoising in the first stage, a second one carries out additional dynamic noise reduction. In this second stage a gain factorisation unit controls the intensity of filtering dependent on the SNR. Figure 9.2 depicts the components of the two noise reduction cycles: First, a framing takes place. Then, the linear spectrum is estimated per frame, and the power spectral density (PSD) is smoothed along the time axis in the PSD Mean block. An audio activity detection (or VAD in the special case of speech) discriminates between audio and noise, and thus the estimated spectrum of the audio frames and noise are used in the computation of the frequency domain Wiener filter coefficients. To obtain a Mel-warped frequency domain Wiener filter, the linear Wiener filter coefficients are smoothed along the frequency axis using a Mel-filterbank [1]. The Mel-warped Inverse DCT unit (Mel IDCT) determines the impulse response of the Wiener filter prior to the input signal's filtering. The signal then passes through a second noise reduction cycle using this impulse response. Finally, the DC offset removal block eliminates the constant component of the filtered signal.

The Wiener filter approach in the AFE algorithm has the advantage that noise reduction is carried out on the frame-level. Further, the Wiener filter parameters are adapted to the current SNR. This allows to handle non-stationary noise. Important is, however, an exact audio activity detection (or VAD). This can be particularly demanding in the case of negative SNR levels (cf. e.g., Sect. 10.1.2). Overall, the AFE is a rather complex approach sensible to errors and inaccuracies within the individual estimation and transformation steps [1].

9.1.2 Unsupervised Spectral Subtraction

USS's [10] spectral subtraction scheme bases on a two-mixture model approach of noisy audio. It aims to distinguish audio and background noise at the magnitude spectogram level. A probability distribution is used to model audio and noise. For the modelling of background noise on silent parts of the time-frequency plane, one usually assumes white Gaussian behaviour for the real and imaginary parts [30, 31]. This corresponds to a Rayleigh probability density function $f_N(m)$ for noise in the magnitude domain:

Fig. 9.1 Feature extraction in the AFE according to ETSI ES 202 050 V1.1.5

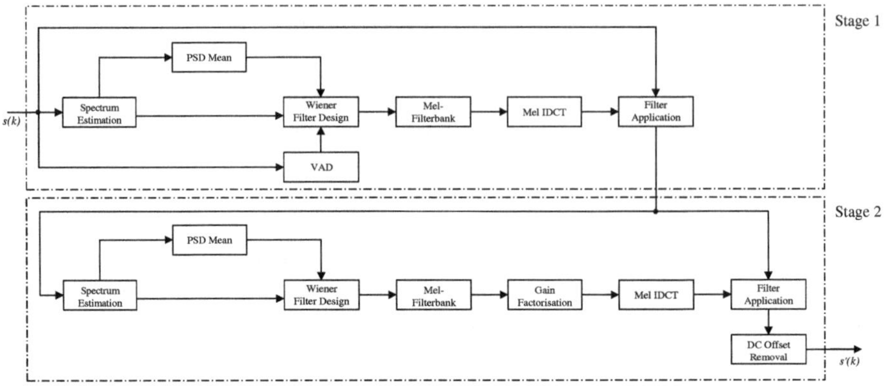

Fig. 9.2 Two-stage Wiener filtering for noise reduction in the AFE according to ETSI ES 202 050 V1.1.5

$$f_N(m) = \frac{m}{\sigma_N^2} e^{-\frac{m^2}{2\sigma_N^2}} \tag{9.1}$$

For the two-mixture model, only an audio 'activity' model modelling large magnitudes is needed besides the Rayleigh silence model. For the audio PDF $f_S(m)$ a threshold δ_S is defined with respect to the noise distribution $f_N(m)$ such that only magnitudes $m > \delta_S$ are modelled. In [10], a threshold $\delta_S = \sigma_N$ is used where σ_N is the mode of the Rayleigh PDF. Consequently, magnitudes below σ_N are assumed as background noise. Two additional constraints are needed for $f_S(m)$:

- The derivative $f_S'(m)$ of the 'activity' PDF may not be zero if m is just above δ_S; otherwise the threshold δ_S is meaningless as it could be set to an arbitrarily low value.
- With m towards infinity the decay of $f_S(m)$ should be lower than the decay of the Rayleigh PDF to guarantee $f_S(m)$ modelling large amplitudes.

The 'shifted Erlang' PDF with $h = 2$ [32] fulfils these two criteria. It can thus be used to model large amplitudes assumed to be audio of interest:

$$f_S(m) = 1_{m>\sigma_N} \cdot \lambda_S^2 \cdot (m - \sigma_N) \cdot e^{-\lambda_S(m-\sigma_N)} \tag{9.2}$$

with $1_{m>\sigma_N} = 1$ if $m > \sigma_N$ and $1_{m>\sigma_N} = 0$ otherwise.

The overall PDF for the spectral magnitudes of the noisy audio signal is

$$f(m) = P_N \cdot f_N(m) + P_S \cdot f_S(m), \tag{9.3}$$

where P_N is the prior for 'silence' and background noise, and P_S is the prior for 'activity' and audio of interest. The parameters of the derived PDF $f(m)$ summarised in the set

$$\Lambda = \{P_N, \sigma_N, P_S, \lambda_S\} \tag{9.4}$$

are independent of time and frequency, and can be trained by the EM algorithm (cf. Sect. 7.3.1) [33]. In the expectation step, posteriors are estimated as

$$p(\text{sil}|m_{f,t}, \Lambda) = \frac{P_N \cdot f_N(m_{f,t})}{P_N \cdot f_N(m_{f,t}) + P_S \cdot f_S(m_{f,t})} \tag{9.5}$$

$$p(\text{act}|m_{f,t}, \Lambda) = 1 - p(\text{sil}|m_{f,t}, \Lambda). \tag{9.6}$$

For the Maximisation step, the moment method is used: An update σ_N employing all data takes place before all data with values above the new σ_N help to update λ_S. Two update equations describe the method as follows:

$$\hat{\sigma}_N = \frac{\left[\sum_{f,t} m_{f,t}^2 \cdot p(\text{sil}|m_{f,t}, \Lambda)\right]^{\frac{1}{2}}}{\left[2\sum_{f,t} p(\text{sil}|m_{f,t}, \Lambda)\right]^{\frac{1}{2}}} \tag{9.7}$$

$$\hat{\lambda}_S = \frac{\sum_{m_{f,t} > \hat{\sigma}_N} (m_{f,t} - \hat{\sigma}_N)^{-1} \cdot p(\text{act}|m_{f,t}, \Lambda)}{\sum_{m_{f,t} > \hat{\sigma}_N} p(\text{act}|m_{f,t}, \Lambda)}. \tag{9.8}$$

Subsequent to the training of all mixture parameters $\Lambda = \{P_N, \sigma_N, P_S, \lambda_S\}$ USS with the parameter σ_N as floor value is applied:

$$m_{f,t}^{USS} = \max\left(1, \frac{m_{f,t}}{\sigma_N}\right) \tag{9.9}$$

Flooring to a non-zero value is required for MFCC or similar features, as zero magnitude values after spectral subtraction can result in unfavourable dynamics. Overall, USS is a simple and efficient preprocessing method that allows for unsupervised EM fitting on observed data. As a downside, it requires reliable estimation of an audio magnitude PDF which is rather challenging. With the PDFs not depending on frequency and time, USS only handles stationary noises. Further, it only models large magnitudes of the audio of interest. Low audio magnitudes thus cannot be distinguished from background noise.

9.2 Feature Enhancement

In feature enhancement, enhancement takes place after the extraction of features to reduce a potential mismatch between test and training conditions. Popular methods include CMS [11], MVN [12], HEQ [9], and the Taylor Series approach [34] able to cope with the non-linear effects of noise. There are some further methods tailored to specific types of features, such as in the cepstrum-domain, where a feature compensation algorithm to decompose audio of interest and noise is introduced in [35]. To enhance noisy MFCCs, a SLDM can also be used to model the dynamics of audio of interest and those of additive noise by a LDM [13]. An observation model then describes how audio and noise produce the noisy observations to reconstruct the features of clean audio. An extension [36] includes time-dependencies among the discrete state variables of the SLDM. Further, a state model for the dynamics of noise can help to model non-stationary noise sources [37]. Finally, incremental on-line adaptation of the feature space is possible as by feature space maximum likelihood linear regression (FMLLR) [38]. Again, we will now take a detailed look at selected popular approaches.

9.2.1 Feature Normalisation

9.2.1.1 Cepstral Mean Subtraction

To ease the influence of noise and transmission channel transfer functions in cepstral features, CMS [11, 39] provides a simple approach. Its basic principle of mean subtraction can also be applied to practically any other audio LLD. Often, the noise can be considered as comparably stationary when opposed to the rapidly changing characteristics of the audio signal of interest. Thus, a subtraction is carried out of the long-term average cepstral or other feature vector

$$\underline{\mu} = \frac{1}{T}\sum_{t=1}^{T} \underline{x}_t \tag{9.10}$$

from the observed noise corrupted feature vector sequence of length T:

$$\underline{X} = \{\underline{x}_1, \underline{x}_2, \ldots, \underline{x}_t, \ldots, \underline{x}_T\} \tag{9.11}$$

By that, a new estimate $\tilde{x}_t$ of the signal in the feature domain results:

$$\underline{\tilde{x}}_t = \underline{x}_t - \underline{\mu}, \quad 1 \leq t \leq T \tag{9.12}$$

The subtraction of the long-term average is particularly interesting in the cepstral domain. Since the audio spectrum is multiplied by the channel transfer function (cf. Sect. 6.2.1.4), by the logarithm application in the MFCC calculation, this multiplication turns into an addition, and this part can be eliminated by subtraction of the cepstral mean from all input vectors. A disadvantage of CMS, as opposed to HEQ, is the disability to treat non-linear noise effects.

9.2.1.2 Mean and Variance Normalisation

The subtraction of the mean per feature vector component corresponds to an equalisation of the first moment of the vector sequence probability distribution. If noise has also an influence on the variance of the features, according variance normalisation of the vector sequence can be applied and by that an equalisation of the first two moments. This is known as MVN. The processed feature vector is obtained by

$$\underline{\tilde{x}}_t = \frac{\underline{x}_t - \underline{\mu}}{\underline{\sigma}}. \tag{9.13}$$

The division by the vector $\underline{\sigma}$ of the standard deviations per feature vector components is computed out element-by-element. The new feature vector's components have zero mean and unity variance.

9.2.1.3 Histogram Equalisation

HEQ is a popular technique in digital image processing [40] where it helps raise the contrast of images and alleviates the influence of the lighting conditions. In audio processing, HEQ can improve the temporal dynamics of noise-affected feature vector components. HEQ extends the principle of CMS and MVN to all moments of the probability distribution of the feature vector components [9, 41], and by that compensates non-linear distortions caused by noise.

In HEQ, the histogram of each feature vector component is mapped onto a reference histogram. The underlying assumption is that noise influence can be described as a monotonic partly reversible feature transformation. With success depending on meaningful histograms, HEQ requires several frames for their reliable estimation. A key advantage lending to HEQ's independence of the noise characteristics is that no assumptions are made on the statistical properties (e.g., normality) of the noise process.

For HEQ, a transformation

$$\tilde{x} = F(x) \tag{9.14}$$

needs to be found for the conversion of the PDF $p(x)$ of an audio feature into a reference PDF $\tilde{p}(\tilde{x}) = p_{ref}(\tilde{x})$. If x is a unidimensional variable with PDF $p(x)$, a transformation $\tilde{x} = F(x)$ modifies the probability distribution, such that the new distribution of the obtained variable $\tilde{x}$ can be expressed as

$$\tilde{p}(\tilde{x}) = p(G(\tilde{x}))\frac{\partial G(\tilde{x})}{\partial \tilde{x}} \tag{9.15}$$

with $G(\tilde{x})$ as the inverse transformation corresponding to $F(x)$. For the cumulative probabilities based on the PDFs, let us consider:

$$
\begin{aligned}
C(x) &= \int_{-\infty}^{x} p(x')dx' \\
&= \int_{-\infty}^{F(x)} p(G(\tilde{x}'))\frac{\partial G(\tilde{x})}{\partial \tilde{x}'}d\tilde{x}' \\
&= \int_{-\infty}^{F(x)} \tilde{p}(\tilde{x}')d\tilde{x}' \\
&= \tilde{C}(F(x))
\end{aligned}
\tag{9.16}
$$

By that, the transformation converting the distribution $p(x)$ into the 'target' distribution $\tilde{p}(\tilde{x}) = p_{ref}(\tilde{x})$ can be expressed as

$$\tilde{x} = F(x) = \tilde{C}^{-1}[C(x)] = C_{ref}^{-1}[C(x)], \tag{9.17}$$

where $C_{ref}^{-1}(\dots)$ is the inverse cumulative probability function of the reference distribution [1]. Further, $C(\dots)$ is the feature's cumulative probability function. To obtain the transformation per feature vector component, a 'rule of thumb' is to use 500 uniform intervals between $\mu_i - 4\sigma_i$ and $\mu_i + 4\sigma_i$ for the derivation of the histograms. μ_i and σ_i are the mean and standard deviation of the ith feature vector element. A Gaussian probability distribution with zero mean and unity variance can be used per element as a reference probability distribution, then, however, ignoring higher moments.

From the feature normalisation strategies discussed above, CMS is the simplest. Together with MVN, it is used most frequently. MVN usually leads to better results at slightly increased computational effort. However, these two techniques both provide a linear transformation. This is different for HEQ, which is able to compensate non-linear effects, but requires sufficient audio frames for good results. HEQ further corrects only monotonic transformations. This can cause an information loss, given that random noise behaviour renders the needed transformation non-monotonic.

9.2.2 Model Based Feature Enhancement

In model based audio enhancement one usually models audio and noise individually plus how these two produce the observation. Then, the features are enhanced to benefit the audio of interest by use of these models. An example is a SLDM to model the dynamics of clean audio of interest [13] that will next be introduced by the mentioned three models for noise, audio, and the combination.

Fig. 9.3 LDM for the modelling of noise

9.2.2.1 Modelling of Noise

Noise is modelled by a simple LDM with the system equation

$$\underline{x}_t = \underline{A}\,\underline{x}_{t-1} + \underline{b} + \underline{g}_{t'}, \tag{9.18}$$

where the matrix $\underline{A}$ and the vector $\underline{b}$ simulate the noise process's evolution over time. Further, $\underline{g}_t$ is a Gaussian noise source that drives the system. A graphical model representation of this LDM is given in Fig. 9.3. In this and the following visualisations in this section, squares again indicate observations. With LDMs being time-invariant, they can model signals such as coloured stationary Gaussian noises. The LDM is expressed by

$$p(\underline{x}_t|\underline{x}_{t-1}) = \mathcal{N}(\underline{x}_t; \underline{A}\,\underline{x}_{t-1} + \underline{b}, \underline{C}) \tag{9.19}$$

$$p(\underline{x}_{1:T}) = p(\underline{x}_1)\prod_{t=2}^{T} p(\underline{x}_t|\underline{x}_{t-1}), \tag{9.20}$$

where $\mathcal{N}(\underline{x}_t; \underline{Ax}_{t-1} + \underline{b}, \underline{C})$ is a multivariate Gaussian with the mean vector $\underline{Ax}_{t-1} + \underline{b}$ and the covariance matrix $\underline{C}$, and T is the input sequence's length.

9.2.2.2 Modelling of Audio of Interest

The SLDM models the audio signal of interest passing through states as in a HMM. It further enforces a continuous state transition in the feature space conditioned on the state sequence. This more complex dynamic model has a hidden state variable s_t at each time t. Like this, $\underline{A}$ and $\underline{b}$ depend on the state variable s_t:

$$\underline{x}_t = \underline{A}(s_t)\underline{x}_{t-1} + \underline{b}(s_t) + \underline{g}_t. \tag{9.21}$$

Likewise, the possible state sequences $s_{1:T}$ describe a non-stationary LDM, as $\underline{A}$ and $\underline{b}$ change with time as do the audio features. In Fig. 9.4 the SLDM is shown as graphical model. As one sees, time dependencies are assumed between the continuous

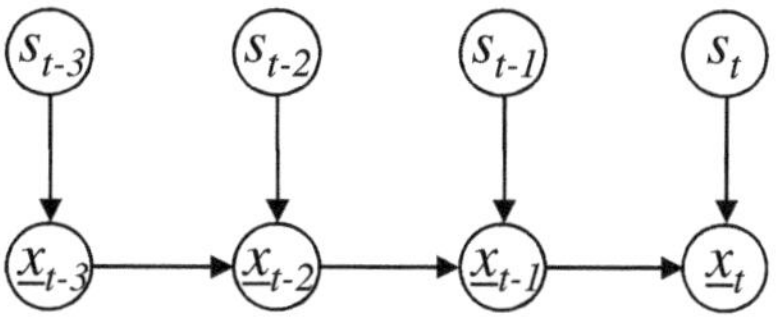

Fig. 9.4 SLDM for the modelling of audio of interest

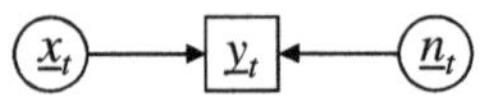

Fig. 9.5 Observation model for noisy audio

variables $\underline{x}_t$, but not between the discrete state variables s_t [13]. An extension in [36] includes time dependencies between the hidden state variables, similar as in enhancing a GMM to a HMM. A SLDM as in Fig. 9.4 is described by

$$p(\underline{x}_t, s_t|\underline{x}_{t-1}) = \mathcal{N}(\underline{x}_t; \underline{A}(s_t)\underline{x}_{t-1} + \underline{b}(s_t), \underline{C}(s_t)) \cdot p(s_t) \tag{9.22}$$

$$p(\underline{x}_{1:T}, s_{1:T}) = p(\underline{x}_1, s_1) \prod_{t=2}^{T} p(\underline{x}_t, s_t|\underline{x}_{t-1}). \tag{9.23}$$

The EM algorithm can be used for the learning of the parameters of the SLDM, namely $\underline{A}(s)$, $\underline{b}(s)$, and $\underline{C}(s)$. If one sets the number of states to one the SLDM turns into a LDM to compute the parameters $\underline{A}$, $\underline{b}$, and $\underline{C}$ required for the noise modelling LDM.

9.2.2.3 Observation Model

The observation model describes the relationship of the noisy observation $\underline{y}_t$ and the hidden audio and noise features. In Fig. 9.5, the graphical model representation of such a model is given by the zero variance observation model with SNR inference as in [42]. It is assumed that audio of interest $\underline{x}_t$ and noise $\underline{n}_t$ mix linearly in the time domain. In the cepstral domain, for example, this corresponds to a non-linear mixing.

9.2.2.4 Posterior Estimation and Enhancement

To reduce the computational complexity of the posterior estimation, an approximation is given by the restriction of the search space size by the generalised pseudo-Bayesian (GPB) algorithm [43]. It neglects distinct state histories with differences more than r frames in the past. Thus, with T as the sequence length, the inference complexity reduces from S^T to S^r where $r \ll T$. In the GPB algorithm, one 'collapses', 'predicts', and 'observes' for each of the audio frames. Estimates of the moments of $\underline{x}_t$ representing the de-noised audio features are computed based on the Gaussian posterior as calculated during the 'observation' in the GPB algorithm. In this process, clean features are assumed to be the Minimum Mean Square Error (MMSE) estimate $E[\underline{x}_t|\underline{y}_{1:t}]$. SLDM feature enhancement can lead to outstanding results including the case of coloured Gaussian noise and negative SNR. This comes by the effort of modelling noise. The audio model's linear dynamics model the the smooth time evolution of typical audio of interest such as speech, music, or certain sound types. The switching states express the piecewise stationarity typical in such audio. However, noise frames are assumed to be independent over time. As a consequence, non-stationary noises are not modelled adequately. Even with the restrictions made in the GPB algorithm, feature enhancement by SLDM is computationally more demanding than the techniques discussed above. Further, as in the AFE (cf. Sect. 9.1), accurate audio activity detection is required to provide correct estimation of the noise LDM.

9.3 Model Architectures

The most frequently used data-driven model representation of audio are HMMs [14]. Beyond the so far described optimisation options along the chain of Intelligent Audio Analysis, extending HMM topologies to more general DBN layouts can also help to increase noise robustness [15, 17, 44]. Generative models such as HMMs assume conditional independence of the audio feature observations, thus ignoring long-range dependencies as given in most audio of interest [45]. To overcome this, Conditional Random Fields (CRF) [46–48] model a sequence by an exponential distribution given the observation sequence. The HCRF [15, 49] further includes hidden state sequences for the estimation of the conditional probability of a class over an entire sequence. Another interesting option is to model the raw audio signal in the time domain [16]. For example, SAR-HMM [16] provide good results in clean audio conditions. To cope with noise, these can be extended to a Switching Linear Dynamical System (SLDS) [17] to model the dynamics of the raw audio signal and the noise. These alternatives will now be shortly presented.

9.3.1 *Conditional Random Fields*

As mentioned above, CRF [46] use an exponential distribution to model a sequence given its observation and by that also non-local dependencies among states and observations. Further, unnormalised transition probabilities are possible. Owing to the ability to enforce a Markov assumption as in HMMs, dynamic programming is applicable for inference. CRFs were also shown beneficial as LM [50].

9.3.2 *Hidden Conditional Random Fields*

An extension to HCRF is needed to make the CRF paradigm suited for general audio recognition tasks. This comes, as CRF provide a class prediction per observation and frame of a time sequence rather than for an entire sequence. HCRF overcome this by adding hidden state sequences [49]. Reports of superiority over HMM in the Intelligent Audio Analysis domain include the recognition of phones [15] and non-linguistic vocalisations [51] or the segmentation of meeting speech [52]. A particular strength is the possibility to use arbitrary functions for the observations without complication of the parameter learning.

The HCRF models the conditional probability of a class c, given the sequence of observations $\underline{X} = \underline{x}_1, \underline{x}_2, \ldots, \underline{x}_T$:

$$p(c|\underline{X}, \underline{\lambda}) = \frac{1}{z(\underline{X}, \underline{\lambda})} \sum_{Seq \in c} e^{\underline{\lambda}\, \underline{f}(c, Seq, \underline{X})}, \tag{9.24}$$

where $\underline{\lambda}$ is the parameter vector and $\underline{f}$ the 'vector of sufficient statistics', and $Seq = s_1, s_2, \ldots, s_T$ is the hidden state sequence run through during the computation of this conditional probability. The probability is normalised by the 'partition function' $z(\underline{X}, \underline{\lambda})$ to ensure a properly normalised probability [15]:

$$z(\underline{X}, \underline{\lambda}) = \sum_{c} \sum_{Seq \in c} e^{\underline{\lambda}\, \underline{f}(c, Seq, \underline{X})}. \tag{9.25}$$

The vector $\underline{f}$ determines the probability to model. With a suited $\underline{f}$ a left-right HMM can be imitated [15]. Let us now now restrict the HCRF to a Markov chain, but without the requirements of the transition probabilities to sum to one and real probability densities for the observations. In analogy to a HMM a parametrisation by transition scores $a_{i,j}$ and observation scores $b_j(\underline{x}_t)$ can then be reached with the parameters $\underline{\lambda}$, where and i and j are states of the model (cf. Sect. 7.3.2). Forward and backward recursions (cf. Sect. 7.3.1) as for a HMM can then further be used.

9.3.3 Audio Modelling in the Time Domain

Modelling of the raw signal in the time domain is a sparsely pursued option, but can offer easy explicit noise modelling [16]. We will look at SAR-HMMs to this end first, and then at the extension to SLDS.

9.3.3.1 Switching Autoregressive Hidden Markov Models

The SAR-HMM models the audio signal of interest as an autoregressive (AR) process. The non-stationarity is realised by switching between different AR parameter sets [17] by a discrete switch variable s_t similar to the HMM states. At a time step t—referring to the sample-level in this case—, exactly one out of S states is occupied. The state at time step t depends exclusively on its predecessor with the transition probability $p(s_t|s_{t-1})$. The sample v_t at this time step is assumed as a linear combination of its R preceding samples superposed by a Gaussian distributed 'innovation' $\eta(s_t)$. $\eta(s_t)$ and the AR weights $c_r(s_t)$ are the parameter set given by the state s_t:

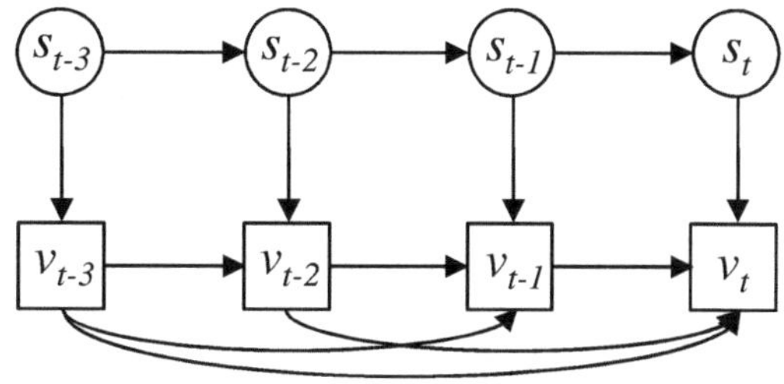

Fig. 9.6 SAR-HMM as DBN structure

$$v_t = -\sum_{r=1}^{R} c_r(s_t) v_{t-r} + \eta(s_t) \quad \text{with} \quad \eta \sim \mathcal{N}(\eta; 0, \sigma^2(s_t)). \tag{9.26}$$

There, $\eta(s_t)$ models variations from pure autoregression rather than an independent additive noise process. The joint probability of a sequence of length T is

$$p(s_{1:T}, v_{1:T}) = p(v_1|s_1)p(s_1) \prod_{t=2}^{T} p(v_t|v_{t-R:t-1}, s_t) p(s_t|s_{t-1}). \tag{9.27}$$

Figure 9.6 visualises the SAR-HMM as DBN structure. Switching of the different AR models is forcedly 'slowed down' by introducing an constant K. The model then needs to remain in a state for an integer multiple of time steps. This is needed, as considerably more sample values usually exist than features on the frame level.

The EM algorithm can be used for learning of the AR parameters. Based on the forward-backward algorithm (cf. Sect. 7.3.1) the distributions $p(s_t|v_{1:T})$ are learnt. The fact that an observation v_t depends on R predecessors makes the backward pass more complicated than in the case of an HMM. A 'correction smoother' [53] can thus be applied such that the backward pass calculates the posterior $p(s_t|v_{1:T})$ by 'correcting' the forward pass's output.

9.3.3.2 Autoregressive Switching Linear Dynamical Systems

With the extension of the SAR-HMM to an AR-SLDS, a noise process can explicitly be modelled [17]. The observed audio sample v_t of interest is then modelled as a noisy version of a hidden clean sample that is obtained from the projection of a hidden vector $\underline{h}_t$ with the dynamic properties of a LDS:

$$\underline{h}_t = \underline{A}(s_t)\underline{h}_{t-1} + \underline{\eta}_t^{\mathcal{H}}, \quad \text{with} \quad \underline{\eta}_t^{\mathcal{H}} \sim \mathcal{N}\left(\underline{\eta}_t^{\mathcal{H}}; 0, \underline{\Sigma}_{\mathcal{H}}(s_t)\right). \tag{9.28}$$

The transition matrix $\underline{A}(s_t)$ describes the dynamics of the hidden variable that depends on the state s_t at time step t. A Gaussian distributed hidden 'innovation' variable $\underline{\eta}_t^{\mathcal{H}}$ models variations from 'pure' linear state dynamics. As for η_t in Eq. (9.26)

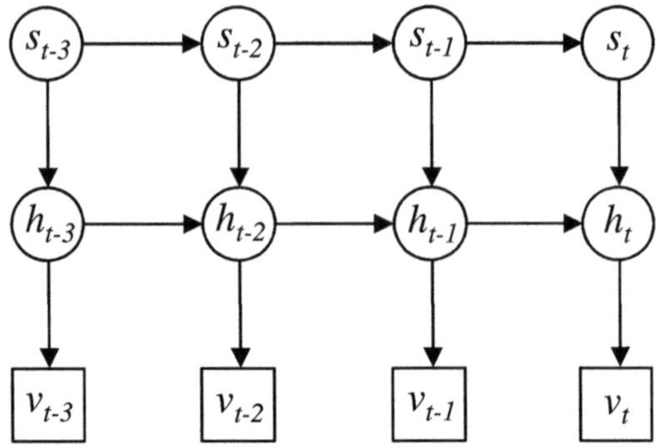

Fig. 9.7 AR-SLDS as DBN structure

in the case of the SAR-HMM, $\underline{\eta}_t^{\mathcal{H}}$ is not modelling an independent additive noise source. For the determination of the observed sample at time step t, the vector $\underline{h}_t$ is projected onto a scalar v_t:

$$v_t = \underline{B}\,\underline{h}_t + \eta_t^{\mathcal{V}}, \quad \text{with} \quad \eta_t^{\mathcal{V}} \sim \mathcal{N}(\eta_t^{\mathcal{V}}; 0, \sigma_{\mathcal{V}}^2), \tag{9.29}$$

where $\eta_t^{\mathcal{V}}$ models independent additive white Gaussian noise (AWGN) assumed to modify the hidden clean sample $\underline{Bh}_t$. The DBN structure of the SLDS that models the hidden clean signal and an independent additive noise is found in Fig. 9.7.

The parameters $\underline{A}(s_t)$, $\underline{B}$ and $\underline{\Sigma}_{\mathcal{H}}(s_t)$ of the SLDS can be chosen to mimic the SAR-HMM (cf. Sect. 9.3.3.1) for the case $\sigma_{\mathcal{V}} = 0$ [17]. Likewise, if $\sigma_{\mathcal{V}} \neq 0$ a noise model is included but no training of a new model is needed. With determination of the exact parameters of the AR-SLDS having a complexity of $\mathcal{O}(S^T)$, the Expectation Correction (EC) approximation [54] provides an elegant reduction to $\mathcal{O}(T)$.

In practice, the AR-SLDS is particularly suited to cope with white noise disturbance, as the variable $\eta_t^{\mathcal{V}}$ incorporates an AWGN model. It is, however, usually inferior to frame-level feature-based HMM approaches in clean conditions. This may be explained by the difference of the approach to human perception which is not performed in the time-domain. In coloured noise environment the AR-SLDS usually also leads to lower performance than frame-level feature modelling as by SLDMs. A limitation for practical use is the high computational requirement, even with the EC algorithm: As an example, for audio at 16 kHz, T is 160 times higher than for a feature vector sequence operated on 100 FPS.

Obviously, further model architectures exist that were not shown here, but are well suited to cope with noises, in particular also for non-stationary noise. An example are the LSTM networks as shown in Sect. 7.2.3.4 [55, 56].

References

1. Schuller, B., Wöllmer, M., Moosmayr, T., Rigoll, G.: Recognition of noisy speech: a comparative survey of robust model architecture and feature enhancement. EURASIP J. Audio Speech Music Process. (Article ID 942617), 17 (2009)
2. de la Torre, A., Fohr, D., Haton, J.: Compensation of noise effects for robust speech recognition in car environments. In: Proceedings of International Conference on Spoken Language Processing (2000)
3. Moreno, P.: Speech recognition in noisy environments. Ph.D. thesis, Carnegie Mellon University, Pittsburgh (1996)
4. Hermansky, H.: Perceptual linear predictive (PLP) analysis of speech. J. Acoust. Soc. Am. **87**, 1738–1752 (1990)
5. Junqua, J., Wakita, H., Hermansky, H.: Evaluation and optimization of perceptually-based ASR front-end. IEEE Trans. Speech Audio Process. **1**, 329–338 (1993)
6. Hermansky, H., Morgan, N., Bayya, A., Kohn, P.: RASTA-PLP speech analysis technique. In: Proceedings of International Conference on Acoustics, Speech, and Signal Processing, vol. 1, pp. 121–124 (1992)
7. Kingsbury, B., Morgan, N., Greenberg, S.: Robust spech recognition using the modulation spectrogram. Speech Commun. **25**, 117–132 (1998)
8. Kim, N.: Nonstationary environment compensation based on sequential estimation. IEEE Signal Process. Lett. **5**, 57–59 (1998)
9. de la Torre, A., Peinado, A.M., Segura, J.C., Perez-Cordoba, J.L., Benitez, M.C., Rubio, A.J.: Histogram equalization of speech representation for robust speech recognition. IEEE Trans. Speech Audio Process. **13**(3), 355–366 (2005)
10. Lathoud, G., Magimia-Doss, M., Mesot, B., Boulard, H.: Unsupervised spectral subtraction for noise-robust ASR. In: Proceedings of Automatic Speech Recognition and Understanding, pp. 189–194 (2005)
11. Rahim, M., Juang, B., Chou, W., Buhrke, E.: Signal conditioning techniques for robust speech recognition. In: Proceedings of IEEE Signal Processing Letters, vol. 3, pp. 107–109 (1996)
12. Viikki, O., Laurila, K.: Cepstral domain segmental feature vector normalization for noise robust speech recognition. Speech Commun. **25**, 133–147 (1998)
13. Droppo, J., Acero, A.: Noise robust speech recognition with a switching linear dynamic model. In: Proceedings of the 2004 IEEE International Conference on Acoustics, Speech and Signal Processing, vol. 1, pp. 953–956 (2004)
14. Rabiner, L.R.: A tutorial on hidden markov models and selected applications in speech recognition. In: Proceedings of the IEEE, vol. 77, pp. 257–286 (1989)
15. Gunawardana, A., Mahajan, M., Acero, A., Platt, J.C.: Hidden conditional random fields for phone classification. In: Proceedings of Interspeech, pp. 1117–1120 (2005)
16. Ephraim, Y., Roberts, W.: Revisiting autoregressive hidden Markov modeling of speech signals. In: IEEE Signal Processing Letters, vol. 12, pp. 166–169 (2005)
17. Mesot, B., Barber, D.: Switching linear dynamical systems for noise robust speech recognition. IEEE Trans. Audio Speech Lang. Process. **15**, 1850–1858 (2007)
18. Sankar, A., Stolcke, A., Chung, T., Neumeyer, L., Weintraub, M., Franco, H., Beaufays, F.: Noise-resistant feature extraction and model training for robust speech recognition. In: Proceedings of the 1996 DARPA CSR, Workshop (1996)
19. Macho, D., Mauuray, L., Noe, B., Cheng, Y., Ealey, D., Jouvet, D., Kelleher, H., Pearce, D., Saadoun, F.: Evaluation of a noise-robust DSR front-end on Aurora databases. In: Proceedings of the International Conference on Spoken Language Processing, pp. 17–20 (2002)
20. Gauvain, J., Lee, C.: Maximum a posteriori estimation for multivariate Gaussian mixture observations of Markov chains. IEEE Trans. Speech Audio Process. **2**, 291–298 (1994)
21. Wang, Z., Schultz, T., Waibel, A.: Comparison of acoustic model adaptation techniques on non-native speech. In: Proceedings of International Conference on Acoustics, Speech, and Signal Processing, vol. 1, pp. 540–543 (2003)

22. He, X., Chou, W.: Minimum classification error linear regression for acoustic model adaptation of continuous density HMMs. In: Proceedings of International Conference on Multimedia and Expo, vol. 1, pp. 397–400 (2003)
23. Szymanski, L., Bouchard, M.: Comb filter decomposition for robust ASR. In: Proceedings of Interspeech, pp. 2645–2648 (2005)
24. Rifkin, R., Schutte, K., Saad, M., Bouvrie, J., Glass, J.: Noise robust phonetic classification with linear regularized least squares and second-order features. In: Proceedings of International Conference on Acoustics, Speech, and Signal Processing (2007)
25. Raj, B., Turicchia, L., S.-N. B., Sarpeshkar, R.: An FFT-based companding front end for noise-robust automatic speech recognition. In: European Association for Signal Processing Journal on Audio, Speech, and Music Processing, volume 2007 (2007)
26. Hirsch, H.G., Pierce, D.: The AURORA experimental framework for the performance evaluation of speech recognition systems under noise conditions. Challenges for the Next Millenium, Automatic Speech Recognition (2000)
27. ETSI. ETSI ES 202 050 V1.1.5—Speech Processing, Transmission and Quality Aspects (STQ), Distributed speech recognition, Advanced front-end feature extraction algorithm, Compression algorithms (2007)
28. Lathoud, G., Doss, M., Boulard, H.: Channel normalization for unsupervised spectral subtraction, In: Proceedings of Automatic Speech Recognition and Understanding (2005)
29. Vaseghi, S., Milner, B.: Noise compensation methods for Hidden Markov model speech recognition in adverse environments. IEEE Trans. Speech Audio Process. **5**, 11–21 (1997)
30. Martin, R., Breithaupt, C.: Speech enhancement in the DFT domain using Laplacian speech priors. In: Proceedings of International Workshop on Acoustic Echo and Noise, Control (2003)
31. Ephraim, Y., Malah, D.: Speech enhancement using a minimum-mean square error short-time spectral amplitude estimator. IEEE Trans. Speech Audio Process. **32**, 1109–1121 (1984)
32. Grinstead, C., Snell, J.: Introduction to probability. American Mathematical Society, Rhode Island (1997)
33. Dempster, A., Laird, N.M., Rubin, D.B.: Maximum likelihood from incomplete data via the EM algorithm. J. Roy. Stat. Soc. **39**, 1–38 (1977)
34. Moreno, P., Raj, B., Stern, R.: A vector Taylor series approach for environment-independent speech recognition. In: Proceedings of International Conference on Acoustics, Speech, and Signal Processing, vol. 2, pp. 733–736 (1996)
35. Kim, H., Rose, R.: Cepstrum-domain acoustic feature compensation based on decomposition of speech and noise for ASR in noisy environments. IEEE Trans. Speech Audio Process. **11**, 435–446 (2003)
36. Deng, J., Bouchard, M., Yeap, T.H.: Noisy speech feature estimation on the Aurora2 database using a switching linear dynamic model. J. Multimedia **2**, 47–52 (2007)
37. Windmann, S., Haeb-Umbach, R.: Modeling the dynamics of speech and noise for speech feature enhancement in ASR. In: Proceedings of International Conference on Acoustics, Speech, and, Signal Processing, pp. 4409–4412 (2008)
38. Li, Y., Erdogan, H., Gao, Y., Marcheret, E.: Incremental on-line feature space MLLR adaptation for telephony speech recognition. In: Proceedings of International Conference on Spoken Language Processing, pp. 1417–1420 (2002)
39. Jankowski, C., Vo, H.-D., Lippmann, R.: A comparison of signal processing front ends for automatic word recognition. IEEE Trans. Speech Audio Process. **3**, 286–293 (1995)
40. Kim, J., Kim, L., Hwang, S.: An advanced contrast enhancement using partially overlapped sub-block histogram equalization. IEEE Trans. Circuits Syst. Video Technol. **11**, 475–484 (2001)
41. Hilger, F., Ney, H.: Quantile based histogram equalization for noise robust large vocabulary speech recognition. IEEE Trans. Audio Speech Lang. Process. **14**, 845–854 (2006)
42. Droppo, J., Deng, L., Acero, A.: A comparison of three non-linear observation models for noisy speech features. In: Proceedings of Eurospeech, vol. 2003, pp. 681–684 (2003)
43. Bar-Shalom, Y., Li, X.: Estimation and Tracking: Principles, Techniques, and Software. Artech House, Norwood (1993)

44. Ganapathiraju, A., Hamaker, J., Picone, J.: Applications of support vector machines to speech recognition. IEEE Trans. Signal Process. **52**, 2348–2355 (2004)
45. Bilmes, J.A.: Maximum mutual information based reduction strategies for cross-correlation based joint distributional modeling. In: Proceedings of ICASSP, pp. 469–472. Seattle, Washington (1998)
46. Lafferty, J., McCallum, A., Pereiar, F.: Conditional random fields: probabilistic models for segmenting and labeling sequence data. In: Proceedings of International Conference on, Machine Learning, pp. 282–289 (2001)
47. Sha, F., Pereira, F.: Shallow parsing with conditional random fields. In: NAACL '03: Proceedings of the 2003 Conference of the North American Chapter of the Association for Computational Linguistics on Human Language Technology, Association for Computational Linguistics. Morristown, NJ, USA. pp. 134–141 (2003)
48. Pinto, D., McCallum, A., Wei, X., Croft, W.: Table extraction using conditional random fields. In: Proceedings of the 26th annual international ACM SIGIR conference on Research and development in, information retrieval, pp. 235–242 (2003)
49. Quattoni, A., Collins, M., Darrell, T.: Conditional random fields for object recognition. In: Advances in Neural Information Processing Systems, vol. 17, pp. 1097–1104 (2005)
50. Roark, B., Saraclar, M., Collins, M., Johnson, M.: Discriminative language modeling with conditional random fields and the perceptron algorithm. In: Proceedings of Association for, Computational Linguistics, pp. 48–55 (2004)
51. Schuller, B., Eyben, F., Rigoll, G.: Static and dynamic modelling for the recognition of non-verbal vocalisations in conversational speech. In: André, E., Dybkjaer, L., Neumann, H., Pieraccini, R., Weber, M. (eds.) Perception in Multimodal Dialogue Systems: 4th IEEE Tutorial and Research Workshop on Perception and Interactive Technologies for Speech-Based Systems. PIT 2008, Kloster Irsee, Germany, 16–18 June 2008, Proceedings of Lecture Notes on Computer Science (LNCS), vol. 5078, pp. 99–110. Springer, Berlin (2008)
52. Reiter, S., Schuller, B., Rigoll, G.:Hidden conditional random fields for meeting segmentation. In: Proceedings 8th IEEE International Conference on Multimedia and Expo, ICME 2007, pp. 639–642, Beijing, China (2007)
53. Rauch, H., Tung, G., Striebel, C.: Maximum likelihood estimates of linear dynamic systems. In: Journal of American Institiute of Aeronautics and Astronautics vol. 3, pp. 1445–1450 (1965)
54. Barber, D.: Expectation correction for smoothed inference in switching linear dynamical systems. J. Mach. Learn. Res. **7**, 2515–2540 (2006)
55. Hochreiter, S., Schmidhuber, J.: Long short-term memory. Neural Comput. **9**(8), 1735–1780 (1997)
56. Fernandez, S., Graves, A., Schmidhuber, J.: An application of recurrent neural networks to discriminative keyword spotting. In: Proceedings of Internet Corporation for Assigned Names and Numbers 2007. vol. 4669, pp. 220–229. Porto, Portugal (2007)

Part III
Intelligent Audio Analysis Applications

In this part applications of Intelligent Audio Analysis in the three domains speech, music, and sound will be presented by selected examples.

Chapter 10
Applications in Intelligent Speech Analysis

Speech is an arrangement of notes that will never be played again.

—Francis Scott Fitzgerald

Speech is broadly considered as being the most natural communication form for humans [1]. Obviously, there are manifold applications opening up for general technical and computer systems, once they are able to recognise speech as well as humans do—be it for interaction purposes with humans [2], mediation purposes between humans [3], or speech retrieval [4]. In fact, spoken language may even become a communication medium among technical systems in the future, e.g., if humans shall be able to follow their conversation.

In this vein, the following sections present application examples selected from the author's recent research on intelligent speech analysis starting with the linguistic side of spoken language. As outlined in the introduction of this book, work of many other authors could have been chosen—the choice of examples from the author's work was simply made to foster consistent application of methods as described previously. We will first have a look at robust recognition of isolated words in severe noise conditions, and then move to the recognition of spontaneous conversational speech. This will be followed by the assessment of 'non-linguistic' human vocalisation such as laughter. Then, the 'paralinguistic' side is highlighted aiming at the automatic recognition of diverse speaker states and traits such as emotion, sleepiness or height of a speaker. This includes analysis of linguistic and acoustic parameters. Linguistic analysis is exemplified in isolation first, then a series of acoustic analyses is presented for various states and traits. For illustration of present-day performances, evaluation results are given on standard data-sets.

B. W. Schuller, *Intelligent Audio Analysis*, Signals and Communication Technology, DOI: 10.1007/978-3-642-36806-6_10, © Springer-Verlag Berlin Heidelberg 2013

10.1 Linguistics: Digits and Spelling

The automatic recognition of speech, enabling a natural and easy to use method of communication between human and machine, is an active area of research as it still suffers from limitations such as the restricted applicability whenever human speech is superposed with background noise [5–7]. Before turning to the recognition of continuous spelling and spontaneous speech, we will first deal with highly noise robust recognition of isolated words, as was presented in [8–11], and [12]. These techniques and their consequent further development led later to the best result for a monaural system in the CHiME Challenge 2011 [13] and later to the best result to-date [14] on the CHiME Challenge 2011 task.

10.1.1 Automotive Digits and Spelling Database

In order to compare the different speech signal preprocessing, feature enhancement, and speech modelling techniques which were presented in Chap. 9 with respect to their recognition performance in various noise scenarios, all of the techniques are tested in a noisy speech recognition experiment which will be outlined in the following. Since the interior of a car is a popular field of application for speech recognisers, allowing hands-free operation of the centre console or text messaging, car noises produced during driving are of great interest when designing a noise robust speech recognition system [15, 16] and have been decided for.

The digits "zero" to "nine" as well as the letters "A" to "Z" from the TI 46 Speaker Dependent Isolated Word Corpus [17] are used as speech database to exemplify noisy digit and spelling recognition similar to the Aurora tasks [18], but tailored for the application in the automotive environment by using different noise for additive overlay. The database contains utterances from 16 different speakers—8 of them are female and 8 of them male. Following the results presentation in [19], only the words which are spoken by male speakers are used. For every speaker 26 utterances were recorded per word class. Of these, 10 are used for training and 16 for testing. By that, the overall digit training corpus consists of 800 utterances and the digit test set 1 280. For the spelling task, 2 080 utterances are used in a similar fashion for training and 3 328 for testing. Babble and white noise scenarios have been chosen as further examples adding to the main focus of the following analyses that lies on designing a robust speech recogniser for an in-car environment. Thus, emphasis is laid on simulating a wide spectrum of different noise conditions that can occur in the interior of a car. In general, interior noise can be split up into four major groups: wind noise which is generated by air turbulences at the corners and edges of the vehicle and arises equivalently to the velocity, engine noise depending on load and number of revolutions, wheels, driving, and suspension noise influenced by road surface and wheel type, and buzz, squeak and rattle noises generated by pounding or relative movement of interior components of a vehicle [20–22]. Usually, the microphone

Table 10.1 Vehicles considered for noise overlay

Vehicle	Derivative	Class
BMW 5 series	Touring	Executive car
BMW 6 series	Convertible	Executive car
BMW M5	Sedan	Exec. sports car
MINI Cooper	Convertible	Super-mini

Table 10.2 Road surfaces and velocities considered for noise overlay

Surface	Velocity (km/h)	Abbreviation
Big cobbles	30	COB
Smooth city road	50	CTY
Highway	120	HWY

would be mounted in the middle of the instrument panel. Consequently, noises as occurring in the interior of a car have been recorded exactly at the same point. The mouth-to-microphone transfer function had been neglected, since the masking effect of background noise was proven to dominate over convolutional noise. As interior noise masking varies depending on vehicle class and vehicle class derivates [21], speech was superposed by noise of four different vehicles as they are listed in Table 10.1.

Besides the vehicle type, the road surface influences the characteristics of interior noise. Hence, three different surfaces with typical velocities are further considered as shown in Table 10.2. A smooth city road at 50 km/h driving velocity and medium revolution (CTY) provides the lowest excitation. At this profile noise caused by wind, engine, wheels, etc. has its minimum. Higher excitation is measured for a highway drive at 120 km/h (HWY), where wind noise is a multiple higher. The worst case noise scenario is given by a road with big cobbles (COB). At 30 km/h, wind noise can be neglected, but the rough cobble surface results in dominant wheel and suspension noise. Figure 10.1 shows the SNR histograms of the accordingly noisy speech utterances.

In spite of SNR levels below 0 dB, speech in the noisy test sequences is still well audible since the recorded noise samples have most of their spectral energy contained in the frequency band from 0 to 500 Hz (cf. Fig. 10.2). As a result, there is

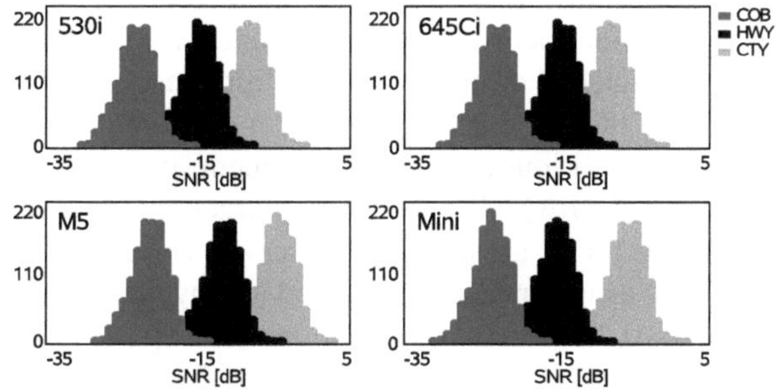

Fig. 10.1 SNR level histograms for noisy speech utterances [10]

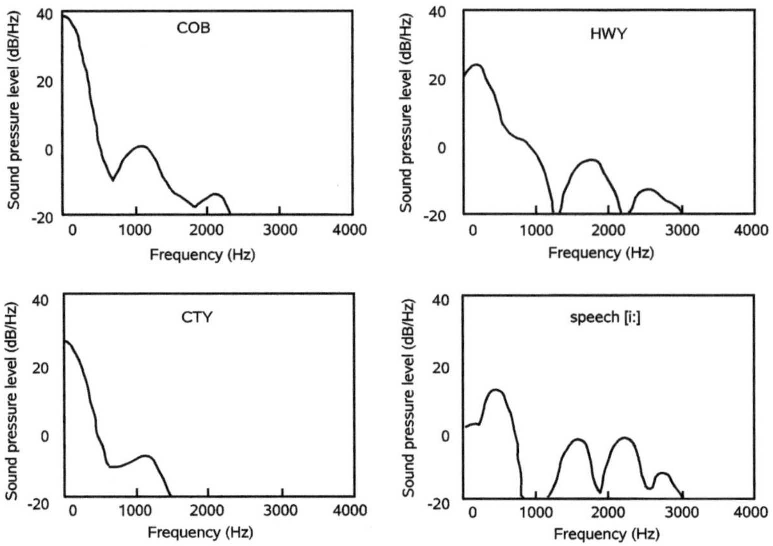

Fig. 10.2 Long-term spectrum of the car noises COB, HWY, CTY (Mini Cooper S) and the spectral characteristics of the vowel [i:] as spoken by a male speaker [10]

little overlap of the spectrum of speech and noise. Extremely low SNR levels for the car noises (see Fig. 10.1) are mainly caused by intense spectral components below the spectrum of human speech (motor drone).

Apart from car noises (CAR), two further noise types are used for the following experiments: First, a mixture of babble and street noise (BAB) at SNR levels 12, 6, and 0 dB recorded in downtown Munich as is present when driving within an urban area with open windows. Furthermore, additive white Gaussian noise (WGN) is considered (SNR levels 20, 10, and 0 dB).

10.1.2 Performance

One model was trained per digit to build an isolated word recogniser. The various strategies for enhancement and robustness as were introduced in Chap. 9 are considered. For HMM and HCRF, each model consists of eight states with a mixture of three Gaussians per state. Clean utterances are used for training to show performance in a non-matched or non-multi-condition test case. As features serve 13 MFCCs or PLPs with their first and second order derivatives. Attempting to remove the effects of noise, the various speech enhancement strategies outlined in Sect. 9.2 are applied: CMS, MVN, HEQ, USS, and AFE extraction. In most of the experiments, the recognition rate for clean speech is observed close to perfection at around 99.9% WA. Note that, as instances are balanced among classes, WA equals UA.

Table 10.3 shows that, for stationary low pass noise like the CAR and BAB noise types, best WA can be reached when enhancing the speech features using a global

Table 10.3 Mean isolated digit recognition rates in [%] UA / WA for different noise types, noise compensation strategies, and features, training on clean data

Strategy	Features	Clean	UA / WA [%]		
			CAR	BAB	WGN
SLDM	MFCC	99.92	99.52	99.29	87.79
HEQ	MFCC	99.92	98.21	96.53	77.50
CMS	PLP	99.84	97.70	97.92	72.67
MVN	MFCC	99.84	94.86	93.32	79.06
CMS	MFCC	99.84	96.96	97.18	72.22
HEQ	PLP	99.92	97.20	95.27	66.51
HCRF/CMS	MFCC	99.76	95.67	94.97	70.06
USS	MFCC	99.05	93.52	92.27	53.19
AFE	MFCC	100.0	87.85	92.84	64.14
None	PLP	99.92	81.06	90.58	67.72
None	MFCC	99.92	75.09	88.37	63.67
AR-SLDS	None	97.37	47.24	78.51	93.32
SAR-HMM	None	98.10	54.26	83.16	41.91

SLDM for speech and a LDM for noise (cf. Sect. 9.2.2). Thereby all clean training sequences were used for global SLDM training. This captures the dynamics of clean speech. The speech model consists of 32 hidden states, and the utterance-specific noise model of a single Gaussian mixture component. It was trained on the first and last 10 frames of the noisy test utterance. To speed up the calculation, the algorithm for speech enhancement was run with history parameter $r = 1$ (cf. Sect. 9.2.2.4). For more demanding recognition tasks like the INTERSPEECH Consonant Challenge [23], SLDM feature enhancement was proven to increase recognition rates for noisy speech: The technique cannot compete with strategies using perfect knowledge of the local SNR of time-frequency components in the spectrogram like oracle masks [24–26], however, compared to the Consonant Challenge HMM baseline recogniser [23], the SLDM approach can improve noisy speech recognition rates by up to 174 % relative [27]. HCRF for the classification of features enhanced by CMS did not result in a better recognition rate as compared to using HMM. For WGN disturbance, the best recognition rate (93.3 % WA, averaged over the different SNR conditions) is reached by the AR-SLDS as was explained in Sect. 9.3.3.2. The noisy speech signal is in this case modelled in the time domain as an AR process. As explained in Sect. 9.3.3.2, the AR-SLDS constitutes the fusion of the SAR-HMM with the SLDS. The AR-SLDS used in the experiment is based on a 10th order SAR-HMM with ten states. This concept is, however, not suited for low pass noise at negative SNR levels in these experiments: For the CAR noise type, only 47.2 % WA are reached, averaged over all car types and driving conditions, for AR-SLDS modelling. A reason for this is the assumption that was made in Eq. (9.29): additive noise is expected to have a flat spectrum. In case of a HMM-based recogniser without feature enhancement, PLP features perform slightly better than MFCC features.

Table 10.4 Mean isolated digit recognition rates (in [%] WA / UA) of a HMM recogniser without feature enhancement for different noise types and training strategies: matched conditions training (MC), mismatched conditions training (MMC), and training with clean data

Training	Clean	CAR	BAB	WGN
Clean data	99.92	75.09	88.37	63.67
MMC	79.42	96.86	98.74	68.51
MC	99.92	99.69	99.73	99.22

Table 10.4 summarises the WA for a HMM recogniser without feature enhancement for three different training strategies: training on clean data, mismatched conditions training, and matched conditions training. In these experiments, mismatched condition training denotes training and testing with the same noise type but at unequal noise conditions (SNR levels and driving conditions, respectively). Matched conditions training stands for exactly identical noise types and noise conditions. If the test sequence is disturbed by noise, mismatched conditions training outperforms a recogniser that had been trained on clean data. However, for clean test sequences the mismatched conditions training significantly downgrades recognition rates, as the noise pattern that had been learnt during the training is missing when testing the recogniser. The results for matched conditions training serve as an upper benchmark for noisy speech recognition performance, because by this strategy one assumes perfect knowledge of the noise properties. Note that, since in the matched conditions experiment one model was trained for every noise condition, this implies knowledge of the noise characteristics and higher memory requirements, as more than one model has to be stored.

The best MFCC feature enhancement methods were further applied in the spelling recognition task as shown in Table 10.5. Again, for noisy test data, SLDM perform better than more 'conventional' techniques such as HEQ.

10.1.3 Summary

In this section evaluation results for the different techniques to improve the performance of ASR in noisy surroundings as were introduced in Chap. 9 were presented for the noisy isolated digit and spelling recognition task. These techniques affect feature extraction, feature enhancement, speech de-coding, and speech modelling.

Table 10.5 Mean spelling recognition rates for different noise types and noise compensation strategies, training on clean data

Strategy	Features	Clean	WA [%]		
			CAR	BAB	WGN
SLDM	MFCC	92.73	82.98	81.59	64.23
HEQ	MFCC	91.85	70.19	69.40	48.20
CMS	MFCC	93.09	73.79	69.78	47.06
none	MFCC	91.04	58.82	66.92	44.30

The use of PLP features as speech representation leads to a relative error reduction of 18.6 % (averaged over all evaluated noise conditions) when compared to 'conventional' MFCC. Furthermore, feature enhancement methods based on spectral subtraction and normalisation like CMS, MVN, USS or HEQ were able to partly remove effects of stationary coloured noises.

As a further approach to enhance speech features, a global SLDM was used. This aims to capture the dynamics of speech enabling a model based speech enhancement through joint speech and noise modelling and prevailed for all car noise types. In fact, the method reached the best WA for the noisy isolated digit recognition task. The usage of HCRF as an alternative model architecture did not outperform HMM. However, embedding a SLDS into a SAR-HMM—modelling the raw signal in the time domain—lead to the best WA in case of speech corrupted with additive WGN. Using noisy training data to build AMs could also improve noise robustness. Mismatched conditions training, which uses training sequences disturbed by a noise type different from that in the test phase, outperformed training on clean data with a relative error reduction of 54.5 %. This shows that multi-condition training is a promising direction. Further, computational complexity and possible fields of application have to be considered when designing a robust speech recogniser. In this respect, AFE and USS are more complex than feature normalisation techniques such as CMS or MVN. But, they are still suited for real-time applications. HEQ and SLDM feature enhancement achieved better recognition rates, however, at the cost of increased computational complexity. Speech-modelling in the time domain as by AR-SLDS requires most computational resources and is therefore at the time not suited for most real-life applications. For stationary noises, the SLDM seems the most promising technique. Yet, it relies on accurate voice activity detection.

Future research effort could be spent on increasing the suitability of promising concepts like SLDM feature enhancement by including discrete state transition probabilities. Another alternative would be finding the optimum compromise between an increment of the history parameter and the computational complexity. Furthermore, the AR-SLDS concept could be optimised for coloured noise when applying AR speech modelling in this context. Further improvements might be achieved by combining the different denoising concepts which were applied in this section.

In a continuous ASR task—as will be considered next—the parameters of a global SLDM as well as the cumulative histogram for the HEQ method could be estimated more precisely due to longer observation sequences than in the so far considered isolated digit or spelling recognition experiments. ASR in noisy environments remains challenging, however, as shown in this section, spending effort on finding accurate techniques for auditory modelling, feature enhancement, speech modelling, and model adaptation can remarkably reduce the performance gap between automatic speech recognition and human perception.

10.2 Linguistics: Spontaneous Speech

Let us now consider spontaneous speech recognition in a real-life setting, as in [28] basing on the related studies published in [11, 28–42]. By that, we will move from isolated to continuous speech recognition independent of the speaker. At the same time, we will consider a setting with real-life noise conditions rather than additive noises. The motivation for this shift is among other reasons given by the fact that ASR is increasingly applied in highly naturalistic human-machine interaction such as with conversational agents [36, 43] or robots. This requires robustly recognising spontaneous, conversational, and by that also disfluent speech. Several strategies to cope with these challenges have been proposed [10, 19]. Of these, most concentrate on improving the signal-processing front-end or computational intelligence back-end side of ASR systems based on HMM. There are, however, also strategies that combine the HMM principle with MLP or RNN [33, 44, 45]. Roughly, one can categorise these into *hybrid* approaches that apply neural networks to generate state posteriors for HMMs, and *Tandem* approaches that use a neural network's output as features to be input into the HMM.

Given co-articulation effects in human speech, modelling of temporal context is essential. For this reason the introduced LSTM networks seem a promising alternative to standard feed-forward networks or RNNs. While temporal context is usually modelled on a higher level by context dependent acoustic models, such as triphone models, and language models, on the feature level only a very limited and inflexible amount of context is modelled. e.g., first and second order regression coefficients of LLDs are added to the feature vector or a fixed number of successive feature frames are 'stacked'. Only few exceptions aimed at modelling of more context, e.g., [46]. Recently, BLSTM networks were shown to be superior to the triphone principle [47], and application of BLSTM phoneme prediction has led to significant performance gains for phoneme classification and keyword spotting [30, 36, 48]. Building on the Tandem technique as was proposed in [35], which uses BLSTM phoneme predictions as additional feature vector components, this section introduces a multi-stream BLSTM-HMM architecture. This architecture models the BLSTM phoneme estimate as a second independent HMM stream of observations to allow for more accurate modelling of observed phoneme predictions. Experiments to show this effect are based on the COSINE corpus [49] which contains noisy conversational speech. With the open-source speech processing toolkit openSMILE [50] this multi-stream technique is implemented in an on-line version in the final SEMAINE[1] system [43]—a multimodal conversational agent.

10.2.1 The COSINE Corpus

All experiments presented in Sect. 10.2.2 are speaker-independent. They were carried out using the 'COnversational Speech In Noisy Environments' (COSINE)

[1] http://semaine-project.eu/

corpus [49]. COSINE is a relatively new database. It contains multi-party conversations that were recorded in real world environments. Speech was captured by a specially crafted wearable recording system in a sort of backpack manner. This allowed the speakers to walk around in the street during the recordings. Participants were asked to speak about anything they liked and to walk to various noisy locations. The corpus thus consists of natural, spontaneous, and highly disfluent speaking styles. The signal is partly masked by indoor and outdoor noise sources including crowds, vehicles, and wind noises. Seven microphones were used simultaneously per speaker. To stick with the precondition of this book to rely on monophonic sources, exclusively speech recorded by a close-talk microphone (Sennheiser ME-3) is exploited in the ongoing.

All ten sessions transcribed at the moment of writing are used. These contain 11.40 h of pairwise conversations and group discussions. The 37 contained speakers are fluent, but not necessarily native English speakers. Each speaker participated in one session exclusively. Their ages range from 18 to 71 years with a median of 21 years. COSINE's test set is used for evaluation (sessions three and ten). This set comprises 1.81 h of speech. Sessions one and eight were chosen as validation set (2.7 h of speech) and the remaining six sessions made up the training set. The vocabulary size is 4.8 k, the out-of-vocabulary rate in the test set is 3.4 %.

10.2.2 Performance

The frame-wise phoneme recognition rate of different network architectures is now presented on the COSINE task as described. It is further compared to a common triphone HMM phoneme recogniser. Then, going from phonemes to words, the accuracy (WA) obtained by the multi-stream system introduced in Sect. 7.4.3 is compared with the performance of a Tandem approach [35]. Again, a baseline is established by a common HMM system that bases only on MFCC features. MFCCs 1–12 are extracted as features for network input in addition to logarithmic energy together with first and second order regression coefficients. To compensate for stationary noise effects, CMS is applied to these features. A HMM system is used to obtain phoneme borders via forced alignment. The following four different network architectures are considered: RNN, BRNN, LSTM networks, and BLSTM networks.

As network topology three hidden layers (per input direction) were chosen for any of these four types. These layers have a size of 78, 128, and 80 hidden units, respectively, and each memory block contains one memory cell. A learning rate of 10^{-5} and a momentum of 0.9 proved optimal for training. To improve the generalisation ability of the networks, zero mean Gaussian noise with standard deviation 0.6 was added to the inputs during training. Prior to the actual training process, weights were uniformly random initialised in the range from -0.1 to 0.1. Input and output gates used tanh activation functions. The forget gates had logistic activation functions.

The standard (CMU) set of 41 different English phonemes is applied. The 41 phonemes include *silence* and *short pause* labels. Once no improvement on the

Table 10.6 Frame-wise phoneme accuracy for BLSTM, LSTM, BRNN, and RNN predictors, and triphone HMMs, and word accuracies obtained for a baseline single-stream HMM, a Tandem system [35], and the proposed multi-stream recogniser ($a = 1.1$) using different network architectures

Accuracy	Phoneme	Word	
[%]	(framewise)	Tandem	Multi-stream
triphone HMMs	56.91	43.36	
RNN	48.91	43.79	46.25
BRNN	50.51	42.59	46.27
LSTM	58.91	44.46	46.45
BLSTM	66.41	45.04	46.50

validation set could be observed for at least 50 epochs, training was stopped. Then, the network was chosen that achieved the best frame-wise phoneme error rate on the validation set.

The second column of Table 10.6 shows the frame-level phoneme accuracies for COSINE's test set obtained in this way. Generally, bidirectional context modelling prevails over unidirectional context modelling and LSTM context modelling outperforms conventional RNNs. The best rate can be achieved with a BLSTM network at 66.41 % WA. The use of bidirectional context in low-latency, responsive, on-line applications is, however, limited or close to impossible. For off-line transcription tasks, and on-line tasks which allow a higher latency, BLSTM networks are perfectly applicable.

When using a triphone HMM system as described below for frame-wise phoneme transcription, the rate is significantly lower at 56.91 % WA—this is in line with [51]. However, triphone HMMs were able to outperform a conventional RNN phoneme predictor (50.51 % and 48.91 % WA for bi- and unidirectional RNNs, respectively).

As explained, BLSTM phoneme estimates are now incorporated as an additional feature stream into a multi-stream HMM framework for the recognition of continuous speech. To this end, each phoneme of the underlying left-to-right HMM system is represented by three emitting states. The initial monophone models consist of one Gaussian mixture for probability density function modelling per state. They were trained using four iterations of embedded Baum-Welch re-estimation. Then, the monophones were mapped to tied-state cross-word triphone models with shared state transition probabilities. This sharing helps to reduce the number of parameters that need to be estimated. Given COSINE's comparably limited size, this is a reasonable standard measure. Two Baum-Welch iterations were executed for re-estimation of the triphone models. Finally, the number of Gaussian mixture components of the triphone models was increased to 16 in four successive rounds of mixture doubling and re-estimation—resembling four iterations in every round. AMs and a back-off bi-gram language model were trained on COSINE's training set. The conditional probability table for the second feature stream was restricted to the 15 most likely phoneme confusions per state. Further, a floor value of 0.01 was used for the remaining confusion likelihoods. As shown in Table 10.6, the word accuracy of the single-stream HMM is 43.36 % WA.

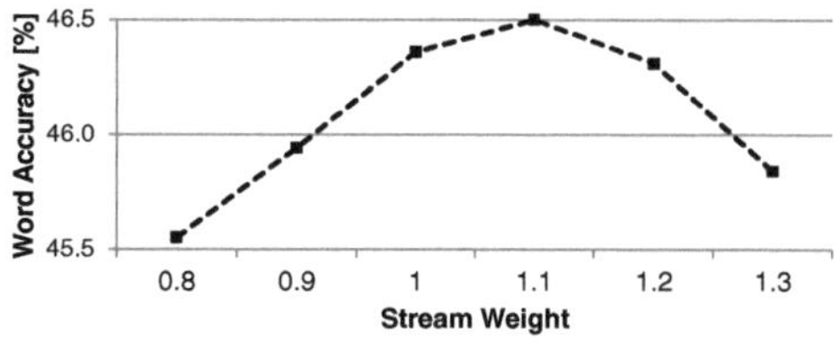

Fig. 10.3 Word accuracies on the COSINE test set using the multi-stream BLSTM-HMM system with different MFCC stream weight parameters a

As was stated in Sect. 7.5.2, word accuracy according to (7.89) is used for measurement of accuracy in the case of continuous speech recognition. In fact, word accuracy is also weighted, as the recall per word—words are the classes in this case—is weighted by the number of instances of this word in the test set. Thus, weighted accuracy (WA) is used as unit.

Using the multi-stream BLSTM-HMM approach outlined in Sect. 7.4.3, the optimal MFCC stream weight parameter a (cf. Eq. (7.84)) can be determined. The best performance on the test set was obtained for $a = 1.1$ (cf. Fig. 10.3).

Table 10.6 depicts word accuracies on COSINE's test set using a Tandem system and the multi-stream approach under variation of the neural network architectures. The multi-stream BLSTM-HMM prevails at overall highest WA of 46.50 %. The usage of bidirectional context, however, implies a non-causal short look-ahead buffer, i.e., recognition is performed with a slight delay. Further, modelling the phoneme confusions of the neural networks as described in Sect. 7.4.3 seemingly results in lower sensitivity to the frame-wise phoneme accuracy: Only little difference is observed for the accuracy of a multi-stream recogniser using BLSTM predictions and a system using RNN-based phoneme estimates.

10.2.3 Summary

In this section, spontaneous continuous speech was recognised in real-life noisy conditions. To best cope with the task, a multi-stream ASR system was introduced. It relies on context-sensitive phoneme estimates generated by a BLSTM network as an additional feature stream to the conventional stream for LLD processing. Evaluation based on the challenging COSINE database of human-to-human conversational speech recorded in originally noisy environments rather than artificially superposing noise afterwards. This implies that noise and speech underlie same reverberation and speakers are directly affected by the noise such as by the Lombard effect of using more vocal effort in louder environment. The shown multi-stream ASR architecture led to higher word accuracies in comparison to a standard single-stream MFCC-based recognition system. It further outperformed a Tandem approach that models MFCC features and BLSTM predictions via Gaussian mixtures in a single observation stream. Explicitly modelling typical phoneme confusions occurring in the BLSTM network helped to improve the results.

Future experiments should include the design of bottle-neck [40, 45] BLSTM networks. Further, the principle of Connectionist Temporal Classification (CTC) [33] could be exploited to let the networks do the phoneme alignments by themselves and thus improve the accuracy of phoneme targets (which in the presented system were obtained by an HMM-based recogniser).

10.3 Non-Linguistics: Vocalisations

Apart from speech—i.e., linguistic entities—, non-linguistic vocalisations are present in spoken language—their computational assessment will now be shown.

Discrimination of speech and non-linguistic vocalisations such as laughter or sighs plays an important role in speech recognition systems dealing with spontaneous speech, such as dialogue systems, call centre loops or automatic transcription of meetings. In contrast to read speech, which conveys only the information contained in the spoken words and sentences, spontaneous speech contains considerably more of such extra-linguistic information—e.g., in the COSINE corpus which was introduced in the previous section. To avoid confusion of non-linguistic information with linguistic information and for higher level natural language understanding, it is vital for an ASR engine to spot the non-linguistic vocalisations and determine their type [52–59].

Several approaches have been proposed in particular for the detection of filled pauses [60] and laughter [61–63]. In this section, we extend this to four different types of non-linguistic vocalisations—laughter, breathing, hesitation (e.g., *"uhm"*) or non-verbal consent (e.g., *"aha"*)—and discriminate them from speech.

Furthermore, it will be shown that features generated by NMF can increase classification accuracy for this task when compared to traditional acoustic feature information—for example MFCCs. To this end, a supervised NMF variant is suggested with pre-computed component spectra from instances of speech and non-linguistic vocalisations. This allows to measure which spectra contribute the most to the signal based on the activations of these components. Previous work in NMF-based ASR uses NMF for speech enhancement applied during pre-processing. In contrast, it is now proposed to use the NMF as data-based feature extractor as was introduced in Sect. 8.3. For sound classification such an approach has been described in [64]. However, for non-linguistic vocalisation classification, this technique was first proposed in [12]. Experimental results are based on the TUM AVIC database [65] (cf. Sect. 5.3.1).

10.3.1 Methodology

The input signal is transformed to the frequency domain. A STFT is applied with a Hamming window, 25 ms window size, and 10 ms frame rate. From the resulting spectrogram, MFCCs 0–12 and their first and second order delta regression coefficients are extracted with 26 Mel-band filter banks. The mean and standard deviation are considered as functionals applied to the LLD. Functionals are applied over the full length of each instance.

As additional functionals, a sub-sampling of the sequence by five equidistant signal frames' feature values from beginning to end of the sequence is carried out. This yields a total of 273 MFCC-based acoustic features per instance. Further, NMF activation features are considered, as were explained in Sect. 8.3.

Classification is carried out by a Support-Vector Machine (SVM) classifier with RBF kernel. This kernel was found to be superior to a linear kernel for this particular task.

10.3.2 Performance

A data set was prepared based on TUM AVIC (cf. Sect. 5.3.1). Data from each of the 21 speakers was assigned to exactly one of the sets training, development, and test in order to evaluate the recognition performance independent of the speaker. The sets were chosen such that the total length of the utterances is evenly distributed across the sets. Furthermore, each set is balanced wrt. the length of male and female utterances, as also is the whole corpus [66].

Based on the manual transcription, the signal parts containing non-linguistic vocalisations of the four classes 'consent', 'laughter', 'hesitation', and 'breathing' were extracted. This was added by the left-over 'speech-only' turns. In total, there are 2 070 instances for training, 1 980 for development, and 2 184 for testing with the following class distribution in the sets training/development/test: breathing (222, 154, 130), consent (83, 88, 177), hesitation (401, 422, 414), laughter (142, 75, 76), and speech (1 222, 1 241, 1 387).

Up-sampling of minority classes was applied to the training set (when testing with the development set) and to the union of training and development set (when testing with the test set). This ensures balance of instances across all classes during the model training.

The suitability of NMF activation features for recognition of non-linguistic vocalisations was evaluated by training with the training set and testing with the development set at first. Component spectra for the extraction of NMF activation features were computed from signals concatenated from all the training utterances separately for the different classes. For the speech class, as in [67], only 10 % of the originally recorded overall speech material was used given the stark contrast in frequency of occurrence of linguistic and non-linguistic vocalisations. Different configurations were tested varying the NMF cost-function and the number of components. Features are linearly scaled to the range $[-1, 1]$.

Next, training and development data was united for training, when testing on the test section of the database. This means that in this step NMF activation features were extracted using spectra that were computed from this larger amount of training

Table 10.7 Recall and UA for four different non-linguistic vocalisations and speech on the test set of the TUM AVIC corpus. Results by SVMs with RBF kernel, trained with different sets of NMF activation features: N70, N90, and N100, corresponding to 70, 90, and 100 NMF components, respectively. NMF cost functions are either Euclidean distance or modified KL divergence

Recall [%]	Euclidean			KL divergence		
	N70	N90	N100	N70	N90	N100
UA	73.0	72.6	72.7	77.7	79.3	79.1
WA	69.5	69.4	70.7	72.5	74.2	74.4
Breathing	95.4	91.5	94.6	87.7	88.5	91.5
Consent	84.2	86.4	89.3	89.3	88.7	85.9
Hesitation	67.6	61.4	62.8	71.0	73.2	76.1
Laughter	51.3	55.3	47.4	71.1	75.0	71.1
Speech	66.7	68.4	69.7	69.5	71.2	71.0

data. 70, 90, and 100 components (N70, N90, N100) were considered and distributed among the classes as follows: For the case of overall 70 components, 20 were assigned each for the speech and laughter classes, and 10 for the remaining three classes (consent, hesitation, breathing). This takes into account the higher spectral diversity of speech and laughter. For 90 components, the number of speech components was doubled (40). For 100 components, an equal distribution of 20 spectra for each of the five classes was chosen. Euclidean distance and modified KL divergence were evaluated as cost-functions. The results for UA are shown in Table 10.7. From these, one can conclude that NMF feature extraction works best when minimising modified KL divergence, outperforming Euclidean distance by up to 7 % absolute. For the N100 feature set this difference is significant at $p \approx 1.1 \cdot 10^{-5}$ in a one-tailed McNemar test. Less influence is observed for the choice of the number of NMF components. Overall, 90 components and minimising modified KL divergence can be recommended as ideal from these experiments.

The effect of combining MFCCs with NMF activation features is elaborated upon in Table 10.8. There, NMF activation features were computed by minimisation of Euclidean distance, which yielded improved—yet not significant—results.

Looking at these results, one can state that NMF activation features alone do not surpass MFCCs in terms of UA. Combining NMF activation features with MFCC-based features, however, increases the recall rate for all classes but hesitation.In particular the M+N100 set leads to the best UA and increases the recall rate for the laughter class by 6.6 % and the consent class by 4.5 % absolute when compared to MFCC-based features alone. The overall difference in UA is significant at $p \approx 10^{-3}$.

10.3.3 Summary

Recognition of non-linguistic vocalisations was shown based on a supervised NMF procedure to compute 'activation features'. In experiments, the method performed

Table 10.8 Recall and UA recall for four different non-linguistic vocalisations and speech on the test set of the TUM AVIC corpus. Results by SVMs with RBF kernel

Recall [%]	MFCC	+N70	+N90	+N100
UA	86.0	87.8	88.0	88.6
WA	82.5	83.4	83.6	83.9
Breathing	93.9	93.9	94.6	94.6
Consent	89.3	91.0	92.1	93.8
Hesitation	86.0	87.0	86.0	85.3
Laughter	81.6	86.8	86.8	88.2
Speech	79.5	80.3	80.6	81.0

Feature sets are varied: 273 MFCC-based features only, and 273 MFCC-based with NMF activation features (+$N70$, +$N90$, +$N100$), corresponding to 70, 90, and 100 NMF components (cf. Table 10.7). NMF activation features base on minimisation of Euclidean distance

considerably well for the discrimination of speech and non-linguistic vocalisations. NMF based on KL divergence gave better results than NMF based on Euclidean distance, and it was demonstrated that NMF activation features can significantly improve performance of MFCC-based recognition.

Upcoming research could elaborate on NMF variants and derivates such as as non-negative matrix deconvolution [68], or various extensions of the cost-functions such as sparseness constraints [67]. This can also be compared to a greater variety of audio features such as the ones proposed in [54]. Features could then also be compared on the LLD level. For the classification, de-correlation and feature selection can be applied.

10.4 Paralinguistics: States and Traits

Once words and other non-linguistic events are recognised, one can aim at their understanding. Here, we will not deal with general understanding of complex and sometimes deeply hierarchically structured intentions such as when dictating mathematical equations [69], but look at the sentiment encoded in text. In Section 10.4.1 the example that was initially presented in [70] and later in more detail in [71] is illustrated.

Beyond the linguistics and non-linguistic vocalisations, the paralinguistic aspects 'how' and by 'whom' things are said are also encoded in the acoustic speech signal. In the sections following Sect. 10.4.1, an overview on benchmark performances for acoustic analysis will be given. These benchmark performances are the results from a series of international research challenges that were held annually at the INTERSPEECH conference since 2009.

We will look at paralinguistics, starting with the short-term affective states as were featured in the INTERSPEECH 2009 Emotion Challenge [72–74] and INTERSPEECH 2010 Paralinguistic Challenge [75] by emotion and interest of a speaker. Then, we will move from such short-term states to long-term traits, i.e., more permanent speaker characteristics, as were also featured in the 2010 Paralinguistic Challenge for age and gender determination. As further speaker trait task and an example of interdependence of speaker traits, we further consider an application where the height of a speaker is inferred from the voice [76] (not included in any of the challenges). The final example of paralinguistics stems from the INTERSPEECH 2011 Speaker State Challenge [77]. The tasks at hand—speaker intoxication and speaker sleepiness—are found somewhat in-between states and traits on a temporal scale, as they are either 'long-term states' or 'short-term traits'.

Looking at applications of such speaker state and trait information, the following are found among the most promising:

First, it seems obvious that speech recognition and interpretation of speakers' intention can benefit from paralinguistic information [78], e.g., when trying to recognise equivocation [79]. The information can also be exploited in the acoustic layer to improve recognition of 'what' has been said, e.g., by adaptation of the acoustic model [39, 80–84].

Next, conversation analysis, mediation, and transmission can benefit from paralinguistics, such as in computer-aided analysis of human-human conversations including the investigation of synchrony in the prosody of married couples [3], specific types of discourse [85] in psychology, or the analysis and summarisation of meetings [86, 87].

Many applications also exist in the public health sector. Hearing-impaired persons can profit, as cochlear implant processors typically alter the spectral cues which are crucial for the perception of paralinguistic information [88]. Children with autism may profit from the analysis of emotional cues as they may have difficulties understanding or displaying them [89, 90].

Also, transmitting paralinguistic information along with other message elements can be used to animate avatars [91], to enrich dictated text messages, or to label calls in voice mailboxes by symbols such as emoticons [92]. Communicative virtual agents and robots should be enriched by social competence [93–96] which requires them to understand paralinguistic information from the voice, face, and gestures.

It is also believed that adapting to callers in a voice portal is of commercial interest [97], including target-group specific advertising. In call centres, also quality management by monitoring agents is being researched [98]. Other applications include serious gaming and fun applications, such as the love detector by Nemesysco Ltd.[2] or the game "Truth or Lies—Someone Will Get Caught" for video consoles that comes with a microphone and claims to detect lies (THQ®Entertainment[3]), further health related applications such as monitoring elderly people living on their own [99] or diseases and speech disorders [100] such as Parkinson's disease [101, 102], autism [103], cancer, cleft lip and palate [104] or dysphonia [105], or further pathological effects [106].

Tutoring systems are another typical field of application, where information on user states such as uncertainty [107], interest, stress, cognitive load [108], or even

[2] http://www.nemesysco.com/

[3] http://www.thq.com/

deception can be employed to adapt the system and the teaching pace [109, 110]; generally, paralinguistic cues are essential for tutors and students to make learning successful [111]. In addition, automatic voice coaching, e.g.,to give better public speeches or simply to intonate appropriately when learning foreign languages, becomes possible [112].

Moreover, there are many security related situations (surgical operation [113], crisis management, and all the tasks connected to aviation and air traffic control) where monitoring of stress level, sleepiness, intoxication, and such, may play a vital role [114]. In addition, counter terrorism or counter vandalism surveillance may be aided by analysing paralinguistic cues such as aggressiveness of potential aggressors [115, 116], or fear of potential victims [117].

Finally, in the field of multimedia retrieval, paralinguistic information is of interest for manifold types of media searches [4, 118].

10.4.1 Sentiment and Opinion

Sentiment analysis and opinion mining have been studied for manifold application scenarios such as product reviews [119–124], the stock market [125], or hotels and travel destinations [119, 122], and film reviews [119, 126, 127]. A particularly difficult task among these is sentiment analysis for film reviews: In [119], a 66 % WA for valence polarity classification are named for film reviews, but 84 % WA for car reviews by identical means of analysis. This may be owed to the discrepancy between the semantic orientation of words describing the elements of a film (i.e., a scene, the plot), and its style or art.

In this light this section introduces valence estimation from text by means of on-line knowledge sources and machine learning. The experiments are reported on a database of over 100 k film reviews collected from the review website Metacritic[4]—the largest to-date. Metacritic's fine-grained review scores as gold standard allow for a regression approach besides binary or ternary valence prediction.

10.4.1.1 Metacritic Database

The database described in this section is likely the largest film review corpus to date. Manifold other do, however, exist, ranging in size from 120 to 2 000 reviews. In [119], 120 film reviews were collected from Epinions.[5] In [127] 11 films were selected from the top 250 list of IMDB.[6] Then, the first 100 reviews were retrieved for each one of them resulting in a total of 16 k sentences and 260 k words. Perhaps the most frequently used film review database was introduced in [126]. In the begin-

[4] http://www.metacritic.com

[5] http://www.epinions.com

[6] http://www.imdb.com

Table 10.9 Metacritic database statistics

[#]	Minimum	Maximum	Average	Standard deviation
Reviews per film	1	65	21.1	10.3
Words	1	104	24.2	12.4
Sentences	1	13	1.3	0.6

ning it contained 752 negative, and 1 301 positive reviews from the Usenet newsgroup *rec.arts.movies.reviews*. Later, other versions were added.[7] The Metacritic film review corpus introduced in the following is larger by far: It comprises a total of 102 622 reviews for 4 901 films. Metacritic[8] compiles reviews for films, video/DVDs, books, music, television series, and computer games from various sources.

Reviews in Metacritic are contained as excerpts of the 'key statement' from the original reviews. Overall, 133 394 sentences, and 2 482 605 words are contained in the database that will simply be referred to as Metacritic database in the ongoing. The average review has 1.3 sentences, with a standard deviation of 0.6. In contrast to other film review database, the reviews thus are short at an average length of 24.2 words (cf. Table 10.9). Its vocabulary comprises 83 328 words. By POS classes, nouns (683 259) come first, followed by verbs (382 822), adjectives (244 825), and adverbs (174 152).

Besides its sheer size, the database features fine-grained score values from 0 to 100 (the higher, the more positive) per review as particular highlight, calculated from the original numeric rating scheme used by each source. These can be assumed reliable in the sense of a ground truth rather than mere gold standard, given that they were assigned by the authors of the reviews. An exception are the cases where no numeric rating by the authors is available—in this case a Metacritic staff member provided these. Further, from the reader's point of view, sentiment expressed can be perceived differently [120, 126]. ConceptNet tries to overcome the problem by letting users vote on the reliability of predicates, which could be used in future approaches.

Metacritic itself provides an additional ternary mapping as can be seen in Table 10.10. There is no balance of instances per class (cf. Table 10.11): Roughly three times as many positive than negative reviews are contained. A partitioning for training and testing is realised by year leading to almost equal size: 49 698 instances are contained in the 'odd' year set, and 52 924 in the 'even' year set.

Table 10.10 Metacritic's mapping of score to valence classes

Score	Valence class	#Reviews
81–100	Positive	15 353
61–80	Positive	38 766
40–60	Mixed	32 586
20–39	Negative	13 194
0–19	Negative	2 723

[7] http://www.cs.cornell.edu/people/pabo/movie-review-data/

[8] http://www.metacritic.com, accessed January 2009.

Table 10.11 Metacritic's statistics by odd and even years of release of reviews

# Reviews	All years	Odd years	Even years
Mixed	32 586 (31.75 %)	15 756 (31.70 %)	16 830 (31.80 %)
Positive	54 119 (52.74 %)	26 410 (53.14 %)	27 709 (52.36 %)
Negative	15 917 (15.51 %)	7 532 (15.16 %)	8 385 (15.84 %)

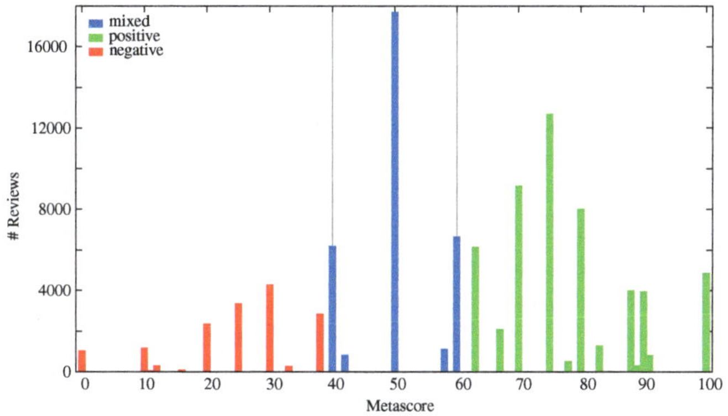

Fig. 10.4 Metacritic's score distribution

Figure 10.4 additionally shows the histogram for the 'continuous' scores S. Visibly, these are also not evenly distributed. Generally, and in particular for the range $40 \leq S \leq 60$ (mixed), one observes spikes for scores S which are integer multiples of ten. With $S > 60$ (positive), the resolution appears more fine-grained. An explanation for this behaviour is the fact that scores were converted from original numerical rating schemes that usually are less fine-grained than Metacritic, such as one to five stars, etc.

10.4.1.2 Performance

Let us now consider the evaluation of the methods introduced in Sect. 6.3 to exploit on-line knowledge sources in comparison to data-based learning by Bag-of-Ngrams (BoNG) features and SVMs or SVR modelling. The optimisation of the methods is limited to sentiment polarity classification. Stemming is achieved by Porter's Snowball Stemmer [128], and OpenNLP[9] is used for text pre-processing. Sentence detection is based on a maximum entropy model to identify end of sentence characters. The stochastic part-of-speech tagger applied bases on maximum entropy, and supports the 48 word-level tags of the Penn Treebank project [129]. The English chunker applied also bases on maximum entropy and a chunking model by CRFs (cf. Sect. 9.3) [130].

[9] http://opennlp.sourceforge.net/

After parameter optimisation, classification results for the ternary problem and by regression for the full 0–100 score range will be shown alongside the out-of vocabulary resolution and attempts of synergistic fusion of knowledge and data.

However, let us begin with a binary classification by excluding the instances of the *mixed* class. This leaves 33 942 instances for training, and 36 094 instances for testing. A development partition is realised as subset of the training data by choosing 'every other odd year', starting at 1, i.e., all years for which $(year - 1) \bmod 4 = 0$. This gives 15 730 instances for evaluation, 18 212 instances for training during development.

To cope with the bias towards positive reviews (cf. Sect. 10.4.1.1) down sampling without replacement is used for the training material. This is the only example in this book of down-sampling instead of up-sampling. The reason is the sheer size of data to handle. After balancing, 15 063 training instances are obtained, from which 8 158 instances are used for training during development.

To start, the parameters c and e of the decay function (cf. Sect. 6.3) are optimised. In direct comparison to the decay function in [124], which is reached by setting $c = 1$ and $e = 1$, WA gains 0.23 % for $c = 1$ and $e = 0.1$. In Fig. 10.5 the WA is visualised depending on c and e. The maximum WA is reaches 70.29 %.

For classification of the BoW and BoNG features serve SMO-trained SVMs with polynomial kernels [131]. After stemming, >62 k word stems are left over from the 83 k vocabulary entries of the Metacritic database. Thus, a minimum term frequency f_{min} with a 'gentle' value of $f_{min} = 2$ is employed to remove infrequent words, taking into account that low-frequency words are likely to be meaningful features for opinionated sentences [132]. Further, 'periodic pruning' is applied to ensure reduction without dropping potentially relevant features: The data set is partitioned with configurable partition size. The pruning discards features that occurred only once after processing of the partitions by the word or N-Gram tokeniser. With a higher partition size—25 % of the data set was chosen as value in the experiments—, the probability to eliminate relevant features is lowered. Next, optimal feature transformation

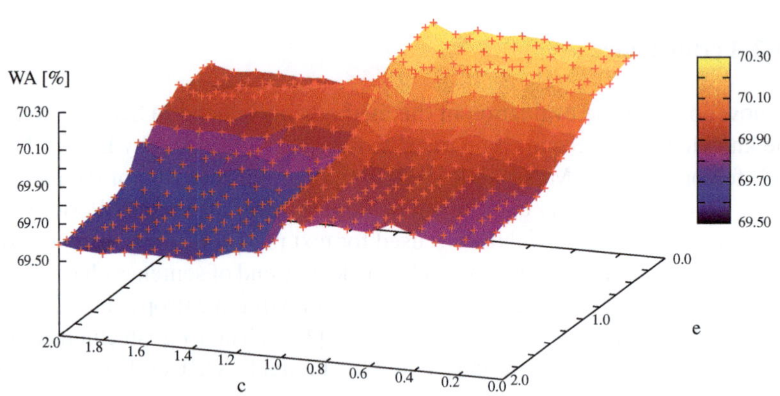

Fig. 10.5 WA throughout optimisation of the decay function parameters c and e [71]

Table 10.12 WA for different BoNG feature types

Feature type	WA [%]
$f_{i,j}$	75.03
Norm($f_{i,j}$)	75.72
TF	75.44
Norm(TF)	75.66
IDF	75.05
Norm(IDF)	76.42
TFIDF	75.45
Norm(TFIDF)	77.14

Table 10.13 WA during N-gram length optimisation from g_{min} to g_{max} for BoNG features

g_{min}	g_{max}	#Features	WA [%]
1	1	18 316	74.58
1	2	96 152	75.95
1	3	151 083	77.14
1	4	171 438	76.41
1	5	177 733	76.92
2	2	77 840	66.72
2	3	132 780	66.54
2	4	153 146	69.62
2	5	159 465	71.66
3	3	54 968	71.59
3	4	75 418	72.33
3	5	81 911	72.61

and normalisation methods were evaluated. Table 10.12 summarises the obtianed WAs for simple N-Gram frequency ($f_{i,j}$), TF, IDF, and normalisation (norm) for N-Grams from one to three terms (i.e., $g_{min} = 1$ and $g_{max} = 3$). It will be shown that this is an optimal choice.

Little difference is observed for the different types of feature representation, of which normalisation combined with TFIDF leads to the best result, and normalisation improves results at any time and in particular in the case of IDF.

Let us now consider the optimal N-Gram length from g_{min} to g_{max} in Table 10.13. As stated, the optimal choice is $g_{min} = 1$ and $g_{max} = 3$. This agrees with [120], where optimal results for product reviews where reached by tri-grams. Yet, the authors of this study found back-off N-Grams to downgrade the accuracies—this is different to the case of Metacritic. In this optimal setting, 12 % of the features are single words, 52 % bi-grams, and 36 % tri-grams. The largest feature space in this evaluation has 177 733 features for $g_{min} = 1$ and $g_{max} = 5$, the smallest one only 18 316 features for $g_{min} = 1$ and $g_{max} = 1$.

Looking next at out-of-vocabulary words (OOV), i.e., such occurring in the test, but not in the training material, these make up 30.3 % of the total vocabulary. Out-of-vocabulary events, i.e., the number of occurrences of OOV words in all reviews,

Table 10.14 OOV statistics for Metacritic for no OOV resolution, stemming, and stemming with on-line knowledge sources-based (OKS) resolution

# Words	Vocabulary	OOV words	OOV events
Baseline	83 328	25 217 (30.3 %)	38 244 (3.0 %)
Stemming	62 212	19 228 (30.9 %)	29 482 (2.3 %)
Stemming and OKS	62 212	14 123 (22.7 %)	22 746 (1.8 %)

Table 10.15 UA and WA for BoNG with SVM and on-line knowledge sources (OKS) for two review classes (positive (+)/negative (−)) on Metacritic

[%]	BoNG	OKS
UA	77.73	60.44
WA	77.37	69.42
$Recall_+$	77.07	77.21
$Recall_-$	78.39	43.67
$Precision_+$	92.18	81.92
$Precision_-$	50.84	36.70

amount to 3.0 % of the total 1 288 384 words in the database (cf. Table 10.14). The delta between OOV words and OOV events is probably owing to proper nouns such as in the title of a film, or the name of actors. OOV words—and by that OOV events—are influenced during stemming: After this step, the OOV words are at a similar level of 30.9 %. However, the OOV events decrease to 2.3 % of all words. As additional solution, OOV words can be substituted by non-OOV 'synonyms' with the help of the on-line knowledge sources ConceptNet and WordNet.

Next, let us compare the BoNG and SVM approach with the on-line knowledge source domain-indpendent one (cf. Sect. 6.3.4.4) on the same test data. The optimal configuration as determined so far is used. In the case of BoNG, $g_{min} = 1$ and $g_{max} = 3$, features are normalised TFIDF, and OOV resolution is applied.

The results are found in Table 10.15, and show a clear advantage for BoNG with SVM with a gap of 7.95 % WA owed to a 34.72 % absolute difference in the recall of negative reviews. An explanation might be the inability of the proposed domain-independent model to cope with negation, assuming negation to be more frequent in negative reviews. In fact, this is a common non-trivial problem for syntax-driven approaches [120, 133]. BoNG features on the other hand model negation, provided that it occurs in proximity of the word to be negated.

We will now turn to three classes of sentiment, which is challenging also, as mixed or neutral reviews are particularly challenging [134, 135]. For the syntax-driven approach, the decision function needs to be extended to handle ternary recognition tasks. This is achieved by a split into two binary tasks: negative plus mixed versus positive and negative versus mixed plus positive. For optimal configuration, these are 'tuned' in isolation, and two decision thresholds τ_- and τ_+ are observed. These thresholds form the basis of an overall valence decision function, where y is the output class label, and S is the score of the sequence of words (cf. Sect. 6.3.3.4):

Table 10.16 UA and WA for BoNG with SVM and on-line knowledge sources (OKS) for three review classes (positive (+)/mixed (0)/negative (−)) on Metacritic

[%]	BoNG	OKS
UA	53.99	38.80
WA	53.71	49.38
$Recall_+$	57.62	68.43
$Recall_0$	43.91	37.32
$Recall_-$	60.43	10.66
$Precision_+$	75.66	58.82
$Precision_0$	42.92	35.74
$Precision_-$	34.70	28.71

$$y = \begin{cases} +1 \text{ if } S > \tau_+ \\ 0 \quad \text{if } \tau_- \leq S \leq \tau_+ \\ -1 \text{ if } S < \tau_- \end{cases} \tag{10.1}$$

With $\tau_- < \tau_+$, a range around $y = 0$ can be chosen for the mixed class. The best constellation observed was reached with $\tau_- = -1.9$, and $\tau_+ = 0.6$. Table 10.16 shows the results for this approach and for BoNG with SVM for the three-class task on the test set.

Just as one might expect, accuracies drop in comparison to handling two classes. BoNG with SVM overall lead to better results and provide more balanced values across the classes from 43.91 % for mixed, up to 60.43 % for negative reviews.

One can now aim at synergistic fusion of the two methods. To this end, again the optimal configuration as determined up to now is chosen for each approach. An early integration on the feature level that preserves the knowledge up to the final decision process is followed first. With the given correlation of the feature streams, this is known to be beneficial [136, 137]. Thus, a super-vector is created by including scores of the knowledge-based approach in the BoNG feature vector prior to SVM classification on this new vector. Table 10.17 shows according results. The WA increases over BoNG 'stand-alone' (cf. Table 10.16) only by 0.13–53.84 %.

The opposite approach of late semantic fusion is a decision based on the predictions per model [136, 137]. A tuning of 'whom to trust when' is thus possible, i.e., it can be modelled which approach is most reliable for which class. The results so far revealed a strength of the knowledge-based method for the recall of positive reviews in the ternary task. This can be emphasised on by according weighting or rules. SVMs are able to provide pseudo-probabilities P in the range of $0 \leq P \leq 1$ per class based on the distance to the hyperplane and the chosen multi-class discrimination strategy. By class, let us denote these pseudo-probabilities in the given ternary case as P_- (negative), P_0 (mixed), and P_+ (positive). Now, with the score S of the knowledge-based approach (cf. Sect. 6.3.4.4), we can influence when to decide for the positive class by setting suited conditions. The SVM decision is decided for if these conditions are not met. A number of such conditions were tested and are summarised alongside the results in Table 10.17. For the knowledge-based score, $S > 0$, and $S > 0.6$—the positive discrimination threshold τ_+ decided for above—were considered, and for the SVMs, $P_+ > 0$, $P_- = 0$, and $((P_+ > 0) \wedge (P_0 > 0))$. As a

Table 10.17 UA and WA plus recalls by early and late fusion for the three review classes (negative (−)/mixed (0)/positive (+)) in Metacritic compared to the baseline (no fusion) with different conditions for the knowledge-based score S, and the SVM class-wise pseudo-probabilities P_-, P_0, P_+ to decide for the positive class by the classifier

[%]	UA	WA	$Recall_-$	$Recall_0$	$Recall_+$
Baseline	53.99	53.71	60.43	43.91	57.62
Early fusion	54.09	53.84	60.67	43.65	57.96
Late fusion					
$(S > 0)$	45.50	55.93	32.65	16.83	86.72
$(S > 0.6)$	46.93	55.95	37.04	20.60	83.14
$(S > 0) \wedge (P_+ > 0)$	52.67	57.77	51.64	29.74	76.64
$(S > 0.6) \wedge (P_+ > 0)$	52.92	57.37	52.74	31.56	74.45
$(S > 0) \wedge (P_- = 0)$	53.72	56.19	60.43	29.74	70.98
$(S > 0.6) \wedge (P_- = 0)$	53.83	55.99	60.43	31.56	69.49
$(S > 0) \wedge (P_+ > 0) \wedge (P_0 > 0)$	53.82	56.83	59.14	29.74	72.58
$(S > 0.6) \wedge (P_+ > 0) \wedge (P_0 > 0)$	53.90	56.54	59.25	31.56	70.89

result, the late fusion significantly outperforms the individual approaches (one-tailed z-test, 0.1 % level).

Finally, to model the 'continuous' values, SVR is chosen for the determination of the Metacritic score value in the range of 0–100. As kernel, a radial basis function with the variance parameter $\gamma = 0.01$ proved optimal on the development set. Given the continuous approximation task, CC and MLE serve as evaluation measure (cf. Sect. 7.5.2). On the test data of Metacritic, the result is a CC of 0.570 and MLE of 14.1, i.e., on average, the regressor is mistaken by 14.1 with respect to the score. An obvious challenge for regression training is the non-even distribution of score values within the Metacritic database (cf. Fig. 10.4).

10.4.1.3 Summary

Two main approaches towards automatic sentiment analysis and opinion mining where discussed in this section—one open-domain approach based on on-line knowledge sources reaching from annotated dictionaries (General Inquirer) to comprehensive semantic networks (ConceptNet), and one based on data. Further, benchmarks were presented for the particular task of film reviews, but the methods can be applied to other sentiment tasks, as will be shown in Sect. 11.7, where song lyrics are analysed in such a way.

The advantage of the on-line knowledge sources-based approach using linguistic methods, dictionaries, and semantic networks, is that no learning material is required. Overall, it led to usable results, but the in-domain data-driven approach based on BoNG features and SVMs reached higher recognition rates. On-line knowledge could thereby be integrated to resolve 40.5 % of the OOV events and slightly improve performance. As another way of combination of the two techniques, a late fusion led to significant gains. The results based on a three-class task, but alternatively regression by SVR was evaluated.

In the future, more specific term categories could be used from General Inquirer. Further, in [138] it was shown that ConceptNet can also be exploited directly for sentiment information. Thus, it could complement General Inquirer in this respect. Also, multi-word terms and complete phrases could be used directly in ConceptNet. Of particular help could also be the addition of a named entity recognition in combination with other types of common knowledge sources such as Wikipedia. Next, out-of-vocabulary resolution could be improved by using more WordNet relations, such as *antonymy*, i.e., opposite meaning, or *hypernymy*, i.e., more general meaning. Finally, OOV N-Grams need to be resolved by a second substitution step after N-Gram creation.

10.4.2 Short-term States: Emotion and Interest

The recognition of a number of short-term states from speech has been addressed so far, of which the following non-exhaustive list names some examples:

- *mode*: speaking style [139] and voice quality [140];
- *emotions* (full-blown, prototypical): [141];
- *emotion-related states or affects*: for example, general [142–144], stress [145], intimacy [146], interest [65, 75], confidence [147], uncertainty [107, 148], deception [149, 150], politeness [151–153], frustration [154–156], sarcasm [157, 158], pain [99].

From these, two examples among most researched candidates have been chosen for illustration of methodology and performances: emotion and interest, both belonging to 'affective' speaker states.

The young field of affect recognition from voice has recently gained considerable interest in the fields of Human-Machine Communication, Human-Robot Communication, and Multimedia Retrieval. Numerous studies have been seen in the last decade trying to improve on features and classifiers [159]. One first cooperative experiment is found in the CEICES initiative [160], where seven sites compared their classification results under exactly the same conditions and pooled their features together for one combined unified selection process. This comparison was not fully open to the public, which was the motivation to create the INTERSPEECH 2009 Emotion Challenge—the first in an ongoing series of challenges on Computational Paralinguistics—which are conducted for strict comparability: all participants use the same database and the same evaluation measures in their experiments. As classes are unbalanced, the primary measure to optimise is UA (unweighted average recall), and secondly WA (weighted average recall by number of instances per class—this is commonly known simply as "accuracy").

10.4.2.1 FAU Aibo Emotion Corpus

One of the major needs of the emotion recognition community ever since—perhaps even more than in many related pattern recognition tasks—is the constant need for data sets. In the early days of the late 1990s, these have not only been few, but also small ($\sim$500 turns) with few subjects ($\sim$10), uni-modal, recorded under studio noise conditions, and acted [161–163]. Further, the spoken content was mostly predefined (e.g., Danish Emotional Speech (DES), EMO-DB, Speech Under Simulated and Actual Stress (SUSAS) databases) [164]. These were seldom made public and few annotators—if any at all—labelled usually exclusively the perceived emotion. Additionally, these were partly not intended for analysis, but for quality measurement of synthesis (e.g., DES, EMO-DB).

Today, there are more diverse emotions covered, more elicited or even spontaneous sets of many speakers, and larger amounts of instances (up to 10 k and more) of more subjects (up to 50), that are annotated by more labellers (4 (AVIC)—17 (VAM, [165])) and partly made publicly available. For acted data, equal distribution among classes is of course easily obtainable. Also transcription is becoming more and more rich: additional annotation of spoken content and non-linguistic interjections (e.g., AVIC, Belfast Naturalistic, FAU AIBO, SmartKom databases [164]), multiple annotator tracks (e.g., VAM), manually corrected pitch contours (FAU AIBO), additional audio tracks under different noise and reverberation conditions (FAU AIBO), phoneme boundaries and manual phoneme labelling (e.g., EMO-DB), different units of analysis, and different levels of prototypicality (e.g., FAU AIBO). At the same time these are partly also recorded under more realistic conditions (or taken from the media). Trying to meet the utmost of these requirements, the FAU AIBO database [166] was chosen for the first Challenge: It is a corpus with recordings of children interacting with Sony's pet robot Aibo. The corpus consists of spontaneous, German speech that is emotionally coloured. The speech is spontaneous, because the children were not told to use specific instructions but to talk to the Aibo like they would talk to a friend. The children were led to believe that the Aibo was responding to their commands, whereas the robot was actually controlled by a human operator. The wizard caused the Aibo to perform a fixed, predetermined sequence of actions; sometimes the Aibo behaved disobediently, thereby provoking emotional reactions. The data was collected at two different schools, Mont and Ohm, from 51 children (age 10–13, 21 male, 30 female; about 9.2 h of speech without pauses). Speech was transmitted with a high quality wireless head set (UT 14/20 TP SHURE UHF-series with microphone WH20TQG) and recorded with a DAT-recorder (sampling rate 48 kHz, quantisation 16 bit, 48 kHz down-sampled to 16 kHz). The recordings were segmented automatically into 'turns' using a pause threshold of 1 s. Five labellers (advanced students of linguistics) listened to the turns in sequential order and annotated each word independently from each other as neutral (default) or as belonging to one of ten other classes of emotion. Since many utterances are only short commands and rather long pauses can occur between words due to Aibo's reaction time, the emotional/emotion-related state of the child can change also within turns. Hence,

Table 10.18 Number of instances for the 2-class emotion problem

#	NEG	IDL	$\sum$
Train	3358	6601	9959
Test	2465	5792	8257
$\sum$	5823	12393	18216

the data is labelled on the word level. If three or more labellers agreed, the label was attributed to the word. All in all, there are 48401 words.

Classification experiments on a subset of the corpus [166] showed that the best unit of analysis is neither the word nor the turn, but some intermediate chunk being the best compromise between the length of the unit of analysis and the homogeneity of the different emotional / emotion-related states within one unit. Hence, manually defined chunks based on syntactic-prosodic criteria [166] are used here (cf. also [167]). The whole corpus consisting of 18216 chunks was used for the 2009 Emotion Challenge.

The two-class problem that was chosen as example for this book consists of the cover classes NEGative (subsuming *angry*, *touchy*, *reprimanding*, and *emphatic*) and IDLe (consisting of all non-negative states). A heuristic approach similar to the one applied in [166] is used to map the labels of the five labellers on the word level onto one label for the whole chunk. Since the whole corpus is used, the classes are highly unbalanced. The frequencies for the two-class problem are given in Table 10.18. Speaker independence is guaranteed by using the data of one school (Ohm, 13 male, 13 female) for training and the data of the other school (Mont, 8 male, 17 female) for testing. In the training set, the chunks are given in sequential order and the chunk id contains the information which child the chunk belongs to. In the test set, the chunks are presented in random order without any information about the speaker. Additionally, the transliteration of the spoken word chain of the training set and the vocabulary of the whole corpus is provided allowing for ASR training and linguistic feature computation.

For the the second task considered as for dealing with short-term user states, namely determination of speaker interest, the TUM AVIC database (cf. Sect. 5.3.1) was used in the follow-up challenge in 2010. It features 2 h of human conversational speech recording (21 subjects), annotated in five different levels of interest. The corpus further features a uniquely detailed transcription of spoken content with word boundaries by forced alignment, non-linguistic vocalisations, single annotator tracks, and the sequence of (sub-)speaker-turns.

10.4.2.2 Methodology

In the past, the main focus was on prosodic features, in particular pitch, durations and intensity [168]. Comparably small feature sets (10–100) were first utilised. In only a few studies, low level feature modelling on a frame level was pursued, usually by HMM or GMM. The higher success of static feature vectors derived by projection of the LLD such as pitch or energy by descriptive statistical functional application such as lower order moments (mean, standard deviation) or extrema is probably justified by the supra-segmental nature of the phenomena occurring with respect to emotional content in speech. In more recent research, also voice quality features such as HNR, jitter, or shimmer, and spectral and cepstral features such as formants and MFCC have become more or less the 'new standard'. At the same time, brute-forcing of features (1 000 up to 50 000), e.g., by analytical feature generation, partly also in combination with evolutionary generation, has become popular. It seems as if this (slightly) outperforms hand-crafted features while the individual worth of automatically generated features seems to be lower [74].

Within expert-based hand-crafted features, perceptually more adequate features have been investigated, reaching from simple log-pitch to Teager energy or more complex features such as articulatory features (e.g., (de-)centralisation of vowels). Further, linguistic features are often added these days, and will certainly also be in the future. However, these demand for robust recognition of speech in the first place.

For the Emotion Challenge, a feature set was provided that should best cover the described gained knowledge. It sticks to the findings in [53] by choosing the most common and at the same time promising feature types and functionals covering prosodic, spectral, and voice quality features. Further, it is limited to a systematic generation of features. For highest transparency, the openSMILE feature extraction (cf. Sect. 6.5) was used. In detail, the 16 LLDs chosen are: ZCR from the time signal, RMS frame energy, pitch frequency (normalised to 500 Hz), HNR by ACF, and MFCC 1–12. For each of these, the delta coefficients are additionally computed. Next, the 12 functionals mean, standard deviation, kurtosis, skewness, minimum and maximum value, relative position, and range as well as two linear regression coefficients with their MSE are applied on a chunk basis as depicted in full detail in the Appendix in Table A.1. Thus, the total feature vector per chunk contains $16 \cdot 2 \cdot 12 = 384$ features.

For the determination of interest, an extended set of features compared to the INTERSPEECH 2009 Emotion Challenge [72] as was described above is used. Speakers' interest determination was featured as one Challenge task in the INTERSPEECH 2010 Paralinguistic Challenge.

The extraction was made by again choosing the open-source Emotion and Affect Recognition toolkit's feature extracting backend openSMILE [169] (cf. Sect. 6.5). This extension intends to better reflect a broader coverage of paralinguistic information assessment [170, 171]. 1 582 acoustic features are obtained in total in this second set by systematic 'brute-force' feature generation in three steps: First, the 38 LLDs shown in the Appendix in Table A.1 are extracted at 100 frames per second with varying window type and size (Hamming, 25 ms for all but pitch with Gaussian, 60 ms) and smoothed by simple moving average low-pass filtering with a window length of three frames. Next, their first order regression coefficients are added. Then, 21 functionals are applied (cf. Table A.1) per instance in the databases. However, 16 zero-information features (e.g., minimum F0, which is always zero) are discarded. Finally, the two single features F0 number of onsets and utterance duration are added.

10.4.2.3 Performance

For provision of baseline results, the two pre-dominant architectures within the field are considered: Firstly, dynamic modelling of LLD as pitch, energy, MFCC, etc. by HMMs (only emotion). Secondly, static modelling using supra-segmental information obtained by statistical functional application to the same LLD on the chunk level. This is done either by classification for emotion or regression in the case of interest.

It was decided to entirely rely on two standard publicly available tools widely used in the community: the Hidden Markov Model Toolkit (HTK)[10] [172] in the case of dynamic modelling, and the WEKA 3 Data Mining Toolkit[11] [131] in the case of static modelling. This ensures easy reproducibility of the results and reduces description of parameters to a minimum: Unless specified, defaults are used.

Constantly picking the majority class for the two-class emotion task of the 2009 Emotion Challenge would result in an accuracy (WA) of 70.1 %, which we consider here, while the chance level for UA is simply 50 %, respectively. As instances are unequally distributed among classes, balancing of the training material to avoid classifier over-fitting is considered. This can be eased by applying the Synthetic Minority Oversampling TEchnique (SMOTE) [173] as data-driven up-sampling. Note that up-sampling does not have any influence in the case of generative modelling: For each class one HMM is trained individually and equal priors are assumed. Table 10.19 depicts these results for the two-class emotion task (classification by linear left-right HMM, one model per emotion, diverse number of states, two Gaussian mixtures, $6 + 4$ Baum-Welch re-estimation iterations, Viterbi) by UA and WA. With increased temporal modelling, i.e., a higher state number, a gradual shift towards preference of NEG is observed in the considered two-class problem case. In Table 10.20 results for this 2-class emotion task are further shown employing the whole feature set and using SVM (SMO learning, linear kernel, pairwise multi-class discrimination). For SVM, an additional pre-processing step is performed: the features are standardised, or z-normalised, i.e.,each feature is normalised to have zero mean and variance one. Table 10.20 shows the influence of these two pre-processing steps and their impact on the target evaluation measure UA. Note that the order of operations is crucial, as the standardisation leads to different results if classes are balanced.

Table 10.21 then depicts the results for the interest baseline. The measures for this task are the Pearson Correlation Coefficient (CC) and the mean linear error

Table 10.19 Baseline results for 2-class emotion by dynamic modelling with HMM

	#States	UA [%]	WA [%]
2-class	1	62.3	71.7
	3	62.9	57.5
	5	66.1	65.3

[10] http://htk.eng.cam.ac.uk/docs/docs.shtml

[11] http://www.cs.waikato.ac.nz/ml/weka/

Table 10.20 Baseline results for 2-class emotion by static modelling with SVM. Diverse pre-processing strategies: training balancing (B) by SMOTE and standardisation (S)

Process		UA [%]	WA [%]
1	2		
–	–	62.7	72.6
S	–	64.9	72.3
B	–	60.5	68.9
S	B	67.6	68.3
B	S	67.7	65.5

Table 10.21 Baseline results for continuous interest by static modelling with unpruned REP-trees (25 cycles) in random-sub-space meta-learning (500 Iterations, sub-space size 5 %)

	CC	MLE
Train versus develop		
Interest	0.604	0.118
Train + develop versus test		
Interest	0.421	0.146

(MLE) as found in other studies (e.g., [65, 165]), where CC is the primary measure. Note that the continuous modelling makes balancing more challenging—none was used for this baseline. Here, a clear downgrade is observed for the apparently more 'challenging' test condition.

10.4.2.4 Summary

The baseline results clearly demonstrate the difficulty of handling not only pre-selected prototypical cases, but all speech that was recorded—just as needed in a working system. The baseline for the emotion task was outperformed by the winners of the challenge and by follow-up work, e.g., in [174] and [175]—the latter is the best result reported by individual groups to date at 70.5 % UA.

The overall best result to date was gained by fusion of the N best participants of the challenge [74]: 71.2 % UA. These results demonstrate that fusion of several engines can slightly improve the results, however, at a rather high overall computational effort. For interest, the baseline was outperformed first in [176], and later in [177] by the use of LSTM networks at 0.459 CC and by inclusion of linguistic cues CC was 0.504 on the test set.

10.4.3 Long-term Traits: Age, Gender, Height

On the opposite end of the temporal scale, we find the long-term traits. As before, let us start by naming most researched candidates in the following list:

- *biological trait primitives* such as height [76, 178], weight, age [75, 179], gender [75, 179];
- *group/ethnicity membership*: race/culture/social class with a weak borderline towards other linguistic concepts, i.e., speech registers such as dialect or nativeness [180];
- *personality traits*: likability [181, 182];
- *personality in general*, 'Big Five' personality traits (openness, conscientiousness, extroversion, agreeableness, and neuroticism) [183–185];
- *speaker idiosyncrasies*, i.e., speaker-ID [186].

As examples, the traits age and gender, as were featured in the INTERSPEECH 2010 Paralinguistic Challenge, and additionally speaker height are discussed in the ongoing. As for age and gender, either mostly prosodic supra-segmental features have been employed, or frame-level features based on MFCCs, and their optimal fusion [187]. For speaker height, very sparse research was carried out so far [178, 188]. The authors in [188] examined the ability of listeners to determine the speaker's height and weight from speech samples and found that especially for male speakers, listeners are able to estimate a speaker's height and weight to a certain degree. A similar study is documented in [189] and deals with the assignment of photographs to voices as well as the estimation of a speaker's age, height, and weight via speech samples. The relationship between formant frequencies and body size was examined in [190]. Especially for female participants, a significant correlation between formant parameters and height could be found. Another study revealed significant negative correlations between F_0, formant dispersion and body shape and weight of male speakers [191].

For the actual experiments, the aGender corpus was provided for age determination in four groups and gender determination in three groups (female, male, and children). It consists of 46 h of telephone speech from 954 speakers. For height determination in centimetres the commonly known TIMIT corpus is picked—though originally intended for automatic speech recognition experimentation, it provides rich speaker trait information and speakers in sufficient number. This meta information includes the speaker trait target task height with the additional speaker information of speaker age, gender, dialect region, education level, and race. Note that the term 'race' stems from the available TIMIT corpus meta-information (cf. also Sect. 11.8).

As feature information the set provided for the INTERSPEECH 2010 Paralinguistic Challenge baseline calculation is used for all three traits. Note, however, that height assessment was not part of the 2010 challenge and is only featured here as additional long-term trait example. Classification and regression of instances in this systematically brute-forced feature space is done with SVM and SVR—a choice motivated by the high popularity of these two variants in the broader field of speaker state and trait assessment [83, 192, 193].

10.4.3.1 aGender Corpus

For the recording of the aGender corpus, an external company was employed to identify possible speakers of the targeted age and gender groups [75, 179]. The subjects received written instructions on the procedure and a financial reward, the calls were free of charge. They were asked to ring up the recording system six times with a mobile phone alternating indoor and outdoor to obtain different recording environments. They were prompted by an automated interactive voice response system to repeat given utterances or produce free content. Between each session a break of one day was scheduled to ensure more variations of the voices. The utterances were stored on the application server as 8 bit, 8 kHz, A-law. To validate the data, the associated age cluster was compared with a manual transcription of the self stated date of birth.

Four age groups—Child (C), Youth (Y), Adult (A), and Senior (S)—were defined. Since children are not subdivided into female and male, this results in seven classes as shown in Table 10.22.

The content of the database was designed in the style of the Speech Dat corpora. Each of the six recording sessions contains 18 utterances taken from a set of utterances listed in detail in [194]. The topics of these were *command words*, *embedded commands*, *month*, *week day*, *relative time description*, *public holiday*, *birth date*, *time*, *date*, *telephone number*, *postal code*, *first name*, *last name*, *yes/no* with according free or pre-set inventory and according 'eliciting' questions as "*Please tell us any date, for example the birthday of a family member*".

In total, 47 h of speech in 65 364 single utterances of 954 speakers were collected. Note that, not all volunteers completed all six calls, and there were cases where some called more often than six times, resulting in different numbers of utterances per speaker. The mean utterance length was 2.58 s. 25 speakers were selected randomly for each of the seven classes as a fixed Test partition (17 332 utterances, 12.45 h) and the other 770 speakers as a Training partition (53 076 utterances, 38.16 h), which was further subdivided into Train (32 527 utterances in 23.43 h of speech of 471 speakers) and Develop (20 549 utterances in 14.73 h of speech of

Table 10.22 Age and gender classes of the aGender corpus, where f and m abbreviate female and male, and x represents children without gender discrimination. The last two columns represent the number of speakers/instances per partition (Train and Develop)

Class	Group	Age	Gender	# Train	# Develop
1	Child	7–14	x	68/4406	38/2396
2	Youth	15–24	f	63/4638	36/2722
3	Youth	15–24	m	55/4019	33/2170
4	Adult	25–54	f	69/4573	44/3361
5	Adult	25–54	m	66/4417	41/2512
6	Senior	55–80	f	72/4924	51/3561
7	Senior	55–80	m	78/5549	56/3826

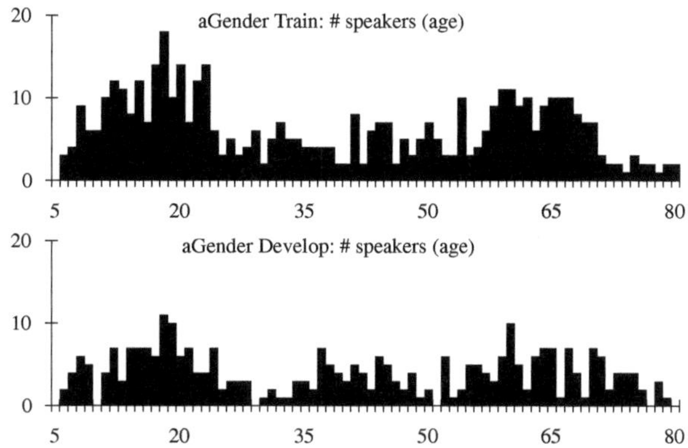

Fig. 10.6 Age (in years) histograms for the train and develop partitions of aGender [75]

299 speakers) partitions. Overall, this random speaker-based partitioning results in roughly 40/30/30 % Train/Develop/Test distribution. Table 10.22 lists the number of speakers and the number of utterances per class in the Train and Develop partitions, Fig. 10.6 depicts the number of speakers as a histogram over their age.

The age group can be handled either as combined age/gender task by classes $\{1, \ldots, 7\}$ as indicated in Table 10.22 or as age group task independent of gender by classes $\{C, Y, A, S\}$. For comparison of results though, only the age group information is used by mapping $\{1, \ldots, 7\} \rightarrow \{C, Y, A, S\}$ as denoted. For gender, the classes $\{f, m, x\}$ have to be classified, as gender discrimination of children is considerably difficult, yet it was again decided to keep all instances (cf. Sect. 10.4.2 for both tasks.

10.4.3.2 TIMIT Database

The TIMIT corpus [195] is well suited for height determination experiments in the sense that it contains a sufficiently high number of speakers—630 in total. This is needed when it comes to speaker trait assessment in order to obtain meaningful and statistically significant results. Each of speaker spoke ten phonetically rich sentences. The fact that these speakers pronounce the same sentences renders the paralinguistic task somewhat text dependent, as for several other databases, e.g., partly the aGender corpus above and in the field of emotion and affective speaker state recognition where the Berlin, the Danish, and the eNTERFACE emotional speech databases or the Speech Under Simulated and Actual Stress database show higher limitation in phonetic content variation [161]. As stated, in addition to featuring sufficient different speakers, TIMIT provides a rich amount of meta-information on its speakers' traits: their age, gender, height, dialect—one out of 8 major American English ones—, their highest education degree, and race. All TIMIT recordings are in 16 bit, 16 kHz.

Figure 10.7 depicts the distribution of height for the speakers in TIMIT. The non-continuous distribution of height in the histrogram is because of TIMIT originally

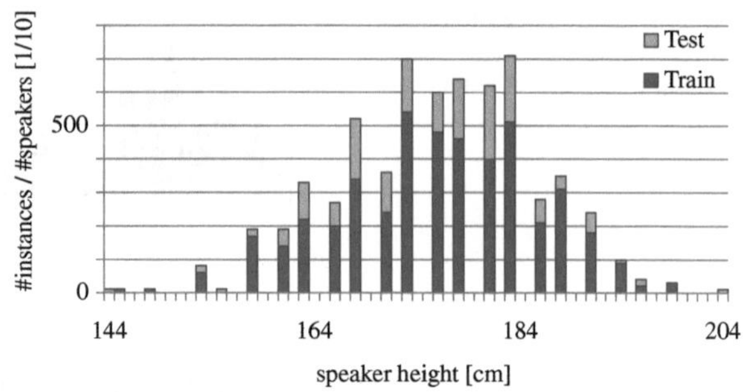

Fig. 10.7 Speaker height distribution in TIMIT's train and test partitions by number of instances and speakers (speaker number is shown by the same bars, but the value has to be divided by ten, as each speaker spoke ten turns)

providing speaker height in the units of feet and inches. For better comparability, though, it was decided to follow the conversion to the SI unit of meters following the result presentation as given in [178].

TIMIT also has a definition of train (462 speakers) and test (168 speakers) partitions to which we stick in the oncoming experiments.

10.4.3.3 Methodology

Due to the size of the aGender corpus, a limited feature set was provided in the Challenge consisting of 450 features This is reached by reducing the number of descriptors from 38 to 29 and that of functionals from 21 to 8 [75, 179]. For height determination, the full set is used.

The Weka toolkit is used [196] for classification and regression. SVM are preferred for age and gender classification experiments; the general Support Vector paradigm further offers SVR for the continuous ordinal task of height. For their training SMO is employed. As kernel function a linear kernel was found optimal in experiments on training exclusively over the different tasks. A kernel complexity of 1 and 0.05 is chosen for classification and regression, respectively. In the case of speaker height determination, additional cases are considered to demonstrate the mutual dependency of speaker traits. To this end, ground truth information on other speaker traits is added as feature information to the acoustic vector in different variations. The use of ground truth information is intentional to show the upper benchmark effect of mutual dependence.

10.4.3.4 Performance

Table 10.23 shows results for the age and gender baselines by UA and WA. Visibly, the 'blind' Test partition shows better results, likely due to the now larger training set. Interestingly, in several cases a 7-group sub-model, separating age groups for gender recognition and vice versa, performs slightly better than direct modelling for the UA. This can be seen as first indication of mutual task dependence.

Table 10.23 Age and gender baseline results obtained by SMO learnt pairwise SVM with linear Kernel

Sub-Ch.	Task	UA [%]	WA [%]
Train versus develop			
–	$\{1, \ldots, 7\}$	44.24	44.40
Age	$\{1, \ldots, 7\} \rightarrow \{C, Y, A, S\}$	47.11	46.17
	$\{C, Y, A, S\}$	46.22	45.85
Gender	$\{1, \ldots, 7\} \rightarrow \{x, f, m\}$	77.28	84.60
	$\{x, f, m\}$	76.99	86.76
Train + develop versus test			
–	$\{1, \ldots, 7\}$	44.94	45.60
Age	$\{1, \ldots, 7\} \rightarrow \{C, Y, A, S\}$	48.83	46.71
	$\{C, Y, A, S\}$	48.91	46.24
Gender	$\{1, \ldots, 7\} \rightarrow \{x, f, m\}$	81.21	84.81
	$\{x, f, m\}$	80.42	86.26

Table 10.24 Selected speaker independent results for height (H) recognition on the TIMIT corpus test partition; contextual information by feature inclusion of age (A), gender (G), American English dialect (D), education level (E), race (R) or all of these (All)

Context	CC	MLE [cm]
–	0.296	7.05
R	0.286	7.09
G	0.299	7.01
A	0.314	6.94
A,G	0.317	6.91
A,R	0.302	7.00
G,R	0.290	7.05
A,G,R	0.304	6.98
All	0.306	7.07

CC, MLE for regression (speaker height in centimetres). 1 582 acoustic features, classification by SVR with linear Kernel, SMO, complexity 0.05

Table 10.24 next depicts results of the speaker height assessment task in strict speaker independence by employing TIMIT's training and test partitions as stated above and exclusively adding speaker contextual meta-information by selected (pairs of) supplementary traits as additional feature(s) to the acoustic vector. Given the case of regression and a continuous ordinal task formulation, CC and MLE are the measures of performance. Gains can be observed by gradual addition of ground-truth supplementary speaker trait information aside of the target task. Little improvement is found for the height recognition task by gender inclusion (1.2 % relative correlation improvement), age inclusion (6.2 %) and combined age and gender inclusion (7.3 %) with the latter being the only significant one. Interestingly, age inclusion helps more for the assessment of height than gender inclusion, even though all speakers can be assumed to have reached their maximal height given their ages above maturity.

At the same time, race meta-information constantly down-grades height assessment in these experiments.

10.4.3.5 Summary

The automatic assessment of speaker's age, gender, and height was shown. Assessing age and gender in combination was observed to prevail over individual assessment. As for the Emotion Challenge, the best participants' results were fused by majority vote. This led to the so far unrivalled upper benchmark of 53.6 % UA for the age class, and 85.7 % UA for the gender classes—again proving the superiority of fusion of multiple engines. When classifying height, information on other traits was added as features. An improvement was observed here as well. There obviously are many other approaches to exploit such knowledge, e.g., by building age, gender or height dependent models for any of the other tasks. This will require further experience in the case of age and height dependent models as to this end reasonable quantisation is required. A further next step will be to find methods to automatically estimate any of these at the same time by mutual exploitation of each other. This can be particularly interesting given different forms of task representation (continuous ordinal or binary) as was chosen here.

Provided that speech databases contain a transcription of the targeted speaker information, the combination of different corpora might result in more accurate results and a versatile applicability of paralinguistic information extraction systems. Thus, cross-corpus evaluations as published for emotion recognition in [161] could be part of future research on combined speaker traits analysis.

Finally, the automatic assessment of certain speaker characteristics such as age potentially also profits from the inclusion of linguistic features in addition to acoustic descriptors. This in turn would require an automatic speech recognition module extracting linguistic information for combined acoustic-linguistic analysis. In the field of emotion recognition [83], recent studies have shown that even though the word accuracies of automatic speech recognisers processing spontaneous, emotional speech are lower than the word accuracies of dictation systems recognising well-articulated, read speech, the inclusion of speech recognisers for linguistic feature generation reliably boosts emotion recognition accuracies. It is of interest whether a similar behaviour can be observed in the case of traits.

10.4.4 Mid-term: Intoxication and Sleepiness

Apart from the short-term related speaker state emotion, mid-term states exist which are not permanent, yet do not change instantly. These comprise, for example:

- *(partly) self-induced*: sleepiness [197], intoxication (e.g., alcoholisation [77, 198, 199]), health state [104], mood (e.g., depression [200]);

- *structural (behavioural, interactional, social) signals*: role in dyads, groups, and the like [201], friendship and identity [202], positive/negative attitude [203], (non-verbal) social signals [204], entrainment [205].

Two such were picked out as a central theme in another follow-up challenge which focused on the crucial application domain of security and safety: the computational analysis of intoxication and sleepiness in speech. Apart from intelligent and socially competent future agents and robots, main applications are found in the medical domain and surveillance in high-risk environments such as driving, steering or controlling [206].

In [207], several differences are shown in the quality of the vocal articulation after a night of sleep deprivation (reduced intonation and a slowing down of the vocal flow); in [208], a reduction of the spontaneous dialogues and performance degradation of the subjects is observed under similar conditions. Generally speaking, these results suggest effects of sleep deprivation on communication, especially with a reduction of the spontaneous verbalisations, trouble finding words, and a degradation of the articulation. Subjects under sleep deprivation produce less details and show less empathy toward a team-mate [209]. Some stressors such as alcohol are likely to influence articulators, which helps to explain intra-speaker and inter-speaker variability [210].

For the experimental evaluation of these tasks, the Alcohol Language Corpus (ALC) and the Sleepy Language Corpus (SLC) with genuine intoxicated and sleepy speech were provided [77]. The first consists of 39 h of speech from 154 speakers in gender balance. It serves to evaluate features and algorithms for the estimation of speaker intoxication in gradual blood alcohol concentration (BAC). The second features 21 h of speech recordings of 99 subjects, annotated in the 10 different levels of sleepiness of the Karolinska Sleepiness Scale (KSS) [211].

The verbal material is of different complexity reaching from sustained vowel phonation to natural communication. In part, the corpora feature detailed speaker meta data, orthographic transcript, phonemic transcript, segmentation, and multiple annotation tracks. Again, both were given with distinct definitions of test, development, and training partitions, with a strict speaker independence as needed in many real-life settings. Two tasks are addressed:

First, the alcoholisation of a speaker is determined as two-class classification task: *alcoholised* for a BAC exceeding 0.5 per mill[12] or *non-alcoholised* for a BAC equal or below 0.5 per mill. The measure of interest is—as before—UA of these two classes to better compensate for imbalance between classes.

Second, the sleepiness of a speaker is determined by a suited algorithm and acoustic features. While the annotation provides sleepiness from 1–10 on the KSS, only two classes are recognised: sleepiness for a level exceeding 7.5 on the KSS, and non-sleepiness for a level equal or below 7.5. Again, the measure is UA of the two classes and a further enlarged standard feature set is used [77].

---

[12] Per mill BAC by volume (standard in most central and eastern European countries; further ways exist, e.g., percent BAC by volume, i.e., the range resembles 0.028 to 0.175 per cent (Australia, Canada, USA), points by volume (GB), per mill by BAC per mass (Scandinavia) or part per million.)

Table 10.25 Partitions of ALC. 'NAL' denotes recordings of non-alcoholised, i.e., BAC per mill in the interval [0; 0.5], and 'AL' recordings of alcolised speakers, i.e., BAC per mill in [0.5; 1.75]

# ALC	NAL	AL	total
Train	3750	1 650	5 400
Develop	2790	1 170	3 960
Test	1620	1 380	3 000
Train + develop	6540	2 820	9 360
Train + develop + test	8160	4 200	12 360

10.4.4.1 ALC Database

A brief description of the ALC project is now given [77]. Details can be found in [199, 210].

ALC comprises 162 speakers (84 male, 78 female) within the age range 21–75, mean age 31.0 years and standard deviation 9.5 years, from five different locations in Germany. Non-native speakers, speakers with a strong dialect as well as non-cooperative speakers were excluded from participation. To obtain a gender balanced set, 154 speakers (77 male, 77 female) are selected randomly; these are further randomly partitioned into gender balanced training, development and test sets according to Table 10.25. Speakers voluntarily underwent a systematic intoxication test supervised by the staff of the Institute of Legal Medicine, Munich. Before the test, each speaker chose the BAC she/he wanted to reach during the intoxication test. Using both Watson- and Widmark formula [210], the amount of required alcohol for each person was estimated and handed to the subject. After consumption, the speaker waited another 20 minutes before undergoing a breath alcohol concentration test (BRAC) and a blood sample test (BAC). However, only the BAC value is considered. The possible range is between 0.28 and 1.75 per mill. Immediately after the tests, the speaker was asked to perform the ALC speech test which lasted no longer than 15 minutes, to avoid significant changes caused by fatigue or saturation/decomposition of the measured blood alcohol level.

At least two weeks later the speaker was required to undergo a second recording in sober condition, which took about 30 minutes. Both tests took place in the same acoustic environment and were supervised by the same member of the staff, who also acted as the conversational partner for dialogue recordings.

The speech signal was recorded with two different microphones of which the headset Beyerdynamic Opus 54.16/3 was used. It is connected to an MAUDIO MobilePre USB audio interface were the analogue signal is converted to digital and transferred to a laptop via USB. Signals are down-sampled to 16 kHz. All speakers are prompted with the same material. Three different speech styles are part of each ALC recording: read speech, spontaneous speech, and command & control.

10.4.4.2 SLC Database

99 participants took part in six partial sleep deprivation studies for the recording of the Sleepy Language Corpus (SLC) [75, 212]. The mean age of subjects was 24.9 years, with a standard deviation of 4.2 years and a range of 20–52 years. The recordings took place in a realistic car environment or in lecture-rooms. Audio was recorded with a sampling rate of 44.1 kHz, then down-sampled to 16 kHz; quantisation is 16 bit, the microphone-to-mouth distance was 0.3 m.

The speech data consists of different tasks as follows: isolated vowels (sustained vowel phonation, sustained loud vowel phonation, and sustained smiling vowel phonation), read speech from "Die Sonne und der Nordwind" (the story of 'the North Wind and the Sun' in German, widely used within phonetics, speech pathology, and alike), commands and requests (10 simulated driver assistance system commands/requests in German, e.g.,"Ich suche die Friesenstrasse" ('I am looking for the Friesen street'), and four simulated pilot-air traffic controller communication statements), and a description of a picture and a regular lecture.

A well established, standardised subjective sleepiness questionnaire measure, the Karolinska Sleepiness Scale, was used by the subjects (self-assessment) and additionally by the two experimental assistants (observer assessment, given by assessors who had been formally trained to apply a standardised set of judging criteria). In the version used, scores range from 1–10: extremely alert (1), very alert (2), alert (3), rather alert (4), neither alert nor sleepy (5), some signs of sleepiness (6), sleepy but no effort to stay awake (7), sleepy and some effort to stay awake (8), very sleepy with great effort to stay awake / struggling against sleep (9), extremely sleepy, cannot stay awake (10). Given these verbal descriptions, scores greater than 7.5 appear to be most relevant from a practical perspective as they describe a state in which the subject feels unable to stay awake.

For training and classification purposes, the recordings (mean KSS = 5.9, standard deviation KSS = 2.2) were divided into two classes: not sleepy ('NSL') and sleepy ('SL') samples with a threshold of 7.5 (approx. 94 samples per subject; in total 9 277 samples). A more detailed description of the data can be found in [197, 212, 213].

The available turns were divided into males (m) and females (f) per study. Then, the turns from male and from female subjects were split speaker-disjunctive, in ascending order of subject ID, into training (roughly 40 %), development (30 %), and test (30 %) instances. This subdivision not only ensures speaker-independent partitions, but also provides for stratification by gender and study setup (environment and degree of sleep deprivation). Out of the 99 subjects, 36 (20 f, 16 m) were assigned to the training, 30 (17 f, 13 m) to the development, and 33 (19 f, 14 m) to the test set. All turns which include linguistic cues for the sleepiness level (e.g.,"Ich bin sehr müde"—"I'm very tired") were removed from the test set—188 in total. The distribution of instances is given in Table 10.26.

Table 10.26 Partitions of SLC. 'NSL' denotes recordings of non-sleepy, i.e., KSS in the interval [1;7.5], and 'SL' recordings of sleepy speakers, i.e., KSS]7.5;10]

#SLC	NSL	SL	Sum
Train	2 125	1 241	3 366
Develop	1 836	1 079	2 915
Test	1 957	851	2 808
Train + develop	3 961	2 320	6 281
Train + develop + test	5 918	3 171	9 089

10.4.4.3 Methodology

For detection of intoxication and sleepiness in the INTERSPEECH 2011 Speaker State Challenge, an extended set of features with respect to the INTERSPEECH 2009 Emotion Challenge (384 features) (cf. Section 10.4.2) [72] and INTERSPEECH 2010 Paralinguistic Challenge (1 582 features) (cf. Section 10.4.3) [75] was given. Features were again extracted with the openSMILE feature extractor [50] (see also Sect. 6.5).

The feature set consists of 4 368 features comprising features known as relevant for these tasks [214, 215] built from three sets of LLDs and one corresponding set of functionals applied on the recording level for each LLD set. The LLD sets are given in Table A.1 in the Appendix. A major novelty concerning LLD compared to the previous challenge set is the auditory spectrum derived loudness measure and the use of RASTA-style filtered auditory spectra instead of Mel-spectra, as well as a slightly extended set of statistical spectral descriptors (such as entropy, variance, etc.).

Further, a base set of 33 functionals is introduced as shown in Table A.1 in the Appendix. Additions include the use of linear predictive coding coefficients and linear prediction gain as functionals, as well as the standard deviation of the intra-peak distances. In the set of functionals applied to the spectral and energy related LLD, the standard deviation of the segment lengths are further additions. Also, a new algorithm for splitting the contour into segments was used. Previously this was based on delta thresholding, where a new segment was started when the signal rose by a pre-defined relative (to the signal's range) amount in a short time frame. Here, a new segment boundary is given each time the LLD's value (after simple moving average filtering with 3 frames width) crosses $(\min + 0.25 \cdot \text{range})$ and $(\min + 0.75 \cdot \text{range})$.

To the 54 energy and spectral LLD and their first order deltas, the base functional set and the mean, max, min, and the standard deviation of the segment length are applied, resulting in 3 996 features. To the 5 pitch and voice quality LLD and their first order deltas, the base functional set as well as the quadratic mean and the rise and fall durations of the signal are applied only to voiced regions (probability of voicing greater 0.7). This adds another 360 features. Another 12 features are obtained by applying a small set of six functionals to the F0 contour (including non-voiced regions where F0 is set to 0) and its first order derivative as also shown in Table A.1 in the Appendix. Please note that, segments in this case correspond to continuous voiced regions, i.e., where F0 is > 0.

10.4.4.4 Performance

The WEKA data mining toolkit was again used for classification [196] with linear kernel SVM trained with the SMO algorithm. For parameter adjustment, optimisation of the complexity on the development partition per corpus is considered. Thereby, the complexity influences the number of Support Vectors for the hyperplane construction. Further, WEKA's implementation of SMOTE [173] is applied again, as was done for the INTERSPEECH 2009 Emotion Challenge baseline (cf. Sect. 10.4.2), to balance instances in the respective learning partitions. If training and development partitions are united, SMOTE is applied subsequently to the unification. The results of the SVM complexity optimisation when training on the train partitions of ALC and SLC and testing on the respective development partitions are shown in Fig. 10.8a for ALC, and Fig. 10.8b for SLC in terms of UA. In these figures we further take a look at evaluation of the former feature sets of the 2009 and 2010 challenges in comparison to the one provided for this challenge.

As can be seen, the new feature set prevails throughout all conditions on these tasks. Based on the optimal complexity as found on the development partitions, Table 10.27 shows baseline results for intoxication (left) and sleepiness (right) by UA and WA. As the distribution among classes is not balanced, the main measure is again UA as earlier stated. Results are given for training on the train partition and testing on the development partition, as well as for training on the unification of the training and development partitions and testing on the test partition.

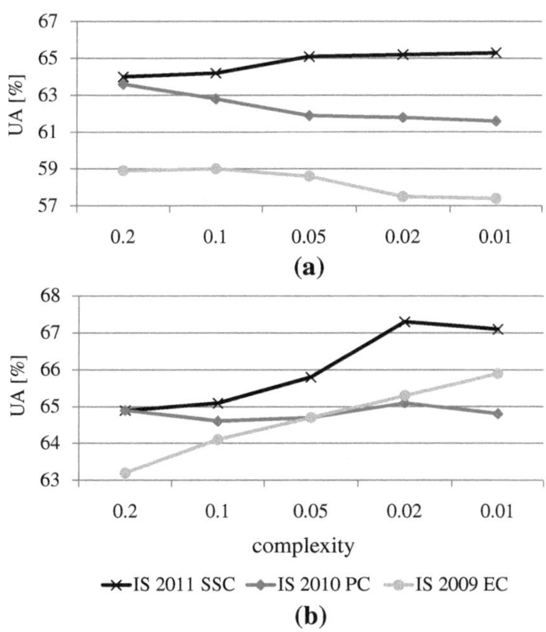

Fig. 10.8 Optimisation of SVM complexity by UA on the development partitions of the ALC and SLC corpora when training on the training partitions after SMOTE. Three different feature sets are evaluated (cf. Table 10.27) [77]. **a** ALC **b** SLC

Table 10.27 Intoxication and sleepiness baseline results by UA and WA. SMO learnt pairwise SVM with linear Kernel, complexity optimised on development partition to 0.01 (intoxication) and 0.02 (sleepiness)

[%]	Intoxication		Sleepiness	
Features	UA	WA	UA	WA
Train versus develop				
IS 2009 EC	57.4	65.3	65.3	64.2
IS 2010 PC	61.6	66.1	65.1	66.4
IS 2011 SSC	65.3	69.2	67.3	69.1
Train + develop versus test				
IS 2009 EC	60.3	60.2	68.0	72.4
IS 2010 PC	63.2	62.6	70.2	72.8
IS 2011 SSC	65.9	66.4	70.3	72.9

SMOTE on (united) learning instances. Feature sets IS 2009 EC, IS 2010 PC, and IS SSC 2011 correspond to the official sets of the Challenges (Emotion [72], Paralinguistic [75, 179], and Speaker State [77] held at INTERSPEECH in the respective years)

10.4.4.5 Summary

The automatic recognition of speakers' intoxication and sleepiness was shown. As for the previous challenges, majority voting of the best participants' results lead to the overall best results of UA 72.2 % (intoxication), and UA 72.5 % (sleepiness).

In [216], however, it was shown how intoxication recognition performance can be further boosted by incorporating not a single speech-chunk, but a series of such. This makes sense, as we are dealing with temporally 'more permanent' speaker states. In addition, focus on specific linguistic entities such as tongue breakers was shown to be beneficial. It seems promising to further elaborate on this findings for other more permanent states and traits.

One of the other most promising future directions seems to be the coupling of tasks—all these are somewhat influencing each other, and it seems intuitive to assess for example age and sleepiness or emotion and gender together rather than in isolation. Further ideas for future research and a summary of recent trends is also found in [217].

References

1. Shriberg, E.: Spontaneous speech: how peoply really talk and why engineers should care. In: Proceedings of Eurospeech, pp. 1781–1784. Lisbon (2005)
2. Schuller, B., Ablameier, M., Müller, R., Reifinger, S., Poitschke, T., Rigoll, G.: Speech communication and multimodal interfaces. In: Kraiss, K.-F. (ed.) Advanced Man Machine Interaction. Signals and Communication Technology. Chapter 4, pp. 141–190. Springer, Berlin (2006)
3. Lee, C.-C., Black, M., Katsamanis, A., Lammert, A., Baucom, B., Christensen, A., Georgiou, P., Narayanan, S.: Quantification of prosodic entrainment in affective spontaneous spoken

interactions of married couples. In: Proceedings of Interspeech, pp. 793–796, Makuhari (2010)
4. Schuller, B., Wöllmer, M., Eyben, F., Rigoll, G.: Retrieval of paralinguistic information in broadcasts. In: Maybury, M.T. (ed.) Multimedia Information Extraction: Advances in Video, Audio, and Imagery Extraction for Search, Data Mining, Surveillance, and Authoring. Chapter 17, pp. 273–288. Wiley, IEEE Computer Society Press (2012)
5. Moreno, P.: Speech recognition in noisy environments. PhD thesis, Carnegie Mellon University, Pittsburgh (1996)
6. Kim, D., Lee, S., Kil, R.: Auditory processing of speech signals for robust speech recognition in real-world noisy environments. IEEE Trans. Speech Audio Process. **7**, 55–69 (1999)
7. Rose, R.: Environmental robustness in automatic speech recognition. In: COST278 and ISCA Tutorial and Research Workshop on Robustness Issues in Conversational, Interaction (2004)
8. Schuller, B., Wöllmer, M., Moosmayr, T., Rigoll, G.: Robust spelling and digit recognition in the car: switching models and their like. In: Proceedings 34. Jahrestagung für Akustik, DAGA. DEGA, pp. 847–848. Dresden, March 2008
9. Schuller, B., Wöllmer, M., Moosmayr, T., Ruske, G., Rigoll, G.: Switching linear dynamic models for noise robust in-car speech recognition. In: Rigoll, G. (ed.) Pattern Recognition: 30th DAGM Symposium Munich, Germany. Proceedings of Lecture Notes on Computer Science (LNCS), vol. 5096, pp. 244–253. Springer, Berlin 10–13 June 2008
10. Schuller, B., Wöllmer, M., Moosmayr, T., Rigoll, G.: Recognition of noisy speech: a comparative survey of robust model architecture and feature enhancement. EURASIP J. Audio Speech Music Process. **2009**(Article ID 942617), 17 (2009)
11. Wöllmer, M., Eyben, F., Schuller, B., Sun, Y., Moosmayr, T., Nguyen-Thien, N.: Robust in-car spelling recognition: a tandem blstm-hmm approach. In: Proceedings INTERSPEECH 2009, 10th Annual Conference of the International Speech Communication Association, pp. 1990–9772. ISCA, Brighton, Sept 2009
12. Schuller, B., Weninger, F., Wöllmer, M., Sun, Y. Rigoll, G.: Non-negative matrix factorization as noise-robust feature extractor for speech recognition. In Proceedings 35th IEEE International Conference on Acoustics, Speech, and Signal Processing, ICASSP 2010, pp. 4562–4565. IEEE, Dallas, March 2010
13. Weninger, F., Geiger, J., Wöllmer, M., Schuller, B., Rigoll, G.: The munich 2011 chime challenge contribution: Nmf-blstm speech enhancement and recognition for reverberated multisource environments. In: Proceedings Machine Listening in Multisource Environments, CHiME 2011, Satellite Workshop of Interspeech, pp. 24–29. ISCA, Florence, Sept 2011
14. Weninger, F., Wöllmer, M., Geiger, J. Schuller, B., Gemmeke, J., Hurmalainen, A., Virtanen, T., Rigoll, G.: Non-negative matrix factorization for highly noise-robust asr: to enhance or to recognize? In: Proceedings 37th IEEE International Conference on Acoustics, Speech, and Signal Processing, ICASSP 2012, pp. 4681–4684. IEEE, Kyoto, March 2012
15. de la Torre, A., Fohr, D., Haton, J.: Compensation of noise effects for robust speech recognition in car environments. In: Proceedings of International Conference on Spoken Language Processing (2000)
16. Langmann, D., Fischer, A., Wuppermann, F., Haeb-Umbach, R., Eisele, T.: Acoustic front ends for speaker-independent digit recognition in car environments. In: Proceedings of Eurospeech, pp. 2571–2574 (1997)
17. Doddington, G., Schalk, T.: Speech recognition: turning theory to practice. In: IEEE Spectrum, pp. 26–32 (1981)
18. Hirsch, H.G., Pierce, D.: The AURORA experimental framework for the performance evaluation of speech recognition systems under noise conditions. Challenges for the Next Millenium, Automatic Speech Recognition (2000)
19. Mesot, B., Barber, D.: Switching linear dynamical systems for noise robust speech recognition. IEEE Trans. Audio Speech Lang. Process. **15**, 1850–1858 (2007)
20. Schuller, B., Rigoll, G., Grimm, M., Kroschel, K., Moosmayr, T., Ruske, G.: Effects of in-car noise-conditions on the recognition of emotion within speech. In: Proceedings 33. Jahrestagung für Akustik, DAGA 2007, pp. 305–306. DEGA, Stuttgart, March 2007

21. Grimm, M., Kroschel, K., Harris, H., Nass, C., Schuller, B., Rigoll, G., Moosmayr, T.: On the necessity and feasibility of detecting a driver's emotional state while driving. In: Paiva, A., Picard, R.W., Prada, R. (eds.) Affective Computing and Intelligent Interaction: Second International Conference, pp. 126–138. ACII 2007, Lisbon, Portugal, September 12–14, 2007. Proceedings of Lecture Notes on Computer Science (LNCS)Springer, vol. 4738/2007. Berlin/Heidelberg (2007)
22. Schuller, B.: Speaker, noise, and acoustic space adaptation for emotion recognition in the automotive environment. In: Proceedings 8th ITG Conference on Speech Communication, vol. 211, p. 4. ITG-Fachbericht, Aachen, Germany, ITG, VDE-Verlag (2008)
23. Cooke, M., Scharenborg, O.: The interspeech 2008 consonant challenge. In: Proceedings of the 2008 IEEE International Conference on Acoustics, Speech and, Signal Processing (2008)
24. Borgström, B., Alwan, A.: HMM-based estimation of unreliable spectral components for noise robust speech recognition. In: Proceedings of the 2008 IEEE International Conference on Acoustics, Speech and, Signal Processing (2008)
25. Jancovic, P., Münevver, K.: On the mask modeling and feature representation in the missing-feature ASR: evaluation on the consonant challenge. In: Proceedings of Interspeech (2008).
26. Gemmeke, J., Cranen, B.: Noise reduction through compressed sensing. In: Proceedings of Interspeech (2008).
27. Schuller, B., Wöllmer, M., Moosmayr, T., Rigoll, G.: Speech recognition in noisy environments using a switching linear dynamic model for feature enhancement. In: Proceedings INTERSPEECH 2008, 9th Annual Conference of the International Speech Communication Association, incorporating 12th Australasian International Conference on Speech Science and Technology, SST 2008, pp. 1789–1792, Brisbane, Australia, ISCA/ASSTA, ISCA (2008)
28. Wöllmer, M., Eyben, F., Schuller, B., Rigoll, G.: A multi-stream asr framework for blstm modeling of conversational speech. In: Proceedings 36th IEEE International Conference on Acoustics, Speech, and Signal Processing, ICASSP 2011, pp. 4860–4863. Prague, Czech Republic, IEEE, IEEE (2011)
29. Wöllmer, M., Eyben, F., Graves, A., Schuller, B., Rigoll, G.: A tandem blstm-dbn architecture for keyword spotting with enhanced context modeling. In: Proceedings ISCA Tutorial and Research Workshop on Non-Linear Speech Processing, p. 9. NOLISP 2009, Vic, Spain. ISCA, ISCA (2009)
30. Wöllmer, M., Eyben, F., Keshet, J., Graves, A., Schuller, B., Rigoll, G.: Robust discriminative keyword spotting for emotionally colored spontaneous speech using bidirectional lstm networks. In: Proceedings 34th IEEE International Conference on Acoustics, Speech, and Signal Processing, ICASSP 2009, pp. 3949–3952. Taipei, Taiwan, IEEE, IEEE (2009)
31. Wöllmer, M., Eyben, F., Schuller, B., Rigoll, G.: Robust vocabulary independent keyword spotting with graphical models. In: Proceedings 11th Biannual IEEE Automatic Speech Recognition and Understanding Workshop, ASRU 2009, pp. 349–353. Merano, Italy, IEEE, IEEE (2009)
32. Wöllmer, M., Sun, Y., Eyben, F., Schuller, B.: Long short-term memory networks for noise robust speech recognition. In: Proceedings INTERSPEECH 2010, 11th Annual Conference of the International Speech Communication Association, pp. 2966–2969. Makuhari, Japan, ISCA, ISCA (2010)
33. Wöllmer, M., Eyben, F., Schuller, B., Rigoll, G.: Spoken term detection with connectionist temporal classification: a novel hybrid ctc-dbn decoder. In: Proceedings 35th IEEE International Conference on Acoustics, Speech, and Signal Processing, ICASSP 2010, pp. 5274–5277. Dallas, TX, IEEE, IEEE (2010)
34. Wöllmer, M., Eyben, F., Graves, A., Schuller, B., Rigoll, G.: Improving keyword spotting with a tandem blstm-dbn architecture. In: Sole-Casals, J., Zaiats, V. (eds.) Advances in Non-Linear Speech Processing: International Conference on Nonlinear Speech Processing, NOLISP 2009, Vic, Spain, 25–27 June 2009. Revised Selected Papers of Lecture Notes on Computer Science (LNCS), vol. 5933/2010, pp. 68–75. Springer (2010)
35. Wöllmer, M., Eyben, F., Schuller, B., Rigoll, G.: Recognition of spontaneous conversational speech using long short-term memory phoneme predictions. In: Proceedings INTERSPEECH

2010, 11th Annual Conference of the International Speech Communication Association, pp. 1946–1949. Makuhari, Japan, ISCA, ISCA (2010)
36. Wöllmer, M., Eyben, F., Graves, A., Schuller, B., Rigoll, G.: Bidirectional lstm networks for context-sensitive keyword detection in a cognitive virtual agent framework. Cogn. Comput. Spec. Issue Non-Linear Non-Conv. Speech Proces. **2**(3), 180–190 (2010)
37. Wöllmer, M., Schuller, B.: Enhancing spontaneous speech recognition with blstm features. In: Travieso-González, C.M., Alonso-Hernández, J. (eds.) Advances in Nonlinear Speech Processing, 5th International Conference on Nonlinear Speech Processing, NoLISP 2011, Las Palmas de Gran Canaria, Spain, 7–9 November 2011. Proceedings of Lecture Notes in Computer Science (LNCS), vol. 7015/2011, pp. 17–24. Springer (2011)
38. Wöllmer, M., Marchi, E., Squartini, S., Schuller, B.: Multi-stream lstm-hmm decoding and histogram equalization for noise robust keyword spotting. Cogn. Neurodyn. **5**(3), 253–264 (2011)
39. Wöllmer, M., Schuller, B., Batliner, A., Steidl, S., Seppi, D.: Tandem decoding of children's speech for keyword detection in a child-robot interaction scenario. In: ACM Transactions on Speech and Language Processing. Special Issue on Speech and Language Processing of Children's Speech for Child-machine Interaction Applications, vol. 7, Issue 4, p. 22 (2011)
40. Wöllmer, M., Schuller, B., Rigoll, G.: A novel bottleneck-blstm front-end for feature-level context modeling in conversational speech recognition. In: Proceedings 12th Biannual IEEE Automatic Speech Recognition and Understanding Workshop, ASRU 2011, pp. 36–41. Big Island, HY, IEEE, IEEE (2011)
41. Wöllmer, M., Schuller, B., Rigoll, G.. Feature frame stacking in rnn-based tandem asr systems—learned vs. predefined context. In: Proceedings INTERSPEECH 2011, 12th Annual Conference of the International Speech Communication Association, pp. 1233–1236. Florence, Italy, ISCA, ISCA (2011)
42. Wöllmer, M., Marchi, E., Squartini, S., Schuller, B.: Robust multi-stream keyword and non-linguistic vocalization detection for computationally intelligent virtual agents. In: Liu, D., Zhang, H., Polycarpou, M., Alippi, C., He, H. (eds.) Proceedings 8th International Conference on Advances in Neural Networks, ISNN 2011, Guilin, China, 29.05.–01.06.2011. Part II of Lecture Notes in Computer Science (LNCS), vol. 6676, pp. 496–505. Springer, Berlin/Heidelberg (2011)
43. Schröder, M., Bevacqua, E., Cowie, R., Eyben, F., Gunes, H., Heylen, D., ter Maat, M., McKeown, G., Pammi, S., Pantic, M., Pelachaud, C., Schuller, B., de Sevin, E., Valstar, M., Wöllmer, M.: Building autonomous sensitive artificial listeners. IEEE Trans. Affect. Comput. **3**(2):165–183 (2012)
44. Aradilla, G., Vepa, J., Bourlard, H.: An acoustic model based on Kullback-Leibler divergence for posterior features. In: Proceedings of the ICASSP, pp. 657–660. Honolulu, HI (2007)
45. Grezl, F., Fousek, P.: Optimizing bottle-neck features for LVCSR. In: Proceedings of the ICASSP, pp. 4729–4732. Las Vegas, NV (2008)
46. Hermansky, H., Fousek, P.: Multi-resolution RASTA filtering for TANDEM-based ASR. In: Proceedings of the European Conference on Speech Communication and Technology, pp. 361–364. Lisbon, Portugal (2008)
47. Graves, A., Fernandez, S., Schmidhuber, J.: Bidirectional LSTM networks for improved phoneme classification and recognition. In: Proceedings of ICANN, pp. 602–610. Warsaw, Poland (2005)
48. Fernandez, S., Graves, A., Schmidhuber, J.: An application of recurrent neural networks to discriminative keyword spotting. In: Proceedings of Internet Corporation for Assigned Names and Numbers 2007, vol. 4669, pp. 220–229. Porto, Portugal (2007)
49. Stupakov, A., Hanusa, E., Bilmes, J., Fox, D.: COSINE—a corpus of multi-party conversational speech in noisy environments. In: Proceedings of the ICASSP, Taipei, Taiwan (2009)
50. Eyben, F., Wöllmer, M., Schuller, B.: Opensmile—the munich versatile and fast open-source audio feature extractor. In: Proceedings of the 9th ACM International Conference on Multimedia, MM 2010, pp. 1459–1462. Florence, Italy, ACM, ACM (2010)

51. Graves, A., Schmidhuber, J.: Framewise phoneme classification with bidirectional LSTM and other neural network architectures. Neural Netw. **18**(5—-6), 602–610 (2005)
52. Campbell, N.: On the use of nonverbal speech sounds in human communication. In: Proceedings of the COST 2102 Workshop, pp. 117–128. Vietri sul Mare, Italy (2007)
53. Schuller, B., Batliner, A., Seppi, D., Steidl, S., Vogt, T., Wagner, J., Devillers, L., Vidrascu, L., Amir, N., Kessous, L., Aharonson, V.: The relevance of feature type for the automatic classification of emotional user states: low level descriptors and functionals. In: Proceedings INTERSPEECH 2007, 8th Annual Conference of the International Speech Communication Association, pp. 2253–2256. Antwerp, Belgium. ISCA, ISCA (2007)
54. Schuller, B., Eyben, F., Rigoll, G.: Static and dynamic modelling for the recognition of non-verbal vocalisations in conversational speech. In: André, E., Dybkjaer, L., Neumann, H., Pieraccini, R., Weber, M. (eds.) Perception in Multimodal Dialogue Systems: 4th IEEE Tutorial and Research Workshop on Perception and Interactive Technologies for Speech-Based Systems, pp. 99–110. PIT 2008, Kloster Irsee, Germany, 16–18 June 2008. Proceedings of Lecture Notes on Computer Science (LNCS), vol. 5078/2008. Springer, Berlin/Heidelberg (2008)
55. Batliner, A., Steidl, S., Eyben, F., Schuller, B., Laughter in child-robot interaction. In: Proceedings Interdisciplinary Workshop on Laughter and other Interactional Vocalisations in Speech, Laughter, Berlin. February, Germany (2009)
56. Eyben, F., Petridis, S., Schuller, B., Tzimiropoulos, G., Zafeiriou, S., Pantic, M.: Audiovisual classification of vocal outbursts in human conversation using long-short-term memory networks. In: Proceedings 36th IEEE International Conference on Acoustics, Speech, and Signal Processing, ICASSP 2011, pp. 5844–5847. Prague, Czech Republic, IEEE, IEEE (2011)
57. Batliner, A., Steidl, S., Eyben, F., Schuller, B.: On laughter and speech laugh, based on observations of child-robot interaction. In: Trouvain, J., Campbell, N. (eds.) The Phonetics of Laughing, p. 23. Saarland University Press, Saarbrücken (2012)
58. Prylipko, D., Schuller, B., Wendemuth, A.: Fine-tuning hmms for nonverbal vocalizations in spontaneous speech: a multicorpus perspective. In: Proceedings 37th IEEE International Conference on Acoustics, Speech, and Signal Processing, ICASSP 2012, pp. 4625–4628, Kyoto, Japan, IEEE, IEEE (2012)
59. Eyben, F., Petridis, S., Schuller, B., Pantic, M.: Audiovisual vocal outburst classification in noisy acoustic conditions. In: Proceedings 37th IEEE International Conference on Acoustics, Speech, and Signal Processing, ICASSP 2012, pp. 5097–5100. Kyoto, Japan, IEEE, IEEE (2012)
60. M. Goto, K. Itou, and S. Hayamizu. A real-time filled pause detection system for spontaneous speech recognition. In: Proceedings of the Eurospeech, pp. 227–230. Budapest, Hungary (1999)
61. Truong, K.P., van Leeuwen, D.A.: Automatic detection of laughter. In: Proceedings of the Interspeech, pp. 485–488. Lisbon, Portugal (2005)
62. Campbell, N., Kashioka, H., Ohara, R.: No laughing matter. In: Proceedings of the Interspeech, pp. 465–468. Lisbon, Portugal (2005)
63. Knox, M.T., Mirghafori, N.: Automatic laughter detection using neural networks. In: Proceedings INTERSPEECH 2007, 8th Annual Conference of the International Speech Communication Association, pp. 2973–2976. Antwerp, Belgium, ISCA, ISCA (2007)
64. Cho, Y.-C., Choi, S., Bang, S.-Y.: Non-negative component parts of sound for classification. In: Proceedings of the ISSPIT, pp. 633–636. Darmstadt, Germany (2003)
65. Schuller, B., Müller, R., Eyben, F., Gast, J., Hörnler, B., Wöllmer, M., Rigoll, G., Höthker, A., Konosu, H.: Being bored? recognising natural interest by extensive audiovisual integration for real-life application. Image Vis. Comput. Special Issue on Visual and Multimodal Analysis of Human Spontaneous Behavior **27**(12), 1760–1774 (2009)
66. Schuller, B., Weninger, F.: Discrimination of speech and non-linguistic vocalizations by non-negative matrix factorization. In: Proceedings 35th IEEE International Conference on Acoustics, Speech, and Signal Processing, ICASSP 2010, pp. 5054–5057. Dallas, TX, IEEE, IEEE (2010)

67. Schmidt, M.N., Olsson, R.K.: Single-channel speech separation using sparse non-negative matrix factorization. In: Proceedings of the Interspeech, pp. 2–5. Pittsburgh, Pennsylvania (2006)
68. Smaragdis, P.: Discovering auditory objects through non-negativity constraints. In: Proceedings of the SAPA, Jeju, Korea (2004)
69. Schuller, B.: Automatisches verstehen gesprochener mathematischer formeln. Technische Universität München, Munich, Germany, October, Diploma thesis (1999)
70. Schuller, B., Schenk, J., Rigoll, G., Knaup, T.: "the godfather" vs. "chaos": comparing linguistic analysis based on online knowledge sources and bags-of-n-grams for movie review valence estimation. In: Proceedings of 10th International Conference on Document Analysis and Recognition, ICDAR 2009, pp. 858–862. Barcelona, Spain, IAPR, IEEE (2009)
71. Schuller, B., Knaup, T.: Learning and knowledge-based sentiment analysis in movie review key excerpts. In: Esposito, A., Esposito, A.M., Martone, R., Müller, V., Scarpetta, G. (eds.) Toward Autonomous, Adaptive, and Context-Aware Multimodal Interfaces: Theoretical and Practical Issues: Third COST 2102 International Training School, Caserta, Italy, 15–19 March 2010, Revised Selected Papers, Lecture Notes on Computer Science (LNCS), vol. 6456/2010, 1st edn, pp. 448–472. Springer, Heidelberg (2011)
72. Schuller, B., Steidl, S., Batliner, A.: The interspeech 2009 emotion challenge. In: Proceedings INTERSPEECH 2009, 10th Annual Conference of the International Speech Communication Association, pp. 312–315. Brighton, UK, ISCA, ISCA (2009)
73. Schuller, B., Steidl, S., Batliner, A., Jurcicek, F.: The interspeech 2009 emotion challenge—results and lessons learnt. Speech and Language Processing Technical Committee (SLTC) Newsletter (2009)
74. Schuller, B., Batliner, A., Steidl, S., Seppi, D.: Recognising realistic emotions and affect in speech: state of the art and lessons learnt from the first challenge. Speech Commun. Special Issue on Sensing Emotion and Affect—Facing Realism in Speech Processing. **53**(9/10), 1062–1087 (2011)
75. Schuller, B., Steidl, S., Batliner, A., Burkhardt, F., Devillers, L., Müller, C, Narayanan, S.: The interspeech 2010 paralinguistic challenge. In: Proceedings INTERSPEECH 2010, 11th Annual Conference of the International Speech Communication Association, pp. 2794–2797. Makuhari, Japan, ISCA, ISCA (2010)
76. Schuller, B., Wöllmer, M., Eyben, F., Rigoll, G., Arsić, D.: Semantic speech tagging: towards combined analysis of speaker traits. In: Brandenburg, K., Sandler, M. (eds.) Proceedings AES 42nd International Conference, pp. 89–97. AES, Audio Engineering Society, Ilmenau (2011)
77. Schuller, B., Batliner, A., Steidl, S., Schiel, F., Krajewski, J.: The interspeech 2011 speaker state challenge. In: Proceedings INTERSPEECH 2011, 12th Annual Conference of the International Speech Communication Association, pp. 3201–3204. Florence, Italy, ISCA, ISCA (2011)
78. Chen, A.: Perception of paralinguistic intonational meaning in a second language. Lang. Learn. **59**(2), 367–409 (2009)
79. Bello, R.: Causes and paralinguistic correlates of interpersonal equivocation. J. Pragmat. **38**(9), 1430–1441 (2006)
80. Fernandez, R., Picard, R.W.: Modeling drivers' speech under stress. Speech Commun. **40**, 145–159 (2003)
81. Athanaselis, T., Bakamidis, S., Dologlou, I., Cowie, R., Douglas-Cowie, E., Cox, C.: ASR for emotional speech: Clarifying the issues and enhancing performance. Neural Netw. **18**, 437–444 (2005)
82. Steidl, S., Batliner, A., Seppi, D., Schuller, B.: On the impact of children's emotional speech on acoustic and language models. EURASIP J. Audio, Speech, Music Process. Special Issue on Atypical Speech **2010**(Article ID 783954), 14 (2010)
83. Wöllmer, M., Schuller, B., Eyben, F., Rigoll, G.: Combining long short-term memory and dynamic bayesian networks for incremental emotion-sensitive artificial listening. IEEE J. Sel. Topics Signal Process. Special Issue on Speech Processing for Natural Interaction with Intelligent Environments **4**(5), 867–881 (2010)

84. Wöllmer, M., Klebert, N., Schuller, B.: Switching linear dynamic models for recognition of emotionally colored and noisy speech. In: Proceedings 9th ITG Conference on Speech Communication, ITG-Fachbericht, vol. 225. Bochum, Germany, ITG, VDE-Verlag (2010)
85. Romanyshyn, N.: Paralinguistic maintenance of verbal communicative interaction in literary discourse (on the material of W. S. Maugham's novel "Theatre"). In: Experience of Designing and Application of CAD Systems in Microelectronics—Proceedings of the 10th International Conference, CADSM 2009, pp. 550–552. Polyana-Svalyava, Ukraine (2009)
86. Kennedy, L., Ellis, D.: Pitch-based emphasis detection for characterization of meeting recordings. In: Proceedings of the ASRU, pp. 243–248. Virgin Islands (2003)
87. Laskowski, K.: Contrasting emotion-bearing laughter types in multiparticipant vocal activity detection for meetings. In: Proceedings of the ICASSP, pp. 4765–4768. Taipei, Taiwan, IEEE (2009)
88. Massida, Z., Belin, P., James, C., Rouger, J., Fraysse, B., Barone, P., Deguine, O.: Voice discrimination in cochlear-implanted deaf subjects. Hear. Res. **275**(1–2), 120–129 (2011)
89. Demouy, J., Plaza, M., Xavier, J., Ringeval, F., Chetouani, M. Prisse, D., Chauvin, D., Viaux, S., Golse, B., Cohen, D., Robel, L.: Differential language markers of pathology in autism, pervasive developmental disorder not otherwise specified and specific language impairment. Res. Autism Spectr. Disord. **5**(4), 1402–1412 (2011)
90. Mower, E., Black, M., Flores, E., Williams, M., Narayanan, S.: Design of an emotionally targeted interactive agent for children with autism. In: Proceedings of the IEEE International Conference on Multimedia and Expo (ICME 2011), pp. 1–6. Barcelona, Spain (2011)
91. de Sevin, E., Bevacqua, E., Pammi, S., Pelachaud, C., Schröder, M., Schuller, B.: A multimodal listener behaviour driven by audio input. In: Proceedings International Workshop on Interacting with ECAs as Virtual Characters, satellite of AAMAS 2010, p. 4. Toronto, Canada, ACM, ACM (2010)
92. Biever, C.: You have three happy messages. New Sci. **185**(2481), 21 (2005)
93. Martinez, C.A., Cruz, A.: Emotion recognition in non-structured utterances for human-robot interaction. In: IEEE International Workshop on Robot and Human Interactive, Communication, pp. 19–23 (2005)
94. Batliner, A., Steidl, S., Nöth, E.: Associating children's non-verbal and verbal behaviour: body movements, emotions, and laughter in a human-robot interaction. In: Proceedings of ICASSP, pp. 5828–5831. Prague (2011)
95. Delaborde, A., Devillers, L.: Use of non-verbal speech cues in social interaction between human and robot: emotional and interactional markers. In: AFFINE'10—Proceedings of the 3rd ACM Workshop on Affective Interaction in Natural Environments, Co-located with ACM Multimedia 2010, pp. 75–80. Florence, Italy (2010)
96. Schröder, M., Cowie, R., Heylen, D., Pantic, M., Pelachaud, C., Schuller, B.: Towards responsive sensitive artificial listeners. In: Proceedings 4th International Workshop on Human-Computer Conversation, p. 6. Bellagio, Italy (2008)
97. Burkhardt, F., van Ballegooy, M., Englert, R., Huber, R.: An emotion-aware voice portal. In: Proceedings of the Electronic Speech Signal Processing ESSP, pp. 123–131 (2005)
98. Mishne, G., Carmel, D., Hoory, R., Roytman, A., Soffer, A.: Automatic analysis of call-center conversations. In: Proceedings of the CIKM'05, pp. 453–459. Bremen, Germany (2005)
99. Belin, P., Fillion-Bilodeau, S., Gosselin, F.: The montreal affective voices: a validated set of nonverbal affect bursts for research on auditory affective processing. Behav. Res. Meth. **40**(2), 531–539 (2008)
100. Schoentgen, J.: Vocal cues of disordered voices: an overview. Acta Acustica United Acustica **92**(5), 667–680 (2006)
101. Rektorova, I., Barrett, J., Mikl, M., Rektor, I., Paus, T.: Functional abnormalities in the primary orofacial sensorimotor cortex during speech in parkinson's disease. Mov. Disord **22**(14), 2043–2051 (2007)
102. Sapir, S., Ramig, L.O., Spielman, J.L., Fox, C.: Formant centralization ratio: a proposal for a new acoustic measure of dysarthric speech. J. Speech Lang. Hear. Res. **53** (2009)

103. Oller, D.K., Niyogic, P., Grayd, S., Richards, J.A., Gilkerson, J., Xu, D., Yapanel, U., Warrene, S.F.: Automated vocal analysis of naturalistic recordings from children with autism, language delay, and typical development. In: Proceedings of the National Academy of Sciences of the United States of America (PNAS), vol. 107. (2010)
104. Maier, A., Haderlein, T., Eysholdt, U., Rosanowski, F., Batliner, A., Schuster, M., Nöth, E.: PEAKS—a system for the automatic evaluation of voice and speech disorders. Speech Commun. **51**, 425–437 (2009)
105. Malyska, N., Quatieri, T., Sturim, D.: Automatic dysphonia recognition using bilogically inspired amplitude-modulation features. In: Proceedings of the IEEE International Conference on Acoustics, Speech and Signal Processing (ICASSP), vol. I, pp. 873–876. Prague (2005)
106. Dibazar, A., Narayanan, S.: A system for automatic detection of pathological speech. In: Proceedings of Conference Signals, Systems, and Computers, Asilomar, CA (2002)
107. Litman, D., Rotaru, M., Nicholas, G.: Classifying turn-level uncertainty using word-level prosody. In: Proceedings of the Interspeech, pp. 2003–2006. Brighton, UK (2009)
108. Boril, H., Sadjadi, S., Kleinschmidt, T., Hansen, J.: Analysis and detection of cognitive load and frustration in drivers' speech. In: Proceedings of the Interspeech 2010, pp. 502–505. Makuhari, Japan (2010)
109. Litman, D., Forbes, K.: Recognizing emotions from student speech in tutoring dialogues. In: Proceedings of ASRU, pp. 25–30. Virgin Island (2003)
110. Ai, H., Litman, D., Forbes-Riley, K., Rotaru, M., Tetreault, J., Purandare, A.: Using system and user performance features to improve emotion detection in spoken tutoring dialogs. In: Proceedings of the Interspeech, pp. 797–800. Pittsburgh (2006)
111. Price, L., Richardson, J.T.E., Jelfs, A.: Face-to-face versus online tutoring support in distance education. Stud. High. Edu. **32**(1), 1–20 (2007)
112. Pfister, T., Robinson, P.: Speech emotion classification and public speaking skill assessment. In: Proceedings of the International Workshop on Human Behaviour Understanding, pp. 151–162. Istanbul, Turkey (2010)
113. Schuller, B., Eyben, F., Can, S., Feussner, H.: Speech in minimal invasive surgery—towards an affective language resource of real-life medical operations. In: Devillers, L., Schuller, B., Cowie, R., Douglas-Cowie, E., Batliner, A. (eds.) Proceedings 3rd International Workshop on EMOTION: Corpora for Research on Emotion and Affect, satellite of LREC 2010, pp. 5–9. Valletta, Malta. ELRA, European Language Resources Association (2010)
114. Ronzhin, A.L.: Estimating psycho-physiological state of a human by speech analysis. Proc. SPIE Int. Soc. Opt. Eng. **5797**, 170–181 (2005)
115. Schuller, B., Wimmer, M, Arsić, D., Moosmayr, T., Rigoll, G.: Detection of security related affect and behaviour in passenger transport. In: Proceedings INTERSPEECH 2008, 9th Annual Conference of the International Speech Communication Association, incorporating 12th Australasian International Conference on Speech Science and Technology, SST 2008, pp. 265–268. Brisbane, Australia. ISCA/ASSTA, ISCA (2008)
116. Kwon, H., Berisha, V., Spanias, A.: Real-time sensing and acoustic scene characterization for security applications. In: 3rd International Symposium on Wireless Pervasive Computing, ISWPC 2008, Proceedings, pp. 755–758 (2008)
117. Clavel, C., Vasilescu, I., Devillers, L., Richard, G., Ehrette, T.: Fear-type emotion recognition for future audio-based surveillance systems. Speech Commun. **50**(6), 487–503 (2008)
118. Boril, H., Sangwan, A., Hasan, T., Hansen, J.: Automatic excitement-level detection for sports highlights generation. In: Proceedings of the Interspeech 2010, pp. 2202–2205. Makuhari, Japan (2011)
119. Turney, P.D.: Thumbs up or thumbs down? Semantic orientation applied to unsupervised classification of reviews. In: Proceedings of the 40th Annual Meeting of the Association for Computational Linguistics (ACL), pp. 417–424. Philadelphia (2002)
120. Dave, K., Lawrence, S., Pennock, D.M.: Mining the peanut gallery: opinion extraction and semantic classification of product reviews. In: Proceedings of the 12th international conference on World Wide Web, pp. 519–528. Budapest, Hungary, ACM (2003).

121. Yi, J., Nasukawa, T., Bunescu, R., Niblack, W.: Sentiment analyzer: extracting sentiments about a given topic using natural language processing techniques. In: Proceedings of the Third IEEE International Conference on Data Mining, pp. 427–434 (2003)
122. Popescu, A., Etzioni, O.: Extracting product features and opinions from reviews. In: Proceedings of the Conference on Human Language Technology and Empirical Methods in Natural Language Processing, pp. 339–346. Association for Computational Linguistics Morristown, NJ, USA (2005)
123. B. Liu, M. Hu, and J. Cheng. Opinion observer: analyzing and comparing opinions on the web. In: WWW '05: Proceedings of the 14th international conference on World Wide Web, pp. 342–351. New York, NY, ACM (2005)
124. Ding, X., Liu, B., Yu, P.S.: A holistic lexicon-based approach to opinion mining. In: WSDM '08: Proceedings of the International Conference on Web Search and Web Data Mining, pp. 231–240, New York, NY, USA, ACM (2008)
125. Das, S.R., Chen, M.Y.: Yahoo! for amazon: sentiment parsing from small talk on the web. In: Proceedings of the 8th Asia Pacific Finance Association Annual Conference (2001)
126. Pang., B., Lee, L., Vaithyanathan, S.: Thumbs up? Sentiment classification using machine learning techniques. In: Proceedings Conference on Empirical Methods in Natural Language Processing (EMNLP), pp. 79–86. Philadelphia, PA (2002)
127. Zhuang, L., Jing, F., Zhu, X.-Y.: Movie review mining and summarization. In: Proceedings of the 15th ACM International Conference on Information and Knowledge Management (CIKM '06), pp. 43–50, New York, NY, USA, ACM (2006)
128. Porter, M.F.: An algorithm for suffix stripping. Program **3**(14), 130–137 (October 1980)
129. Marcus, M., Marcinkiewicz, M., Santorini, B.: Building a large annotated corpus of english: the Penn Treebank. Comput. Linguist. **19**(2), 313–330 (1993)
130. Sha, F., Pereira, F.: Shallow parsing with conditional random fields. In: NAACL '03: Proceedings of the 2003 Conference of the North American Chapter of the Association for Computational Linguistics on Human Language Technology, pp. 134–141. Morristown, NJ, USA. Association for Computational Linguistics (2003)
131. Witten, I.H., Frank, E.: Data Mining: Practical Machine Learning Tools and Techniques, 2nd edn. Morgan Kaufmann, San Francisco (2005)
132. Wiebe, J., Wilson, T., Bell, M.: Identifying collocations for recognizing opinions. In: Proceedings of the ACL-01 Workshop on Collocation: Computational Extraction, Analysis, and Exploitation, pp. 24–31 (2001)
133. Wilson, T., Wiebe, J., Hoffmann, P.: Recognizing contextual polarity in phrase-level sentiment analysis. In: HLT '05: Proceedings of the conference on Human Language Technology and Empirical Methods in Natural Language Processing, pp. 347–354. Morristown, NJ, USA, Association for Computational Linguistics (2005)
134. Turney, P.D., Littman, M.L.: Measuring praise and criticism: Inference of semantic orientation from association. ACM Trans. Inf. Syst. **21**(4), 315–346 (October 2003)
135. Esuli, A., Sebastiani, F.: Determining term subjectivity and term orientation for opinion mining. In: Proceedings of the 11th Conference of the European Chapter of the Association for Computational Linguistics (EACL '06), Trento, Italy (2006)
136. Lizhong, W., Oviatt, S., Cohen, P.R.: Multimodal integration—a statistical view. IEEE Trans. Multimed. **1**, 334–341 (1999)
137. Wöllmer, M., Al-Hames, M., Eyben, F., Schuller, B., Rigoll, G.: A multidimensional dynamic time warping algorithm for efficient multimodal fusion of asynchronous data streams. Neurocomputing **73**(1–3), 366–380 (2009)
138. Liu, D.: Automatic mood detection from acoustic music data, pp. 13–17. In: Proceedings International Conference on Music, Information Retrieval (2003)
139. Nose, T., Kato, Y., Kobayashi, T.: Style estimation of speech based on multiple regression hidden semi-markov model. In: Proceedings INTERSPEECH 2007, 8th Annual Conference of the International Speech Communication Association, pp. 2285–2288. Antwerp, Belgium, ISCA, ISCA (2007)

140. Zhang, C., Hansen, J.H.L.: Analysis and classification of speech mode: whispered through shouted. In: International Speech Communication Association—8th Annual Conference of the International Speech Communication Association, Interspeech 2007, vol. 4, pp. 2396–2399 (2007)
141. Scherer, K.R.: Vocal communication of emotion: a review of research paradigms. Speech Commun. **40**, 227–256 (2003)
142. Batliner, A., Schuller, B., Seppi, D., Steidl, S., Devillers, L., Vidrascu, L., Vogt, T., Aharonson, V., Amir, N.: The automatic recognition of emotions in speech. In: Cowie, R., Petta, P., Pelachaud, C. (eds.) Emotion-Oriented Systems: The HUMAINE Handbook, Cognitive Technologies, 1st edn, pp. 71–99. Springer, New York (2010)
143. Batliner, A., Steidl, S., Schuller, B., Seppi, D., Vogt, T., Wagner, J., Devillers, L., Vidrascu, L., Aharonson, V., Kessous, L., Amir, N.: Whodunnit—searching for the most important feature types signalling emotion-related user states in speech. Comput. Speech Lang. Special Issue on Affective Speech in real-life interactions **25**(1), 4–28 (2011)
144. Batliner, A., Steidl, S., Hacker, C., Nöth, E.: Private emotions vs. social interaction—a data-driven approach towards analysing emotions in speech. User Modeling and User-Adapted Interaction. J. Personal. Res. **18**(1–2), 175–206 (2008)
145. Hansen, J., Bou-Ghazale, S.: Getting started with susas: a speech under simulated and actual stress database. In: Proceedings of the EUROSPEECH-97, vol. 4, pp. 1743–1746. Rhodes, Greece (1997)
146. Batliner, A., Schuller, B., Schaeffler, S., Steidl, S.: Mothers, adults, children, pets—towards the acoustics of intimacy. In: Proceedings 33rd IEEE International Conference on Acoustics, Speech, and Signal Processing, ICASSP 2008, pp. 4497–4500. Las Vegas, NV, IEEE, IEEE (2008)
147. Pon-Barry, H.: Prosodic manifestations of confidence and uncertainty in spoken language. In: INTERSPEECH 2008—9th Annual Conference of the International Speech Communication Association, pp. 74–77. Brisbane, Australia (2008)
148. Black, M., Chang, J., Narayanan, S.: An empirical analysis of user uncertainty in problem-solving child-machine interactions. In: Proceedings of the 1st Workshop on Child, Computer and Interaction, Chania, Greece (2008)
149. Enos, F., Shriberg, E., Graciarena, M., Hirschberg, J., Stolcke, A.: Detecting deception using critical segments. In: Proceedings INTERSPEECH 2007, 8th Annual Conference of the International Speech Communication Association, pp. 2281–2284. Antwerp, Belgium, ISCA, ISCA (2007)
150. Bénézech, M.: Vérité et mensonge : l'évaluation de la crédibilité en psychiatrie lgale et en pratique judiciaire. Annales Medico-Psychologiques **165**(5), 351–364 (2007)
151. Nadeu, M., Prieto, P.: Pitch range, gestural information, and perceived politeness in catalan. J. Pragmat. **43**(3), 841–854 (2011)
152. Yildirim, S., Lee, C., Lee, S., Potamianos, A., Narayanan, S.: Detecting politeness and frustration state of a child in a Conversational Computer Game. In: Proceedings of the Interspeech 2005, pp. 2209–2212. Lisbon, Portugal, ISCA (2005)
153. Yildirim, S., Narayanan, S., Potamianos, A.: Detecting emotional state of a child in a conversational computer game. Comput. Speech Lang. **25**, 29–44 (2011)
154. Ang, J., Dhillon, R., Shriberg, E., Stolcke, A.: Prosody-based automatic detection of annoyance and frustration in human-computer dialog. In: Proceedings International Conference on Spoken Language Processing (ICSLP), pp. 2037–2040. Denver, CO, (2002)
155. Arunachalam, S., Gould, D., Anderson, E., Byrd, D., Narayanan, S.S.: Politeness and frustration language in child-machine interactions. In: Proceedings EUROSPEECH, pp. 2675–2678, Aalborg, Denmark, (2001)
156. Lee, C., Narayanan, S., Pieraccini, R.: Recognition of negative emotions from the speech signal. In: Proceedings of the Automatic Speech Recognition and Understanding Workshop (ASRU'01) (2001)
157. Rankin, K.P., Salazar, A., Gorno-Tempini, M.L., Sollberger, M., Wilson, S.M., Pavlic, D., Stanley, C.M., Glenn, S., Weiner, M.W., Miller, B.L.: Detecting sarcasm from paralinguistic

cues: anatomic and cognitive correlates in neurodegenerative disease. NeuroImage **47**(4), 2005–2015 (2009)
158. Tepperman, J., Traum, D., Narayanan, S.: "Yeah Right": sarcasm recognition for spoken dialogue systems. In: Proceedings of the Interspeech, pp. 1838–1841. Pittsburgh, Pennsylvania (2006)
159. Zeng, Z., Pantic, M., Roisman, G.I., Huang, T.S.: A survey of affect recognition methods: audio, visual, and spontaneous expressions. IEEE Trans. Pattern Anal Mach. Intell. **31**(1), 39–58 (2009)
160. Batliner, A., Steidl, S., Schuller, B., Seppi, D., Laskowski, K., Vogt, T., Devillers, L., Vidrascu, L., Amir, N., Kessous, L., Aharonson, V.: Combining efforts for improving automatic classification of emotional user states. In: Proceedings 5th Slovenian and 1st International Language Technologies Conference, ISLTC 2006, pp. 240–245. Ljubljana, Slovenia, October 2006. Slovenian Language Technologies Society (2006)
161. Schuller, B., Vlasenko, B., Eyben, F., Wöllmer, M., Stuhlsatz, A., Wendemuth, A., Rigoll, G.: Cross-corpus acoustic emotion recognition: Variances and strategies. IEEE Trans. Affect. Comput. **1**(2), 119–131 (2010)
162. Schuller, B., Vlasenko, B., Eyben, F., Rigoll, G., Wendemuth, A.: Acoustic emotion recognition: a benchmark comparison of performances. In: Proceedings 11th Biannual IEEE Automatic Speech Recognition and Understanding Workshop, ASRU 2009, pp. 552–557. Merano, Italy, IEEE, IEEE (2009)
163. Stuhlsatz, A., Meyer, C., Eyben, F., Zielke, T., Meier, G., Schuller, B.: Deep neural networks for acoustic emotion recognition: raising the benchmarks. In: Proceedings 36th IEEE International Conference on Acoustics, Speech, and Signal Processing, ICASSP 2011, pp. 5688–5691, Prague, Czech Republic, IEEE, IEEE (2011)
164. Ververidis, D., Kotropoulos, C.: A state of the art review on emotional speech databases. In: 1st Richmedia Conference, pp. 109–119. Lausanne, Switzerland (2003)
165. Grimm, M., Kroschel, K., Narayanan, S.: The Vera am Mittag German audio-visual emotional speech database. In: Proceedings of the IEEE International Conference on Multimedia and Expo (ICME), pp. 865–868. Hannover, Germany (2008)
166. Steidl, S.: Automatic Classification of Emotion-Related User States in Spontaneous Speech. Logos, Berlin (2009)
167. Batliner, A., Seppi, D., Steidl, S., Schuller, B.: Segmenting into adequate units for automatic recognition of emotion-related episodes: a speech-based approach. Adv. Human Comput. Interact. Special Issue on Emotion-Aware Natural Interaction **2010**(Article ID 782802), 15 (2010)
168. Cowie, R., Douglas-Cowie, E., Tsapatsoulis, N., Votsis, G., Kollias, S., Fellenz, W., Taylor, J.: Emotion recognition in human-computer interaction. IEEE Signal Process. Mag. **18**(1), 32–80 (2001)
169. Eyben, F., Wöllmer, M., Schuller, B.: Openear—introducing the munich open-source emotion and affect recognition toolkit. In: Proceedings 3rd International Conference on Affective Computing and Intelligent Interaction and Workshops, ACII 2009, vol. I, pp. 576–581, Amsterdam, The Netherlands, HUMAINE Association, IEEE (2009)
170. Ishi, C., Ishiguro. H., Hagita, N.. Using prosodic and voice quality features for paralinguistic information extraction. In: Proceedings of Speech Prosody 2006, pp. 883–886, Dresden (2006)
171. Müller, C.: Classifying speakers according to age and gender. In: Müller, C. (ed.) Speaker Classification II, vol. 4343. Lecture Notes in Computer Science/Artificial Intelligence. Springer, Heidelberg (2007)
172. Young, S., Evermann, G., Gales, M., Hain, T., Kershaw, D., Liu, X., Moore, G., Odell, J., Ollason, D., Povey, D., Valtchev, V., Woodland, P.: The HTK Book (v3.4). Cambridge University Press, Cambridge (2006)
173. Chawla, N.V., Bowyer, K.W., Hall, L.O., Kegelmeyer, W.P.: Synthetic minority over-sampling technique. J. Artif. Intell. Res. **16**, 321–357 (2002)

174. Steidl, S., Schuller, B., Seppi, D., Batliner, A.: The hinterland of emotions: facing the open-microphone challenge. In: Proceedings of the 3rd International Conference on Affective Computing and Intelligent Interaction and Workshops, ACII 2009, vol. I, pp. 690–697, Amsterdam, The Netherlands, HUMAINE Association, IEEE (2009)
175. Schuller, B., Metze, F., Steidl, S., Batliner, A., Eyben, F., Polzehl, T.: Late fusion of individual engines for improved recognition of negative emotions in speech—learning vs. democratic vote. In: Proceedings of the 35th IEEE International Conference on Acoustics, Speech, and Signal Processing, ICASSP 2010, pp. 5230–5233, Dallas, TX, IEEE, IEEE (2010)
176. Wöllmer, M., Weninger, F., Eyben, F., Schuller, B.: Computational assessment of interest in speech - facing the real-life challenge. Künstliche Intelligenz (German J. Artif. Intell.), Special Issue on Emotion and Computing **25**(3), 227–236 (2011)
177. Wöllmer, M., Weninger, F., Eyben, F., Schuller, B.: Acoustic-linguistic recognition of interest in speech with bottleneck-blstm nets. In: Proceedings of INTERSPEECH 2011, 12th Annual Conference of the International Speech Communication Association, pp. 3201–3204. Florence, Italy, ISCA, ISCA (2011)
178. Mporas, I., Ganchev, T.: Estimation of unknown speaker's height from speech. Int. J. Speech Tech. **12**(4), 149–160 (2009)
179. Schuller, B., Steidl, S., Batliner, A., Burkhardt, F., Devillers, L., Müller, C., Narayanan, S.: Paralinguistics in speech and language—state-of-the-art and the challenge. Comput. Speech Lang. Special Issue on Paralinguistics in Naturalistic Speech and Language **27**(1), 4–39 (2013)
180. Omar, M.K., Pelecanos, J.: A novel approach to detecting non-native speakers and their native language. In: ICASSP, IEEE International Conference on Acoustics, Speech and Signal Processing—Proceedings, pp. 4398–4401. Dallas, Texas (2010)
181. Weiss, B., Burkhardt, F.: Voice attributes affecting likability perception. In: Proceedings of the INTERSPEECH, pp. 2014–2017. Makuhari, Japan (2010)
182. Bruckert, L., Lienard, J., Lacroix, A., Kreutzer, M., Leboucher, G.: Women use voice parameter to assess men's characteristics. Proc. R. Soc. B. **237**(1582), 83–89 (2006)
183. Gocsál, A.: Female listeners' personality attributions to male speakers: the role of acoustic parameters of speech. Pollack Period. **4**(3), 155–165 (2009)
184. Mohammadi, G., Vinciarelli, A., Mortillaro, M.: The voice of personality: mapping nonverbal vocal behavior into trait attributions. In: Proceedings of the SSPW 2010, pp. 17–20, Firenze, Italy (2010)
185. Polzehl, T., Möller, S., Metze, F.: Automatically assessing personality from speech. In: Proceedings—2010 IEEE 4th International Conference on Semantic Computing, ICSC 2010, pp. 134–140. Pittsburgh, PA (2010)
186. Wallhoff, F., Schuller, B., Rigoll, G.: Speaker identification—comparing linear regression based adaptation and acoustic high-level features. In: Proceedings 31. Jahrestagung für Akustik, DAGA 2005, pp. 221–222. Munich, Germany, DEGA, DEGA (2005)
187. Müller, C., Burkhardt, F.: Combining short-term cepstral and long-term prosodic features for automatic recognition of speaker age. In: Interspeech, pp. 1–4,.Antwerp, Belgium (2007)
188. van Dommelen, W., Moxness, B.: Acoustic parameters in speaker height and weight identification: sex-specific behaviour. Lang. Speech **38**(3), 267–287 (1995)
189. Krauss, R.M., Freyberg, R., Morsella, E.: Inferring speakers physical attributes from their voices. J. Exp. Soc. Psychol. **38**(6), 618–625 (2002)
190. Gonzalez, J.: Formant frequencies and body size of speaker: a weak relationship in adult humans. J. Phonetics **32**(2), 277–287 (2004)
191. Evans, S., Neave, N., Wakelin, D.: Relationships between vocal characteristics and body size and shape in human males: an evolutionary explanation for a deep male voice. Biol. Psychol. **72**(2), 160–163 (2006)
192. Grimm, M., Kroschel, K., Narayanan, S.: Support vector regression for automatic recognition of spontaneous emotions in speech. In: International Conference on Acoustics, Speech and Signal Processing, vol. IV, pp. 1085–1088. IEEE (2007)
193. Hassan, A., Damper, R.I.: Multi-class and hierarchical SVMs for emotion recognition. In: Proceedings of the Interspeech, pp. 2354–2357, Makuhari, Japan (2010)

194. Burkhardt, F., Eckert, M., Johannsen, W., Stegmann, J.: A database of age and gender annotated telephone speech. In: Proceedings of the 7th International Conference on Language Resources and Evaluation (LREC 2010), pp. 1562–1565, Valletta, Malta (2010)
195. Fisher, M., Doddington, G., Goudie-Marshall, K.: The DARPA speech recognition research database: specifications and status. In: Proceedings of the DARPA Workshop on Speech Recognition, pp. 93–99 (1986)
196. Hall, M., Frank, E., Holmes, G., Pfahringer, B., Reutemann, P., Witten, I.H.: The WEKA data mining software: an update. SIGKDD Explor. **11**(1) (2009)
197. Krajewski, J., Batliner, A., Golz, M.: Acoustic sleepiness detection—framework and validation of a speech adapted pattern recognition approach. Behav. Res. Meth. **41**, 795–804 (2009)
198. Levit, M., Huber, R., Batliner, A., Nöth, E.: Use of prosodic speech characteristics for automated detection of alcohol intoxination. In: Bacchiani, M., Hirschberg, J., Litman, D., Ostendorf, M. (eds.) Proceedings of the Workshop on Prosody and Speech Recognition 2001Red Bank, NJ, pp. 103–106 (2001)
199. Schiel, F., Heinrich, C.: Laying the foundation for in-car alcohol detection by speech. In: Proceedings of INTERSPEECH 2009, pp. 983–986, Brighton, UK (2009)
200. Ellgring, H., Scherer, K.R.: Vocal indicators of mood change in depression. J. Nonverbal Behav. **20**, 83–110 (1996)
201. Laskowski, K., Ostendorf, M., Schultz, T.: Modeling vocal interaction for text-independent participant characterization in multi-party conversation. In: Proceedings of the 9th SIGdial Workshop on Discourse and Dialogue, pp. 148–155, Columbus (2008)
202. Ipgrave, J.: The language of friendship and identity: children's communication choices in an interfaith exchange. Br. J. Relig. Edu. **31**(3), 213–225 (2009)
203. Fujie, S., Ejiri, Y., Kikuchi, H., Kobayashi, T.: Recognition of positive/negative attitude and its application to a spoken dialogue system. Syst. Comput. Jpn. **37**(12), 45–55 (2006)
204. Vinciarelli, A., Pantic, M., Bourlard, H.: Social signal processing: survey of an emerging domain. Image Vis. Comput. **27**, 1743-1759 (2009)
205. Lee, C.-C., Katsamanis, A., Black, M., Baucom, B., Georgiou, P., Narayanan, S.: An analysis of pca-based vocal entrainment measures in married couples' affective spoken interactions. In: Proceedings of Interspeech, pp. 3101–3104, Florence, Italy (2011)
206. Brenner, M., Cash, J.: Speech analysis as an index of alcohol intoxication—the Exxon Valdez accident. Aviat. Space Environ. Med. **62**, 893–898 (1991)
207. Harrison, Y., Horne, J.: The impact of sleep deprivation on decision making: a review. J. Exp. Psychol. Appl. **6**, 236–249 (2000)
208. Bard, E.G., Sotillo, C., Anderson, A.H., Thompson, H.S., Taylor, M.M.: The DCIEM map task corpus: spontaneous dialogue under SD and drug treatment. Speech Commun. **20**, 71–84 (1996)
209. Caraty, M., Montacie, C.: Multivariate analysis of vocal fatigue in continuous reading. In: Proceedings of Interspeech 2010, pp. 470–473, Makuhari, Japan (2010)
210. Schiel, F., Heinrich, C., Barfüßer, S.: Alcohol language corpus—the first public corpus of alcoholized German speech. Lang. Res. Eval. **46**(3), 503–521 (2012)
211. Akerstedt, T., Gillberg, M.: Subjective and objective sleepiness in the active individual. Int. J. Neurosci. **52**(1–2), 29–37 (May 1990)
212. Krajewski, J., Schnieder, S., Sommer, D., Batliner, A., Schuller, B.: Applying multiple classifiers and non-linear dynamics features for detecting sleepiness from speech. Neurocomputing. Special Issue From neuron to behavior: evidence from behavioral measurements **84**, 65–75 (2012)
213. Krajewski, J., Kröger, B.: Using prosodic and spectral characteristics for sleepiness detection. In: Proceedings of INTERSPEECH 2007, 8th Annual Conference of the International Speech Communication Association, pp. 1841–1844, Antwerp, Belgium, ISCA, ISCA (2007)
214. Chin, S.B., Pisoni, D.B.: Alcohol and Speech. Academic Press Inc, New York (1997)
215. Dhupati, L., Kar, S., Rajaguru, A., Routray, A.: A novel drowsiness detection scheme based on speech analysis with validation using simultaneous EEG recordings. In: Proceedings of

IEEE Conference on Automation Science and Engineering (CASE), pp. 917–921, Toronto, ON (2010)

216. Weninger, F., Schuller, B., Fusing utterance-level classifiers for robust intoxication recognition from speech. In: Proceedings MMCogEmS, : Workshop (Inferring Cognitive and Emotional States from Multimodal Measures), held in conjunction with the 13th International Conference on Multimodal Interaction, ICMI 2011, Alicante, Spain, ACM, ACM (2011)
217. Schuller, B., Weninger, F.: Ten recent trends in computational paralinguistics. In: Esposito, A., Vinciarelli, A., Hoffmann, R., Müller, V.C. (eds.) 4th COST 2102 International Training School on Cognitive Behavioural Systems. Lecture Notes on Computer Science (LNCS), p. 15. Springer, New York (2012)

Chapter 11
Applications in Intelligent Music Analysis

Of all noises, I think music is the least disagreeable.
—Samuel Johnson

As digitised music has conquered the market for more than ten years, advanced techniques of MIR are gaining interest and importance. Caused by the progress in lossy perceptual audio coding (MP3 and similar), broadband internet connections and high volume storage capacities, large music databases have emerged which demand novel handling strategies [1–3].

The increasing popularity of portable music players and music distribution over the internet has made worldwide, instantaneous access to rapidly growing music archives possible. Such archives must be well structured and sorted in order to be user friendly. For example, many users face the problem of having heard a song they would like to buy but not knowing its bibliographic data, i.e., title and artist, which is necessary to find the song in conventional (on-line) music stores. According to Downie in [4], almost three fourths of all MIR queries are of bibliographic nature. The querying person gives information he or she knows about the song, most likely genre, metre, tempo, lyrics or acoustic properties, e.g., tonality, and requires information about title and/or artist. In order to have machines assist in building a song database queryable by attributes such as tempo, metre or genre, intelligent Information Retrieval algorithms are necessary to automatically extract such high level features from raw music data. Hence, many new tasks in private as well as in professional environments have occurred, as for example rhythm recognition [5, 6], genre [7] and mood classification [8], melody extraction [9], chord detection [10] or key determination [11]. Many works exist that describe or give overviews over basic MIR methods, e.g., [5, 12–20].

In this chapter, an overview shall be given on intelligent music analysis. The selected application scenarios cover methods on rhythmic aspects first: The separation of drum-beats [21] is followed by the determination of onsets in music [22, 23], and tempo, metre and ballroom dance style determination [6, 24–27]. Subsequently, analysis of the tonal aspects, reaching from musical key [28] to chords

B. W. Schuller, *Intelligent Audio Analysis*, Signals and Communication Technology,
DOI: 10.1007/978-3-642-36806-6_11, © Springer-Verlag Berlin Heidelberg 2013

[10, 29] are presented. Then, the structure of music is analysed targeting chorus sections [30, 31]. Such extracted information is then used in the assessment of mood in music [8, 32, 33]. Finally, in analogy to speech analysis, it is attempted to assess traits of singers [34–36].

Other examples could have been chosen, such as the full transcription of music on the note event level [37], the recognition of genre [38, 39], music spotting in audio streams [40], query by humming [41, 42], or the recognition of vibrato singing [43], and of course the querying per se [2, 3], to name a few. However, the chosen examples provide a good overview on core topics and allow for a general understanding of the principles and methods involved. Each topic is addressed by selected exemplary data with according test results to provide the reader with a feeling for obtainable state-of-the-art performances under realistic conditions as were outlined in Sect. 11.1.

11.1 Drum-Beat Separation

For the analysis of music, we will first see how the harmonic section and the drum beat in Rock, Pop, or similar music can be separated. This was first shown in [21].

Non-negative Matrix Factorisation (NMF) is known for its suitability in BASS of drums and melodic parts of music recordings [44–46]. An isolation of these parts can serve as enhancement in manifold MIR tasks such as the ones to follow including automatic onset, metre, tempo detection or key and chord labelling and even the recognition of singer traits. Let us thus consider in this section the combination of an NMF based blind music separation into several isolated audio tracks with subsequent classification of obtained isolated NMF components to label them as either rhythmic or melodic.

In [47] drum beat separation based on ICA was introduced. Opposed to this, in [48] it is relied on NMF for separation of sources within transcription of polyphonic music. Remarkable results were reported on piano music. Also, the work in [44] is based on NMF. There, a feature extraction and subsequent classification is already used. The authors report promising results for the separation of drum beats in popular music. Such an approach was later proven beneficial for drum transcription [45, 46] and vocal separation [49].

11.1.1 Methodology

Let us first take a look at different cost functions and parameters. As we remember from Sect. 11.8, given a matrix $\underline{V} \in \mathbb{R}_{\geq 0}^{n \times m}$ and a constant $r \in \mathbb{N}$, NMF computes two matrices $\underline{W} \in \mathbb{R}_{\geq 0}^{n \times r}$ and $\underline{H} \in \mathbb{R}_{\geq 0}^{r \times m}$, such that

$$\underline{V} \approx \underline{W}\,\underline{H}. \tag{11.1}$$

Table 11.1 Considered cost functions for NMF-based drum-beat separation

Cost-function	Denomination
$\Vert (\underline{V} - \underline{W}\,\underline{H}) \Vert_F$	Frobenius norm
$\sum_{i,j} \left(V_{i,j} \log \frac{V_{i,j}}{(\underline{W}\,\underline{H})_{i,j}} - \underline{V}_{i,j} + (\underline{W}\,\underline{H})_{i,j} \right)$	Modified KL divergence

Only an approximate solution exists for $r \ll n, m$. Factorisation is usually realised by iterative algorithms. These aim at minimisation of suited cost-functions as those two shown in Table 11.1.

The Frobenius norm cost-function minimises some form of quadratic error, the 'modified KL divergence' interprets the matrices $\underline{V}$ and $(\underline{WH})$ as probability distributions and minimises their divergence. It is a modification due to the additional term $(\underline{W}\,\underline{H})_{i,j} - \underline{V}_{i,j}$. Besides a measurement of the absolute error, this term ensures non-negativity. Several further cost-functions exist and the available algorithms in principle only differ by their particular cost-function.

In NMF-based instrument separation the monophonic signal's short-time magnitude spectra are considered as linear combinations of several distinct components' spectra. The assumption of non-negativity of each component suffices for this approach. Applying NMF on a signal's magnitude spectrum the resulting columns of $\underline{W}$ and the rows of $\underline{H}$ in Eq. (11.1) can be understood as *spectral components* and their *gains over time*, respectively (cf. Sect. 11.8). By that, the overall contribution of the ith component to the magnitude spectrum of the original signal can be determined as the dyadic product of the ith column of $\underline{W}$ and the ith row of $\underline{H}$. As the magnitude spectrum is factorised, one can transform the separated components back into the time domain based on their magnitude spectra. In addition, the original phase spectrum is needed. Typically, the spectrum of an instrument is modelled by several components. Hence, one can distinguish between *instrument* separation and *component* separation.

The decisive factor in instrument separation now is finding the optimal parameters of the combined STFT and NMF approach. In the case of the STFT these parameters include the window function, size, and overlap. Window functions used in this context are the previously introduced rectangular and Hanning window (cf. Sect. 6.1.2), and the square root of the Hanning window, as used in [44]. The most influential parameter in the choice and design of the window function as for perceptual quality of the resulting factorisation seems to be the window size. In drum-beat separation, typical window sizes are between 40 and 60 ms—a window size of 62.5 ms is, e.g., the equivalent of an eighth note at 120 BPM. Depending on these parameter choices, the STFT may produce larger amounts of data: With a sample rate of 44.1 kHz, a window size of 60 ms, and 50 % overlap, the magnitude- and phase-spectrum matrices for 30 s of music result in a dimension of 1322×1000.

If we imagine a sequence of single different notes, it seems intuitive that each of these notes can be represented by its own spectrum and likewise by a single corresponding component, taking the non-negativity constraint as introduced into account. In [48] the authors thus speak of *events* rather than *components* to emphasise on the singularity. A priori, however, one will rarely know how many components to select, except for a certain subset of music or by preliminary 'manual' analysis of the music to be separated. Should this number of components be chosen larger than 'needed', the contributions of superfluous components' to the whole magnitude spectrum will be nearly zero. However, usage of more components leads to smaller absolute values and likewise to less maximum amplitudes of the separated components. As a rule of thumb, 20 to 30 components for unsupervised instrument separation of popular music are recommended [21].

There are a few peculiarities one should bear on mind when using NMF for instrument separation as follow:

Usage of a-priori information is advisable given the model parameters' significant influence on the (perceptual) quality of the factorisation.

Careful initialisation is mandatory, in particular using NMF as a preprocessing step for feature extraction. This comes as, e.g., random initialisation with small values within $[0.01, 0.02]$, in comparison to values within $[0.1, 0.2]$, often yields results with totally different scale [21]. This difference may significantly influence the extracted features' values. Randomised initialisation of $\underline{W}$ and $\underline{H}$ further leads to slightly different results on each application of NMF. An alternative is targeted initialisation by application-dependent training sets [45] which comes at the need of more a-priori knowledge.

Separation limitations arise in particular when components or events at no time in the signal occur in isolation and the matrices $\underline{W}$ and $\underline{H}$ were initialised randomly. This means that the algorithm cannot separate events occurring exclusively simultaneously. The reason is that the algorithm achieves as good a solution in terms of the cost-function when uniting these events in a single component, unless sparsity constraints are added.

Sub-optimality characterises NMF as there is not guarantee to find a global minimum of the respective cost-function.

Factorisation ambiguity can be shown by looking at the product $\underline{W} \cdot \underline{H}$ compared to $\underline{W} \cdot \underline{A}^{-1} \cdot \underline{A} \cdot \underline{H}$, where $\underline{A}$ is some arbitrary permutation or affine transformation. As a result, the order of the separated components is non-deterministic. For drum-beat separation as shown in the ongoing, however, this has no effect, as the resulting components are classified automatically without assumptions about their order.

11.1.1.1 Component Classification

With the number of components typically being higher than the number of classes to separate (here: drums and harmonics), a classification is needed to put the components into the 'right bag'. In the aimed at drum beat separation, the classes are drum-beat and non-drum-beat or 'harmonic'. However, as stated, 20–30 components are advisable for separation of pop music. Once the components are classified, all components in one 'bag' are superposed to generate the signals corresponding to each class.

The first step for such a classification is thus the extraction of suitable features to characterise 'drums versus rest' as the ones suggested in [44, 47]. Each component is described by a column of the spectral matrix W—the *spectral vector* in the following—and a row of the gains matrix H—the *gains vector* in the following—as obtained by NMF. One could transform the components back into the time domain for feature extraction. However, the features then would be extracted from redundant representations of the components. Rather than that, features can be based on the spectral and gains vectors as all relevant information apart from phase information is contained in these.

In detail, per spectral vector $\underline{x} = (x_1, \ldots, x_N)^T$, corresponding to frequencies $f_1, \ldots, f_N$, these features are extracted: sample standard deviation (by the common unbiased estimator), and 10 MFCCs (with a Mel filter bank that ranges from 20 to 8000 Hz). Further, a number of spectral features are computed: The *spectral centroid* is the weighted mean of frequencies f_i:

$$\frac{\sum_{i=1}^{N} f_i x_i}{\sum_{i=1}^{N} x_i} \tag{11.2}$$

The *roll-off point* is the frequency f_r with

$$r = \min \left\{ k : \sum_{i=1}^{k} x_i^2 \geq 0.95 \sum_{i=1}^{N} x_i^2 \right\}. \tag{11.3}$$

It is the 'point' at which 95 % of the energy of the spectrum are contained at frequencies below this point.

In the data set considered in the evaluation (cf. Sect. 11.1.2), harmonic patterns are characterised by spectral centroids and roll-off points at middle frequencies. For drum patterns, however, these features are less specific.

Next is *noise-likeness* [47] that is based on the assumption that spectra of harmonic components have a limited number of sharp peaks, whereas spectra of percussive components are smoother. It is computed as:

1. Find the local maxima of $\underline{x}$, i.e., leave all components x_i with $x_{i-1} < x_i > x_{i+1}$ and set all other components to zero.
2. Convolve the resulting vector with a Gaussian function ($\mu = 0, \sigma = 83.3$ Hz). For a feasible calculation, we assume that the function is zero outside the interval $[-3\sigma, 3\sigma]$.
3. Noise-likeness is then the (Pearson product-moment) correlation coefficient of the result and the original vector $\underline{x}$.

Figure 11.1 illustrates this procedure.

Then, the *spectral flatness* [47] is the ratio of the geometrical to the arithmetical mean of the vector $\underline{x}$, squared element by element.

For the computation of *spectral dissonance* [47], the dissonance measure for two sinusoids with frequencies f_1 and f_2, $f_1 \leq f_2$, and amplitudes a_1 and a_2, are employed [50]:

Fig. 11.1 Spectra of a drum (*top*) and harmonic component (*bottom*) from David Bowie and Mick Jagger's "*Dancing In The Street*", and their convolution with a Gaussian function

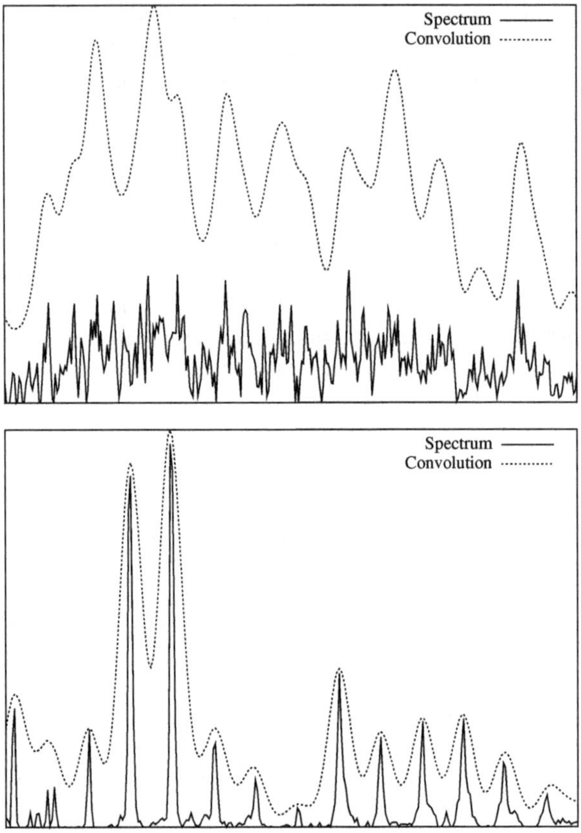

$$d(f_1, f_2, a_1, a_2) = a_1 a_2 \left(e^{-as(f_2 - f_1)} - e^{-bs(f_2 - f_1)} \right) \tag{11.4}$$

with $a = 3.5$, $b = 5.75$ and $s = \frac{0.24}{0.021 f_1 + 19}$.

Then, spectral dissonance of $\underline{x}$ is defined as the sum of pairwise dissonances of all its components:

$$\sum_{i=1}^{N} \sum_{j=1}^{i-1} d(f_j, f_i, x_j, x_i), \tag{11.5}$$

where f_i is the frequency corresponding to index i in the spectral vector.

Further, temporal features are calculated from the gains vectors. Per gains vector $\underline{g} = (g_1, \ldots, g_M)$ sample standard deviation is extracted.

Further, *Percussiveness* [47] measures how accurately $\underline{g}$ can be modelled using instantaneous attacks and linear decays resembling the structure of typical drum patterns. Its computation is similar to the one applied to spectral vectors for the calculation of noise-likeness. The local maxima of $\underline{g}$ are determined and convolved

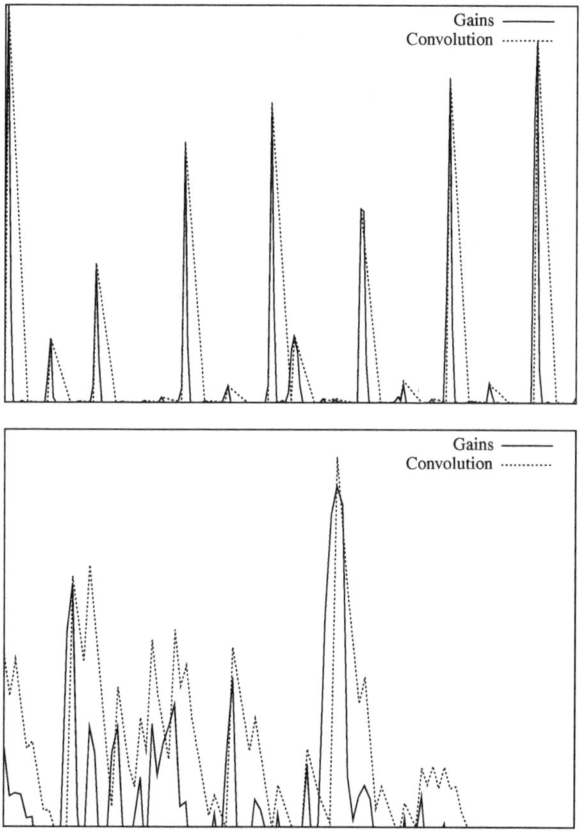

Fig. 11.2 Gains of a drum (*top*) and harmonic component (*bottom*) from David Bowie and Mick Jagger's "*Dancing In The Street*", and their convolution with a linear decay function

with a linear decay function of height 1 and length 200 ms. Percussiveness is then the (Pearson product-moment) correlation coefficient of the convolution and the original vector $\underline{g}$. This is illustrated in Fig. 11.2.

Next, *Periodicity* [44] models drum patterns often being periodic in intervals corresponding to a musical piece's tempo. Autocorrelation values normalised by mean and variance of $\underline{g}$ are computed for delays corresponding to tempi of 30–240 BPM, at intervals of 5 BPM. The maximum of these coefficients is defined as the periodicity.

Finally, *average peak length* and *peak fluctuation* are added, where a peak is 'any area' of $\underline{g}$ that is above a threshold of 20 % of the maximum of $\underline{g}$. Formally, a peak of length l is a set of consecutive indices $\{i, i+1, \ldots, i+l-1\} \subseteq \{1, \ldots, M\}$ such that [21]:

$$g_i, g_{i+1}, \ldots, g_{i+l-1} \geq 0.2 \cdot \max\{g_i\}. \tag{11.6}$$

Once the peaks in $\underline{g}$ are located, the average peak length is given by the sample mean of the peak lengths, and the peak fluctuation by their sample standard deviation

[44]. The music data considered provide evidence that drum components generally have short peaks of similar length. As opposed to this, harmonic components tend to have longer peaks varying more in length.

11.1.1.2 Synthesis

Subsequent to the assignment of components to their 'bags' by automatic classification, time signals for each class can be computed as follows [44]:Per class, the magnitude spectrograms of the components belonging to that class are added, where the magnitude spectrogram of a component is the dyadic product of its spectral and gains vectors.the class spectrograms.Then, a column-wise IDFT is performed on As alluded above,the phase values from the corresponding columns of the phase matrix of the original signal are used in this step. Time signals are finally computed by windowing the columns with the square root of the Hanning function using overlap-add.

11.1.2 Performance

For evaluation the set "20 Years on MTV" (1981–2000, Sony/BMG) was used. It consists of 200 songs, from each of which one data instance of 15–30 s duration is extracted. With the framework as described in Sect. 11.8, spectrograms were computed using the square root of the Hanning function with a window size of 60 ms and 50 % window overlap. For subsequent NMF application, 30 components were used. Out of the 6000 resulting components 344 were manually selected by perceptual quality. Music experts carried out the labelling attaching either the label "Drum" (95 components) or "Harmonic" (249 components) to these. Evaluation with linear kernel SVM of this data is carried out in ten-fold SCV after scaling features in the range $[-1, 1]$. Different feature subsets are considered:

- The "complete" feature set contains all features described above and led to a WA of 95.9 %.
- The "reduced" feature set as proposed in [44] includes standard deviation,10 MFCCs, noise-likeness, spectral centroid and roll-off for spectral vectors, and average peak length, percussiveness, peak fluctuation and periodicity for gains vectors. It led to a slightly improved WA of 96.2 %.

11.1.3 Summary

In this section a separation of music into drum-beat and the harmonic parts was shown. Generally judging, the audible results are well usable in, e.g., DJ applications or music remixing. There are, however, some cases which seem to pose difficulties to the separation and/or the classification procedure. These are 'limit cases' between noisy and harmonic components, such as distorted guitars or cowbells. Further, 'noisy' phonemes in the vocal parts, such as "s", were often assigned to the drum part. Unless Wiener filtering is used in synthesis, harmonic parts may partly appear distorted in the result, which is especially true for vocal parts. Future efforts could combine a targeted initialisation and exploit semi-supervised learning strategies on large music archives to reduce the number of erroneous component classifications: In the presented results roughly every twenty-fifth component was misclassified, and 30 were chosen for decomposition—this means that roughly one component is in the wrong 'bag' on average.

11.2 Onsets

An essential task in intelligent music analysis is the determination of onsets in music. This is another good example of the application of LSTM networks: The MIREX (Music Information Retrieval Evaluation eXchange) 2010 contributions [22, 23], which are based on LSTM-RNN, were able to reach the best result for the audio onset detection task.

Onsets mark the beginning of acoustic events. Locating these onsets is a major part of segmenting music. It therefore serves as basis for many high level MIR tasks such as music transcription. As opposed to studies focusing on beat and tempo detection via the analysis of periodicities (e.g., [25, 51]) exploiting larger chunks of audio, an audio onset detector aims at the detection of single events. These need not follow a periodic pattern. Automatic onset detection (e.g., [52–54]) has reached reasonable robustness for polyphonic music by now. Approaches towards onset detection are, however, often rather specialised or optimised for specific types of onsets such as pitched or percussive onsets. Thus, they may show low generalisation ability for music with mixed types of onsets. To overcome this, either diverse methods are required in synergistic combination or a selector has to be implemented to chose the appropriate onset detector fitting the music.

Most onset detectors are based on a three step model: First, some methods include a preprocessing step to emphasise relevant parts of the audio signal. Then, a reduction is carried out by means of a suited function, to obtain the 'detection function'. This can be considered as the core component. Common reduction functions are summarised later in this section. Finally, the last stage serves to extract the onsets from this detection function. The final step can further be subdivided into post processing such as smoothing and normalising of the detection function, thresholding, and peak picking. Given a fixed number of thresholds, the methods tend to 'insert' onsets in louder parts, or 'delete' onsets in quieter parts. Adaptive thresholds are thus often employed to ease this behaviour. As a last of these sub-steps, the peak picking algorithm identifies the local maxima above the threshold. These correspond to the detected onsets together with their according position in time.

Methods operating in the time domain were frequently seen in early reduction functions. An example is the one in [55] that normalises the loudness of the signal before splitting it into multiple bands via bandpass filters. Onsets are then detected per band as peaks in the first order difference of the logarithm of the amplitude envelope. The band-wise onsets are then combined to determine the final set of detected onsets. Onset detection in the time domain has, however, its short-comings as onsets are often masked in this domain by higher energy signals. Today, many reduction functions thus operate on a spectral audio signal representation. Typical solutions—all based on the STFT of the signal—are:

Spectral difference: The spectral difference (SD) function is the bin-wise difference of two consecutive short-time spectra. Positive differences are then summed up across bins. The L_1-norm [52] or L_2-norm [56] can be used to assess the function. In case of the L_1-norm, the function is referred to as spectral flux (SF). These methods are among the best so far.

High frequency content: Percussive sounds tend to have a high amount of energy in upper frequency bands. This fact can be used by weighting each STFT bin proportionally to its frequency. The sum of the weighted bins is the high frequency content (HFC), and can be used as a detection function. The HFC method is suited for percussive onsets, but less for other types of onsets [56].

Phase deviation: So far, functions were based on the spectral magnitudes. The phase change in a STFT frequency bin can serve as rough estimate of its instantaneous frequency. Should this frequency change, it is likely because of an onset [56]. The mean phase change over all frequency bins helps to reduce 'deletions' of onsets because of phase wrap around. This method is known as the phase deviation (PD) detection function. An extension is the normalised weighted phase deviation (NWPD) [52], that first weights each frequency bin's contribution to the phase deviation by its magnitude and then normalises the result by the sum of the magnitudes.

Complex domain: In this method, also magnitude and phase information are used. It is calculated for the current frame based on the last two predecessors under the assumption of constant amplitude and phase change rate. The sum of the magnitude of the complex differences between the actual values for each frequency bin and the estimated values is then computed as a detection function [57]. The rectified complex domain (RCD) [52] modifies this algorithm by only summing over positive amplitude changes. This is based on the observation that for onset detection increases of the signal amplitude are generally more relevant than decreases.

Pitch detection: Discontinuities and perturbations in the pitch contour can be assumed as indication for onsets [58]. The information on the location of these phenomena can also be used in combination with energy analysis [53].

Probabilistic models: The negative log-likelihood (NLL) method [59] defines two different statistical models for the signal. A sudden change in these models indicates a potential onset. This is known to work well for soft onsets [56].

Automatic classification: Employing a trained machine learning algorithm allows for the design of more general detection functions, such as the one in [60] basing on a convolutional neural network—the winner of the MIREX 2005 audio onset detection evaluation.

Table 11.2 Number of files, onsets, and length distributions for the onset detection data sets

Set	# files	# onsets	min/max/mean length [s]
BRD_o	87	5474	10.0 / 10.0 / 10.0
PNP	1	93	13.1 / 13.1 / 13.1
PP	9	489	2.5 / 60.0 / 10.5
NPP	6	212	1.4 / 8.3 / 4.3
MIX	7	271	2.8 / 15.1 / 8.0

11.2.1 The Bello Database

The onset detector described here is evaluated using the data set introduced by Bello in [56], which consists of 23 sound excerpts with lengths ranging from a few seconds to one minute (cf. Table 11.2) and is divided into pitched percussive (*PP*), pitched non-percussive (*PNP*), non-pitched percussive (*NPP*), and complex music mixes (*MIX*), and includes audio synthesised from MIDI files as well as original recordings.

For effective RNN training, the onset annotations needed to be partly corrected by addition of missing onsets. Further, the temporal annotation precision of onsets in polyphonic pieces was manually improved by an expert musician. Temporal inaccuracies in onset annotations can be assumed to be less severe for rule-based onset detection, as inaccuracies of a couple of frames are levelled out by the detection window during evaluation. Yet, these inaccuracies have a large impact when training neural networks with them. Still, for the sake of fair and comparable evaluation, the original transcriptions are used for scoring.

Additional 87 10 s excerpts of ballroom dance style music (BRD_o in the ongoing) from the ISMIR 2004 tempo induction contest[1] [51] were taken for training (cf. Table 11.2). This data was partly already annotated for ANN training [60].[2] The remaining parts were labelled by an expert musician. Table 11.2 shows the number of files and onsets for the data sets.

11.2.2 Methodology

Using bidirectional LSTM networks, the approach is able to learn the properties of an onset and the relevant context it occurs in. As audio features two STFT magnitude spectra computed from differently sized windows of the audio signal together with their first order differences are extracted. From these features as input, the BLSTM RNN produces an onset activation function as output. This principle is shown in Fig. 11.3, and individual blocks are described now in more detail.

---

[1] http://mtg.upf.edu/ismir2004/contest/tempoContest/node5.html

[2] http://w3.ift.ulaval.ca/~allac88/dataset.tar.gz

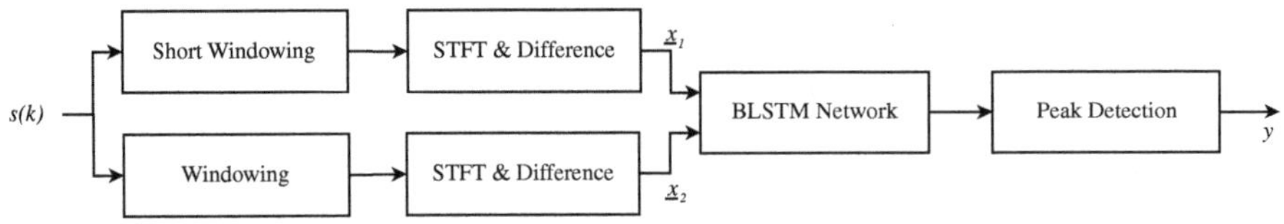

Fig. 11.3 Basic signal flow of the BLSTM RNN based onset detector

11.2.2.1 Feature Extraction

The raw PCM audio signal with a sampling rate of $f_s = 44.1$ kHz is used, and stereo signals are converted to a monaural signal by averaging of the stereo channels. Hamming windowing uses overlapping frames of W samples length with $W = 1024$ or $W = 2048$, and a frame rate of 100 Hz. Note that by this principle, onset annotations are available on a frame level. Then, applying the STFT to the signal values $s(k)$ leads to the complex spectrogram $S(n, m)$, with n being the frame index, and m the frequency bin index. This complex spectrogram is then converted to the power spectrogram:

$$S_{power}(n, m) = |S(n, m)|^2 \tag{11.7}$$

To reduce the dimensionality of the spectra and exploit psychoacoustic knowledge, a filterbank with 40 triangular filters equidistant on the Mel scale, is used to transform the spectrogram $S(n, m)$ to the Mel spectrogram $M(n, m')$. A logarithmic representation helps match human perception of loudness:

$$M_{log}(n, m') = log\left(M(n, m') + 1.0\right) \tag{11.8}$$

Motivated by spectral difference approaches, the positive first order difference $D^+(n, m)$ is calculated by applying a half-wave rectifier function $H(x) = \frac{x+|x|}{2}$ to the difference of two consecutive Mel spectra:

$$D^+(n, m') = H\left(M_{log}(n, m') - M_{log}(n-1, m')\right) \tag{11.9}$$

Overall, the 160 resulting features thus are two log Mel-spectrograms $M_{log}^{23}(n, m')$ and $M_{log}^{46}(n, m')$ computed with window sizes of 23.2 ms and 46.4 ms for $W = 1024$ and $W = 2048$ samples, respectively, and their corresponding positive first order differences $D_{23s}^{+}(n, m')$ and $D_{46s}^{+}(n, m)$.

11.2.2.2 BLSTM Network Stage

As classifier serves a BLSTM RNN with three hidden layers per forward and backward processing—thus six layers in total—, with 20 LSTM units, each. The output layer has two units using the softmax function. The normalised outputs represent the probabilities for the classes 'onset' as used during evaluation and 'no onset' (ignored). This allows for usage of the cross entropy error criterion during network training [61]. Alternative networks were evaluated with a single output and trained using the mean squared output error as criterion, but led to inferior results.

The network was trained with frame-by-frame presentation of the audio sequence. Iterative weight updates were done by standard gradient descent with BPTT. After each training iteration (epoch), the performance was measured on a separate validation set to prevent over-fitting. Once no improvement over 20 epochs had been observed, the training was stopped. Network weights were initialised randomly from a Gaussian distribution with mean 0 and standard deviation 0.1, ensuring non-zero values given by the requirement of the used gradient descent algorithm.

11.2.2.3 Peak Detection Stage

Thresholding and peak detection is now applied to the output activation of the 'onset' class, as follows: For high dynamic ranges, in existing magnitude based reduction functions, adaptive thresholding of the detection function prior to peak picking is mandatory. This comes, as the detection functions amplitude either depends on the one of the signal or on the magnitude of its short time spectrum.

The output activation function of the BLSTM network is not affected by input amplitude variations, similar to phase based reduction functions. The reason is that by the way the networks have been trained its value represents a probability of observing an onset rather than onset strength. This renders intra song adaptive thresholds obsolete. Yet, with varying confidence of the network from song to song the margin between high amplitude peaks (ideally $\approx$1) corresponding to onsets and low amplitude random peaks (ideally $\approx$0) varies. From observation (cf. Fig. 11.4), it seems that especially the random peaks lead to higher activations in the case of low confidence. Thus, a fixed threshold θ is computed per song. This threshold is chosen proportional to the median of the activation function (frames $n = 1 \ldots N$), and constrained to the range from $\theta_{min} = 0.1$ to $\theta_{max} = 0.3$:

$$\theta^* = \lambda \cdot \text{median}\{a_o(1), \ldots, a_o(N)\} \quad (11.10)$$

$$\theta = \min\left(\max\left(0.1, \theta^*\right), 0.3\right) \quad (11.11)$$

with $a_o(n)$ as the output activation function of the BLSTM RNN for the onset class, and the scaling factor λ chosen to maximise the F_1-measure on the validation set. The final onset detection function $o_o(n)$ exclusively contains activation values above this threshold:

$$o_o(n) = \begin{cases} a_o(n) & \text{for } a_o(n) > \theta \\ 0 & \text{otherwise} \end{cases} \quad (11.12)$$

The onsets are likewise represented by the local maxima of $o_o(n)$. With a standard peak search, the final onset function $o(n)$ is:

$$o(n) = \begin{cases} 1 & \text{for } o_o(n-1) \le o_o(n) \ge o_o(n+1) \\ 0 & \text{otherwise} \end{cases} \tag{11.13}$$

11.2.3 Performance

In the literature, an onset is correctly located when detected ±50 ms [52, 56] or ±25 ms [62] around the annotated gold standard onset position. Results with a fixed threshold scaling factor of $\lambda = 50$ are given per set for both of these two tolerance criteria. Note that humans are believed to perceive two onsets as one if they are no more than 30 ms apart [63]. On the Bello set, this would in theory leave 4294 from the 5474 onsets that can be distinguished by humans.

Evaluation on BRD_o and the Bello set bases on eight-fold SCV where six folds are used for training, one for development, and one for testing. Owing to the random initialisation of BLSTM RNN, the eight-fold cross validation is repeated ten times

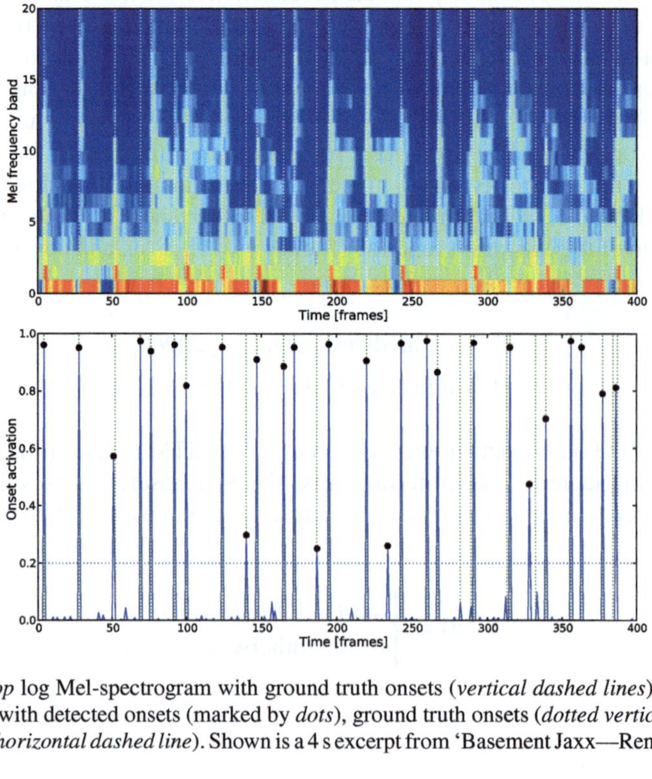

Fig. 11.4 *Top* log Mel-spectrogram with ground truth onsets (*vertical dashed lines*). *Bottom* network output with detected onsets (marked by *dots*), ground truth onsets (*dotted vertical lines*), and threshold θ (*horizontal dashed line*). Shown is a 4 s excerpt from 'Basement Jaxx—Rendez-Vu' [23]

Table 11.3 Results for the Bello data sets *PNP*, *PP*, *NPP*, and *MIX* and the complete set *ALL* by precision (PR), recall (RE), and F_1-measure (F_1)

Tolerance	*PNP*			*PP*			*NPP*			*MIX*			*ALL*		
[%]	PR	RE	F_1	PR	RE	F_1	PR	RE	F_1	PR	RE	F_1	PR	RE	F_1
±50 ms	96.8	96.8	96.8	98.7	98.7	98.7	99.1	99.5	99.3	94.1	89.7	91.8	94.5	92.5	93.5
±25 ms	91.8	95.7	93.7	95.5	98.1	96.8	98.2	99.5	98.9	84.4	86.5	85.5	92.0	90.1	91.1

BLSTM with ±50 ms and ±25 ms detection tolerance

with the same folds, and the means of the output activation functions are used for the final evaluation. Table 11.3 shows the results for each set of onsets and the overall database. Following [52], recall, precision, and F_1-measure are shown for this task. Note that, as the *PNP* data set consists of 93 onsets from a single audio file of string sounds, it should be considered as less representative.

11.2.4 Summary

The principle of onset detection in audio was discussed including a BLSTM RNN onset detector. This detector had achieved results on par with or better than existing results on the same data (wrt. F_1-measure), regardless of the onset type. The average improvement on the whole Bello data set over previous variants introduced in [52, 56] is 1.7 % F_1-measure absolute. Highest gain (3.6 % F_1-measure, absolute) is measured for complex music mixes. This reflects the data driven approach's adaptivity to different musical genres. Good results are obtained even if the onset location tolerance is reduced to ±25 ms instead of ±50 ms.

Follow-up work in [64] and [65] again base on LSTM and standard recurrent neural networks, and further show their high potential. Overall, and given its dominance at MIREX 2010 and MIREC 2011, the BLSTM leads to excellent results for all types of onsets. As particular advantage when compared to 'conventional' onset detection approaches, it can detect multiple types of onsets equally well— given representative and sufficient training data—, which is an important step towards a universal onset detector.

Since the types of onsets the BLSTM method can detect depend on the onsets contained in the training set, future work should investigate whether the approach is suitable for identifying the onset type such as type of instrument or vocal.

11.3 Tempo, Metre, Ballroom Dance Style

Related to the onsets in music are the tempo, and more distantly also metre and ballroom dance styles. All these can be assessed comparably well in fully automatic processing, as was shown in [24, 25] following the later extension presented in [6].

Generally speaking, 'rhythm' describes patterns of changes. In music, a 'beat' corresponds to the perceived pulses which mark off equal durational units and is our basis of comparison for measurements of rhythmic durations. The 'tempo' refers to the beats' 'striking rate', whereas 'metre' represents accent structure of the beats. Considering 'metre', the metrical structure of a musical piece is composed of multiple hierarchical levels [66]. There, the tempo on higher levels is an integer multiple of the one on the lowest level, which is also referred to as 'tatum' level. When we tap along with a song, we do this on the 'pulse' or 'beat' level, which can be referred to as the quarter-note tempo. The 'bar' or 'measure' level corresponds to the unit of a bar in notated music. The relation between measure and beat level then is the metre or 'time signature' of a musical piece.

Current tempo detection algorithms mostly base on periodicity detection: Autocorrelation, resonant filter banks or onset time statistics (cf. Sect. 11.2) are some examples as summarised in [51]. Very few approaches, however, aim at synergistic common or combined assessment of tempo together with related information such as metre or beat-tracking to provide a robust basis for higher level tasks, such as ballroom dance style or genre recognition. Further, few studies introduce data-driven genre and metre recognition [67, 68]. Others [69–71] use rhythmic feature information for specialised tasks such as audio identification.

In this section, an approach for robust data-driven rhythm analysis is discussed. To this end, LLDs modelling rhythmic information are presented that are tailored to classify duple and triple metre and ballroom dance styles. Once these are determined, the information is used to reliably assess the quarter-note tempo and avoid 'octave' errors, i.e., doubling, tripling, halving, etc., of the tempo by mistake.

The determination of tempo, metre, and (on-)beat positions [25] can be roughly divided into two major principles:

The first strategy starts with the location of onsets in the audio (or symbolic notation such as MIDI) as was shown in the last section. Then, the desired determination tasks are based on the analysis of the inter-onset intervals (IOIs) [72–78]. To this end, histogram approaches are found most frequently [13, 75]. There, duration and weight of all possible IOIs are calculated. IOIs are binned by similarity clustering and the clusters are arranged in a histogram. From the weights and the centres of the clusters the tempo of several metrical levels can be estimated. Alternatively, rule-based approaches are employed [13]. Or, exclusively the Tatum pulse, i.e., the fastest tempo present in a piece is computed by choosing the cluster with the centre of the smallest IOI [75]. Then, within a window around each Tatum pulse features are extracted and the Tatum pulses are classified, e.g., by Bayesian methods, with respect to their perceived accentuation. By that, the beat level is detected based on the assumption that beats are more accented than off-beat pulses.

In the second strategy to determine tempo, metre, and (on-)beat positions the order is inverted, i.e., after analysis of tempo and metrical structure onset positions are retrieved. In this case, resonator methods or the related correlation approaches are commonly used. Onset localisation then benefits from the knowledge gained throughout tempo detection [5, 13, 14, 16, 19, 79]. This second strategy tends to lead to more reliable results if the tempo is sufficiently constant over longer segments. It will thus be followed in the ongoing. As in the case of onset detection (cf. Sect. 11.2), the assumption is made that beats, percussion or note onsets, and other rhythmic events are marked by a change in signal amplitude in a few non-linear frequency bands. The starting point thus is the envelopes or the differentials of the envelopes of six frequency bands, however, without peak picking. The 'detection function' (cf. Sect. 11.2) [57] will be the envelope, its differential, or any other function related to perceivable change in the signal $s(k)$.

The beat level tempo we aim at now, can be seen as periodicity in the envelope function. Just as for the detection of pitch periodicity in Sect. 6.2.1.9, auto-correlation can be used here to find the periodicity [19, 80]. The periodic auto-correlation is calculated over a window of 10 s of the envelope function. As in the case of pitch detection, the index of the ACF's highest peak indicates the strongest periodicity. However, this strongest periodicity does not necessarily correspond to the periodicity perceived as dominant [81], which may be influenced by an interval of preferred tapping linked to a supposed resonance between the human perceptual and motor system. Ignoring this fact, however, and using this highest peak as indication of the beat level tempo can give reasonable results if the music of consideration has strong beats in the preferred tapping range. Given, however, that multiple frequency bands were used, their results need to be combined in a meaningful way. A straightforward approach is the addition of the bands' individual ACFs (cf., e.g., Fig. 11.5) leading to the summary ACF (SACF). In the SACF, one then picks the highest peak. Alternatively, one can determine the tempo per band and carry out a majority vote over these decisions—potentially even weighted according to the type of music. An example of ACF application is given in [13] for tempo and in [19] for metre detection.

A related method is the use of a resonant filter bank [5]. Such a bank is made up by resonators tuned to different frequencies or periodicities, respectively. The detection function is input to all resonators. Then, the total output energy is measured per resonator. Similar to the ACF approach, the resonator with the highest output energy best matches the piece's periodicity. Thus, one assumes the beat level tempo to be its resonance frequency. As stated, this is an incomplete, yet 'working' model of the considerably more complex human rhythm perception. In fact, most state-of-the-art

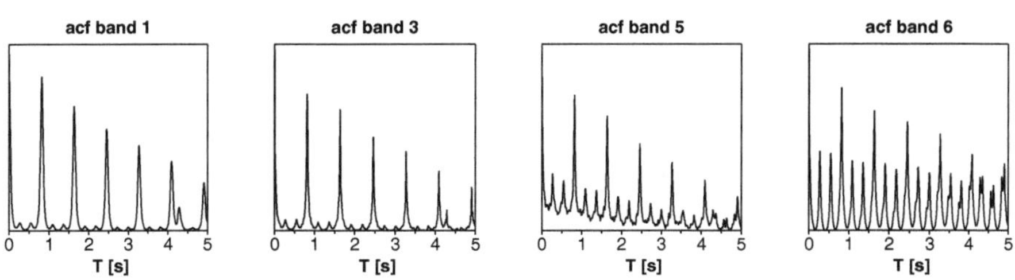

Fig. 11.5 Periodic ACF of band envelope differentials from 10 s of OMD—"*Maid of Orleans*" [6]

systems do not assess beat level tempo fully reliably, and octave errors are a major issue [51], even among several human listeners that tap at different levels.

A detection function is needed by the approaches up to now. Alternatively, a different periodicity detection approach can be followed based on retrieving self-similarities among audio features [14]. To this end, FFT coefficients or MFCCs [82] are computed from short overlapping windows (20–40 ms). Based on these, a vector-by-vector self-similarity matrix $\underline{S}$ is calculated by distance measurement or cross-correlation between vectors at different positions in time. Then, the 'beat-spectrum' [14] B is derived from $\underline{S}$. It is comparable to the ACF or the output of the resonant filter bank discussed before:

$$B(IOI) = \sum_{k=1}^{K} \underline{S}_{k,k+IOI} \tag{11.14}$$

The acoustic features $\underline{S}$ is based upon still influence the performance, but relations between all feature vectors are modelled in this way. However, given the time needed for computation of a self-similarity matrix and the fact that for most music the sensation of the tempo corresponds to a loudness periodicity, a set of sub-band detection functions is considered as sufficient in the ongoing.

Let us now switch to a brief overview on selected metre detection and ballroom dance style recognition methods. Tempo information from various metrical levels is ideally used for metre classification, as in [16], where music is processed on the tatum, pulse, and measure level by comb-filter bank periodicity analysis and probabilistic modelling of dependencies across the metrical levels. Further, the approach can model change of metrical structure within a song for a broad variety of genres, and obtains phase and tempo robustness on the beat level, but not on the measure level. Limiting the kinds of metres helps to reduce the complexity of the task. For ballroom dance music, considering only duple or triple periods on the measure level can be a reasonable [67]. There, a segmentation of the song on the beat level is assumed to be given for subsequent determination of duple or triple metre on the measure level. Per beat-segment, LLDs are extracted and periodic similarities across beats and LLDs are analysed by ACF leading to decision criterion features for metre classification.

Ballroom dance style recognition is a comparably novel task, but experience exists to some degree from the related classification of musical genre as in [83]. In [7] timbral texture, rhythmic and pitch content is modelled in the acoustic features. Rhythmic feature information bases on the ACF of sub-band envelopes. GMMs and k-Nearest-Neighbour (kNN) are compared. In [39] a brute-forced large feature space is the basis for SVM classification. Ballroom dance style recognition based on a data-learnt model is presented in [68]. Different features based on IOI histograms are input to a kNN classifier, and 15 MFCC-like descriptors derived from the IOI histogram lead to the best result.

Table 11.4 Tempo distribution in the BRD set by mean, standard deviation σ, minimum and maximum tempo in BPM by dance style

Tempo [BPM]	Mean σ	Min		Max
All	128.5	38.7	68	208
Cha-Cha-Cha	122.0	6.5	92	136
Foxtrot	114.8	2.1	104	116
Jive	165.9	11.5	124	176
Quickstep	200.7	6.7	153	208
Rumba	97.7	8.3	76	141
Samba	100.7	8.8	68	202
Tango	127.4	3.2	112	136
Viennese Waltz	177.1	2.3	168	186
Waltz	86.2	1.7	72	94

11.3.1 BRD Database

To provide an impression of obtainable results, a set of 1855 pieces of typical Ballroom and Latin dance music sampled from [84] serve as database—the BRD database for short. The detailed list of these is available for reproduction of results.[3] According to the World Dance Council's (WDC) regulation, the five International Standard dances Foxtrot, Quick Step, Tango, Viennese Waltz, and Waltz are covered next to the four most typical International Latin dances Cha-Cha-Cha, Jive, Rumba, and Samba—Paso Doble is left out, as it is hardly taught and seldom danced in most dance schools. Their tempi range from 68 to 208 BPM. 30 s are available per song. The audio was converted from a Real Audio like format to 44.1 kHz PCM. The overall duration of the audio is thus 15.5 h. The distribution of the tempi and the instances across dance styles can be seen in Table 11.4 and respectively later in Table 11.6.

The tempo and dance style labels are taken over from [84]. The ground truth with respect to duple or triple metre is known from the dance style: Waltzes have triple metre as opposed to the rest with duple metre.

11.3.2 Methodology

Let us now take a look at the algorithm used for provision of benchmark results. The approach is (partly) data-driven and performs rhythm analysis based on 82 rhythmic features. Its output is duple or triple metre, quarter-note tempo, and one of nine ballroom dance styles. In order to best cope with the above described octave errors which are challenging even for human listeners, ballroom dance style recognition is integrated into the tempo analysis. Once the ballroom dance style is known, a

[3] http://www.mmk.ei.tum.de/~~sch/brd.txt

tempo range deduced from the dance style can be enforced on the quarter-note tempo detection. This method is very effective in eliminating octave errors.

The envelopes or detection functions of six non-linear frequency bands are fed into comb filters as first introduced in [5] to detect the fastest, i.e., Tatum tempo [19, 75] by highest output energy. The comb-filter bank used in the ongoing is a specialised version. From this information features are derived that describe the distribution of resonances throughout the musical piece of analysis. These allow for the automatic decision upon duple or triple metre, and ballroom dance style classes, which assist the tempo detection algorithm. The tempo is from now on denoted by θ and is specified by a frequency with the unit BPM. The subscript IOI indicates that it is given as IOI period in frames.

Let us next look at the comb filters in detail. It basically adds the signal itself to a delayed version of the signal and is characterised by two parameters: the 'delay' d or period, being the reciprocal value of the filter's resonance frequency, and the gain α. For tempo detection IIR comb filters are used with the output $y(k)$ in the discrete time domain:

$$y(k) = (1-\alpha)\cdot s(k) + \alpha \cdot y(k-d) \tag{11.15}$$

The according transfer function $H(z)$ in the z-domain is:

$$H(z) = \frac{1-\alpha}{1-\alpha \cdot z^{-d}} \tag{11.16}$$

$H(z)$ is depicted in Fig. 11.6 for two exemplary gains α. Optimising α is a crucial factor to achieve best tempo detection. In [5] it is suggested to use a constant half-energy time by using variable gain α depending on d. This was, however, not observed ideal in the oncoming experiments, and a fixed value for α is thus preferred. As small temporary tempo drifts within a musical piece have to be assumed, the gain $\alpha \to 1$

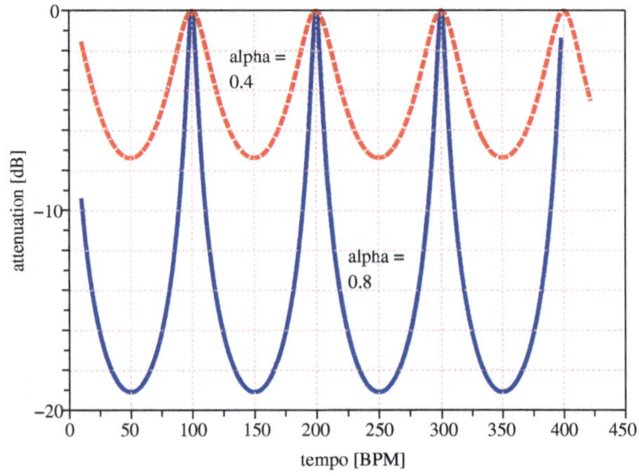

Fig. 11.6 Frequency responses of IIR comb filters for the gains $\alpha = 0.8$ and $\alpha = 0.4$ set for 100 BPM

being optimal in theory cannot be used. Evaluating α in [0.20, 0.99] revealed $\alpha = 0.7$ as best solution.

Comb filter banks are instantiated for a broad ranger to also cover higher metrical layers. Features as obtained by the outputs of the filters describe the distribution of resonances of several metrical layers, and by that the metrical structure. To keep the number of comb filters in reasonable limits, one can exploit the multiple metrical layers present in a musical piece: At first, the Tatum tempo is estimated. Then, potentially present higher metrical levels are assumed to have tempi at integer multiples of this tempo. This is true for a broad range of genres.

For the processing, the audio signal is down sampled to f_s =11.025 kHz and converted into a monophonic signal by stereo-channel addition. The input of length L_i seconds is chunked by Hamming windowing into $N_{frames} = 100 \cdot L_i$ frames of $N_{s,block} = 256$ samples with a frame overlap of 0.57. This resembles a frame rate of 100 FPS. 128 DFT coefficients are then computed per frame. M_{mel} overlapping triangular filters which are equidistant on the Mel-frequency scale as used in speech recognition for the computation of MFCC [82] (cf. Sect. 6.2.1.4) reduce these coefficients to envelope samples of M_{mel} non-linear bands. The reduced number of frequency bands covers the human auditory frequency range. According to [5], the rhythmic structure is entirely preserved in this compact form of representation. The envelope samples $x_{m,k}$ per Mel-frequency band m are logarithmised by:

$$x_{m,k,log} = 10 \cdot log\left(x_{m,k} + 1\right) \tag{11.17}$$

The envelopes $\underline{x}_m$ of the Mel-frequency bands are then low-pass filtered for smoothing. This is realised by convolution with a half-wave raised cosine filter h_{cos}. The length of 15 envelope samples, or 150 ms, respectively has proven a good value—overall, it preserves fast attacks, but filters noise and rapid modulation, similar to human sound sensation:

$$h_{cos}(k) = \cos\left(\frac{\pi k}{15}\right) + 1\,, \quad k \in [1, 15] \tag{11.18}$$

Per low-pass filtered Mel-frequency band envelope m a weighted differential d_m is applied:

$$d_m(k) = \left(x_{m,k} - \overline{x}_{m,k,l}\right) \cdot \overline{x}_{m,k,r} \tag{11.19}$$

For a sample $x_{m,k}$ at position k a moving average is calculated over one window of 10 samples to the left of sample $x_{m,k}$ (left mean $\overline{x}_{m,k,l}$) as well as a second window of 20 samples to the right of sample $x_{m,k}$ (right mean $\overline{x}_{m,k,r}$) [6]. The motivation is that human's perceive note onsets as more intense after a longer phase of lower sound level [85]. Further, note duration and energy are crucial factors in the perceived note accentuation [75].

11.3.2.1 Tatum Features

The Tatum grid is the lowest metrical level, i.e., all onsets are contained in it [75, 86]. By that, it represents the highest tempo and lowest inter-onset-interval in a piece. The Tatum tempo θ_T is assessed by an IIR comb filterbank with 57 filters of gain $\alpha = 0.7$ and delays from $d_{min} = 18$ to $d_{max} = 74$ envelope samples. Tempi from 81 to 333 pulses-per-minute can likewise be captured which is sufficient except for very slow music, where a different range can be chosen, accordingly. The weighted differential d_m of each Mel-frequency band envelope m is input $\underline{u}_m$ to the filters $h_{m,d}$ with delays d, whose output is referred to as $y_{n,d,m}$. The total energy output $\underline{T}'(d - d_{min} + 1)$ over all bands is calculated per filter $h_{m,d}$ by:

$$\underline{T}'(d - d_{min} + 1) = \sum_{m=0}^{M_{mel}} \sum_{n=0}^{N_{frames}} y_{n,d,m} \tag{11.20}$$

This leads to the 'unflattened' Tatum vector $\underline{T}'$ of 57 elements $\underline{T}'(d - d_{min} + 1)$. Figures 11.7 and 11.8 show $\underline{T}'$ for exemplary songs.

From $\underline{T}'$ three peak statistics are derived as follows: T_{ratio}—the ratio of the highest and lowest value, T_{slope}—the fraction of the first over the last value, and $T_{peakdist}$—the mean of the maximum and minimum value normalised by the global mean. These describe the vector's peaks' 'visibility' and flatness as can be seen in Figs. 11.7 and 11.8. Despite their constantly periodic spectral response, the comb filters inherently have higher resonances at higher tempi for less rhythmic content (due to the way the data is distributed and the comb filter outputs are normalised). Thus, a *flattening* of

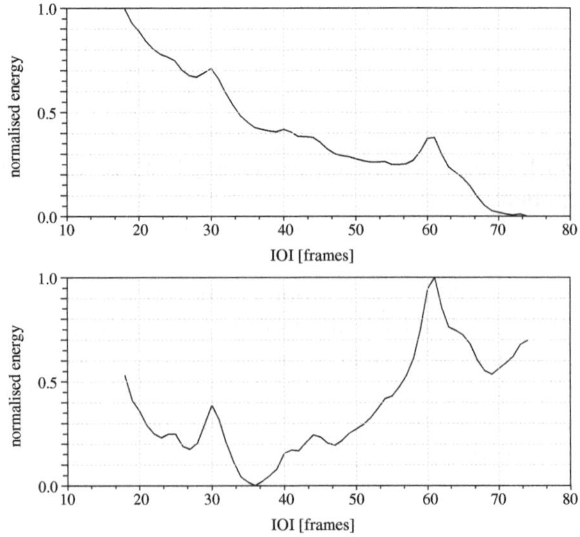

Fig. 11.7 Plots of Tatum vector $\underline{T}'$ (*top*) and flattened Tatum vector $\underline{T}$ (*bottom*) for Celine Dion—"*My Heart Will Go On*"

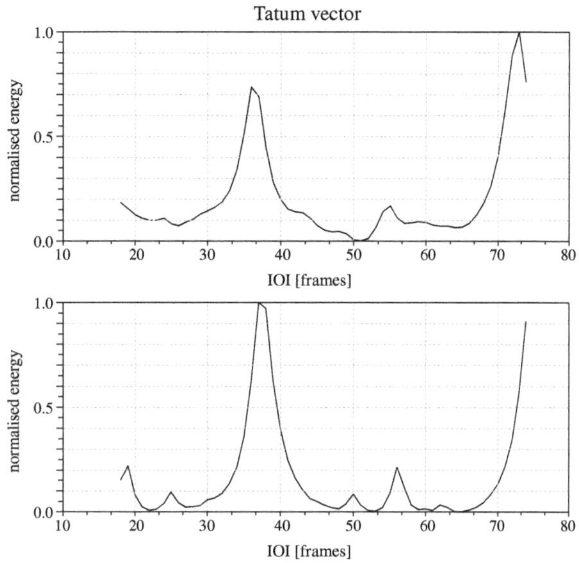

Fig. 11.8 Plots of Tatum vector $\underline{T}'$ for "*Moon River*" (Waltz, triple metre) (*top*) and "*Hit the Road Jack*" (Jive, duple metre) (*bottom*)

the vector by the difference between the means of the first and last six values leads to the flattened Tatum vector $\underline{T}$. From this vector, the two most dominant peaks are located by determining all local minima and maxima at first. Next the height D is calculated per maximum based on its mean minus the adjacent left and right minima. The indices of these two maxima are the Tatum candidates ($\theta_{T1,IOI}$ and $\theta_{T2,IOI}$), and confidences $C_{Ti,IOI}$ are calculated for these:

$$C_{Ti} = D_{Ti} + \underline{T}(\theta_{Ti,IOI}), \quad i \in \{1, 2\}. \tag{11.21}$$

The candidate with higher confidence is chosen as final Tatum tempo θ_T. The IOI period $\theta_{T,IOI}$ of the final Tatum tempo is converted into the final tatum tempo (θ_T) in BPM by:

$$\theta_T = \frac{6000}{\theta_{T,IOI}} \tag{11.22}$$

11.3.2.2 Metre Features

The 63 Tatum features made up by θ_T, θ_{T1}, θ_{T2}, T_{ratio}, T_{slope}, $T_{peakdist}$ and the Tatum vector $\underline{T}$ with 57 elements only model a very small tempo range. To extend to tempo distributions over a broader range within the estimation of the main or beat-level tempo θ_B, and the metrical grouping M, a 'metre vector' $\underline{m}$ is next described. Note, however, that explicit metre information is not contained—the feature vector is rather needed to assess this information. This metre vector captures the distribution of resonances among 19 metrical levels, starting at the Tatum level.

The 19 elements m_i of $\underline{m}$ are normalised score values of the tempo $\theta_T \cdot i$. By that, they provide information on the degree to which the tempo $\theta_T \cdot i$ resonates with the musical piece. For their calculation, an un-normalised score value m'_i is computed, first, by setting up a comb filter bank for each value of $i \in [1, 19]$. Each filter bank consists of $2i + 1$ filters with delays from $(\theta_{T,IOI} \cdot i - i)$ to $(\theta_{T,IOI} \cdot i + i)$ [6]. As in Sect. 11.3.2.1, the total energy output per filter in the bank is calculated. Then, the maximum value is assigned to m'_i. The delay d of the filter with the highest total energy output is stored as adjusted tempo $\theta_{i,IOI}$ belonging to m'_i. The 19 elements m'_i make up the unflattened metre vector $\underline{m}'$, with

$$m'_i = \max_{j \in [-i,+i]} \left(\sum_{m=0}^{M_{mel}} \sum_{n=0}^{N_{frames}} y_{n,\theta_T \cdot i+j,m} \right) \tag{11.23}$$

Figures 11.9 and 11.10 provide according examples. Given same behaviour for higher resonances of higher tempi as was described for the Tatum vector above also for $\underline{m}'$ (cf. Fig. 11.9), this vector is flattened, accordingly, considering the difference $m'_{19} - m'_1$.

This leads to the flattened metre vector $\underline{m}$, simply called metre vector. The requirement of a minimal input length $L_i = d_{max} \cdot 19 \approx 14$ s is needed, as the higher metrical levels correspond to very slow tempi and by that to large comb filter delays.

11.3.2.3 Feature Selection

82 features, including all the 19 metre vector elements m_i and the 63 Tatum features, namely θ_T, θ_{T1}, θ_{T2}, T_{ratio}, T_{slope}, $T_{peakdist}$ plus all 57 elements of the Tatum vector

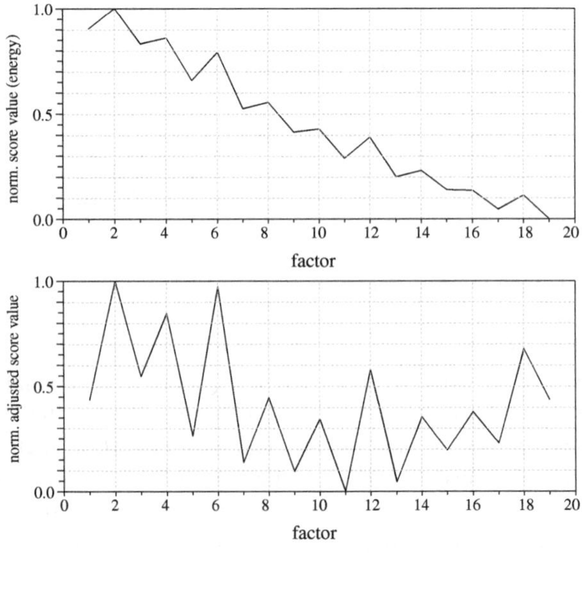

Fig. 11.9 Plots of metre vector $\underline{m}'$ (*top*) for "*Moon River*" (Waltz) and flattened metre vector $\underline{m}$ (*bottom*)

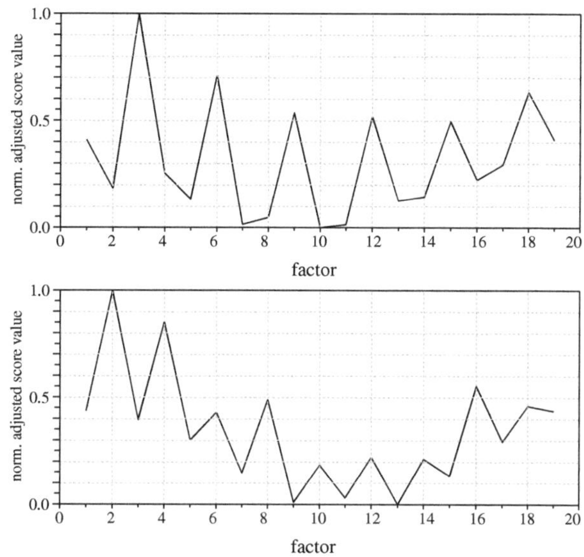

Fig. 11.10 Plots of flattened metre vector $\underline{m}$ for "*Maid Of Orleans*" (3/4 metre, *top*) and "*Hit the Road Jack*" (4/4 metre, *bottom*)

$\underline{T}$ were introduced so far—they form the feature set F_{all}. Given the suitability of linear SVMs with SMO learning in [24, 39], these have been used for classification in the work described here. To determine the most relevant features for metre and ballroom dance style classification from the set F_{all}, a Sequential Floating Forward Search (SFFS) [39] with the target classifier—the SVMs—was carried out once per task. This led to the feature sub-set F_{metre} for metre classification: T_{ratio}, metre vector $\underline{m}$ elements 4, 6, 8, 16, and the Tatum vector $\underline{T}$. Further, the ballroom dance style classification feature sub-set F_{dance} found resembles: metre M, T_{ratio}, T_{slope}, $T_{peakdist}$, metre vector $\underline{m}$ elements 4–6, 8, 11, 12, 14, 15, 19, and the Tatum vector $\underline{T}$ without elements 21 and 29.

11.3.2.4 Recognition

Metre and ballroom dance style are classified by a data-learnt approach, namely SVM. Given the continuous value nature of tempo, one may think of using SVR for tempo assessment. This was tested on the BRD set, but observed as not able to identify a few percent relative BPM deviation. Thus, a hybrid classification and regression approach is considered: The tempo range is divided into few overlapping tempo ranges. A natural choice in the context of ballroom dance style is to use these styles as tempo classes as these are usually limited to a specific tempo range. Such a regulation is officially provided by the International DanceSport Federation's tempo regulation for competitions. In [13, 87], this fact is used the other way around: Tempo ranges are used there to assess the ballroom dance style.

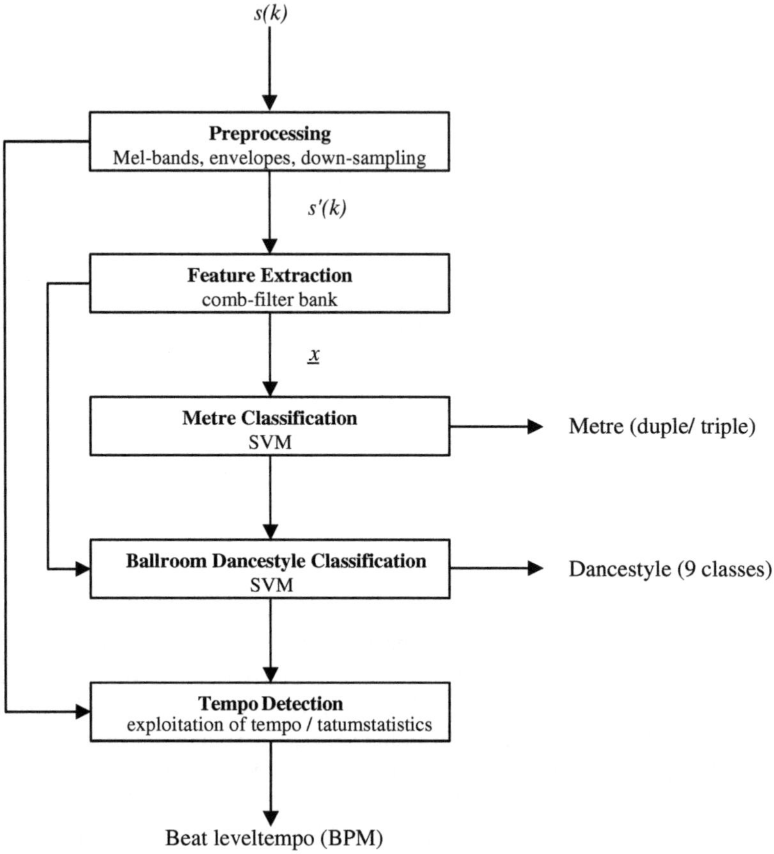

Fig. 11.11 Flow in the described data-driven tempo detection basing on metre and ballroom dance style recognition [6]

Figure 11.11 shows the overall processing flow for metre, ballroom dance style, and quarter-note tempo determination: First, a SVM-model for metre classification is built using the feature sub-set F_{metre} to assign a metre M (duple or triple). Then, the metre M is used as a feature in the set F_{dance} (cf. Sect. 11.3.2.3) for ballroom dance style classification. Finally, determined metre and ballroom dance style are used to assess quarter-note tempo robustly.

From the training data the means $\mu_{q/T}$ and variances ${\sigma^2}_{q/T}$ of the annotated quarter-note tempi and tatum tempi θ_T are calculated per ballroom dance styles. As no annotation for tatum tempo is usually available, the tempo estimated automatically as in the first step (cf. Sect. 11.3.2.1) serves as substitute. Higher WA could be reached given manual annotation also for this tempo.

Then, the tempo of unknown test instances is determined: With the two tatum candidates θ_{T1} and θ_{T2} as extracted in the first step in Sect. 11.3.2.4, the final tatum is decided upon based on the statistics from the training data. The confidence $C_{T1/2}$

(cf. Sect. 11.3.2.1) is replaced by a Gaussian function $G(\theta_{T1/2})$:

$$G(\theta) = \exp\left(-\frac{(\theta - \mu)^2}{2\sigma^2}\right), \tag{11.24}$$

and the parameters μ and σ^2 are set to the values of μ_T and ${\sigma^2}_T$ of the ballroom dance style.

Next, the candidate $\theta_{T1/2}$ that maximises the function $G(\theta_{T1/2})$ is selected as the final tatum tempo θ_{T*}. With this new tatum, a new flattened metre vector $\underline{m}^*$ is calculated, and used for determination of the quarter-note tempo. The elements m_i^* are multiplied by a Gaussian weighting factor $G(\theta_i)$, and the parameters μ and σ^2 in Equation (11.24) are chosen as μ_q and ${\sigma^2}_q$ according to the ballroom dance style. θ_i indicates the tempo the metre vector element m_i^* belongs to (cf. Sect. 11.3.2.2) [6]. Then, the index i_{max} that maximises $m_i^* \cdot G(\theta_i)$ is determined, and the tempo $\theta_{i_{max}}$ according to this index i_{max} is chosen as the detected quarter-note (beat level) tempo θ_q.

11.3.3 Performance

Table 11.5 depicts benchmark WA for the detection with and without prior ballroom dance style recognition. These were computed in a ten-fold SCV as were further results in this section based on data. Thereby, at no time throughout processing test instances' labels are used except for the final comparison if the decision was correct. Tempo tolerance in the evaluation is 3.5 % relative BPM deviation as in [24]. For the case without ballroom dance style recognition a single predefined Gaussian is used for the overall tempo distribution instead of the nine dance style specific Gaussians.

As can be seen in the table, WA is increased by almost 20 % absolute with the prior recognition of the ballroom dance style. With the 'perfect' ballroom dance style as given by the manual annotation, the tempo octave is near always correct. Overall, 88 % of all instances were assigned the correct tempo octave.

With all steps as described in Sect. 11.3.2.4, the performances in Table 11.6 are obtained, which are the best on this data set to-date [67, 68]. There, ballroom dance style recognition is obtained without the quarter-note tempo as feature information.

Table 11.5 WA for tempo detection on the BRD set without (w/o BDS), with prior ballroom dance style recognition (w/ BDS), and using manually annotated 'ground truth' ballroom classes as upper idealistic benchmark (gt BDS)

WA [%]	w/o BDS	w/ BDS	gt BDS
Tempo	88.8	92.4	93.1
Octave	70.0	88.5	93.0

Table 11.6 UA / WA on the BRD set for metre M, quarter-note tempo θ_q, and ballroom dance style (BDS) by genre

[%]	UA	WA	Cha-Cha -Cha	Foxtrot	Jive	Quickstep	Rumba	Samba	Tango	Vien. Waltz	Waltz
#Instances			211	245	138	242	217	188	185	136	293
Metre	97.1	96.9	99.1	97.6	97.8	99.6	90.8	98.9	98.4	97.8	94.2
BDS	88.9	89.1	87.7	95.5	88.4	90.1	77.9	84.0	91.4	91.9	93.2
Tempo	93.0	92.4	97.2	93.9	97.1	96.3	90.3	93.6	94.1	92.6	81.8
Tempo octave	89.0	88.5	94.8	93.5	90.6	87.6	81.6	86.2	93.5	91.2	81.8

Further, only 30 s were available per song. Given longer segments, one can expect results to improve [24].

11.3.4 Summary

In this section we discussed automatic rhythm analysis on real-world music. In particular, a data-learnt approach of sequential combination of ballroom dance style, metre, and tempo recognition was introduced. It could be observed that the information on ballroom dance style highly increased the tempo estimation accuracy. Further, 82 rhythmic features were discussed that will be re-used later in Sect. 11.7 for music mood analysis.

Further efforts could consider complementary features, such as the ones in [19, 68, 71] to augment the rhythm analysis.

11.4 Key

Apart from the rhythm-focused application presented so far, the tonal analysis is of course of interest. Starting from a shallow analysis, let us first consider musical key in music as in [28].

The musical key is decisive for the notes—mostly seven—that 'belong' to the according scale. The key itself usually refers to the 'tonic' chord. The tonic chord is normally perceived as arrival or 'resolution'. Other chords in a musical piece of a certain key create different types and degrees of tension. Keys can be divided into major and minor. However, one major and one according minor key share the same notes that belong to the scale. Pieces or songs in pop and classical music are mostly in one key. However, longer pieces can have several, partly contrasting keys. The underlying rules that manifest the key of a musical piece are non-trivial and varied over the history of western music. Most decisive are the chords appearing in a piece that are usually constructed by the notes of the corresponding scale.

Knowledge of this key is an essential information as a high level-feature in MIR [88]: It indicates more probable semitones in melody extraction or chord detection. Of course there may be occurrences of out-of-key notes, but in-key notes will usually appear considerably more frequently. Other applications include automatic 'sub-bass' addition. These are audible sounds typically in the range of 20–90 Hz popular in clubs or discotheques to increase the use of sub-woofer loudspeakers. Finally, key-matching by transposition can be realised in automatic mixing of music for DJ tools, in radio broadcast or for automatic play-list creation.

Given this relevance, substantial research efforts were made in this direction. Most earlier publications base on synthesised music from MIDI (cf., e.g., [89–91]) or other symbolic representations [92]. Further, often single genres are dealt with (cf., e.g., [93–96]). It has to be stated, though, that even key detection from symbolic music representation is not solved to-date given the complexity of the topic [92]. Most systems base on a chain of frequency analysis including pitch class mapping, feature extraction, and a key detection algorithm. The last stage marks the major difference in approach: It may be based on knowledge such as correlation with templates or more recently increasingly on data [97–99]. As features, PCP variations dominate [100] together with diverse variations such as CHROMA (cf. Sect. 6.2.2.2) based harmonic PCP [89, 90, 101], PCP modulations [102], data-learnt PCPs including scale transitions [103], 'tonal centroid' features [104, 105], constant-quotient profiles [106, 107], overtone removal [108], ANNs for human perception modelling [109], and weighting the contribution of FFT bins by their distance to the closest note [110].

Correlation with templates includes such constructed from monophonic instrument clips, weighted by a combination of the Krumhansl-Schmuckler and Temperley's modified PCPs using a multidimensional tonal representation [91, 111, 112]. Other variants include a decaying spectral impulse train for chromagram modelling with subsequent template based correlation [93], rules learnt from MIDI data [113, 114], and a geometric topology of tonality with an inter-key distance—the Spiral Algorithm and Centre of Effect Generator, also known as FACEG [94, 96, 115]. In this process knowledge on rhythm structure and chord change progressions can be integrated [116].

More recently, HMMs emerge for this task [97, 104, 105, 117] given their ability to incorporate temporal dependencies and key changes [118]. The HMM approaches partly include transient and noise reduction or tuning estimation [95]. As static classifiers, distance-based approaches such as KL divergence or Mahalanobis distance [101] and SVMs [98] are applied. A further overview on methods is given in [119].

From the described methods, none has emerged as single best alternative. Further, different datasets reaching from synthesised to real audio and varying in size, as well as differing evaluation criteria such as accuracy or point systems as in the MIREX challenge, make it difficult to compare the diverse approaches. None the less, some advantages are inherent in the methods, such as HMMs or more general DBNs being able to segment or allowing for inclusion of an LM [10]. SVMs or ANNs provide discriminative learning, generalise well, and can handle larger feature spaces. The charm of using templates without training is obvious: Besides fast classification, almost no storage space is needed. This comes, however, at the price that adaptation to genres or instrumentation is almost impossible.

In the ongoing, we will oppose the two main approaches of a template based correlation model and a data-driven approach and compare features based on music theory and human perception studies (cf. Sect. 6.2.2)

11.4.1 Key Databases

Four datasets will be used to cover the prevailing music genres typically aired reaching from Rock and Pop, to Jazz, and Classical music. Overall, 520 pieces—respectively 35 h and 25 min of playtime, are contained in the 'KEY-ALL' database. The subsets were selected from commercial CDs to ensure availability for reproduction of results. Pieces are limited to constant key. This is, however, no real constraint, as the time window of the methods presented can be shortened, or overlapping analysis can be carried out. In addition, structure analysis such as introduced in Sect. 11.6 can be applied first to subsequently assign the key per found segment such as chorus or verse. The annotation of the key was carried out in 24 keys (12 major and 12 minor) by three professional musicians. No disagreement was found for the pieces in the database.

The CLASSIC dataset contains 89 classical pieces (5 h 38 min) from "*100 Meisterwerke der Klassischen Musik*", a six CD collection of '100 masterpieces of classical music' [28]. 11 pieces were excluded because they contained key changes.

The JAZZ dataset contains the 82 constant key pieces (7 h 52 min) throughout all stages of development in Jazz music from the 106 in the "*Blue Note Jazz History Collection Vol. 1–Vol. 5*". Special challenges of this set include live-recordings and the fact that western music theory is not always applicable in Jazz—for example blue notes and special scales and modes may leave the semitones of a key.

The CHANSON dataset contains the 150 constant key pieces (7 h 35 min) of the 162 French Chansons in the 10 CD collection "*Chansons de France Vol. 2*" or 'Chansons of France volume two'. Particular challenges of this set include low recording quality in the sense of noisiness and tape speed variation similar to 'older' vinyl recordings—studio time was often too pricy and low-budget live recordings were common habit, just as for Jazz music.

Finally, the MTV dataset contains all 200 contemporary popular pieces (14 h 24 min) of the annual top ten songs of the "*MTV-Europe Most Wanted*" from the years 1981–2000. One piece needed to be excluded because of key changes. A challenge in this set is its genre variety including Electronic Dance, Hip Hop, Pop, and Rock. Further, drums and percussion are comparably dominant, traditional instruments partly absent, and intentional detuning or addition of samples present.

Table 11.7 shows the distribution of playtime for 24 keys (twelve major and twelve minor). For 12 and 24 keys this is additionally visualised in Fig. 11.12.

In the experiments described here 12 and 24 keys are considered, whereby always the same instances are handled. In the case of 12 keys, the minor keys are re-labelled

Table 11.7 Distribution of 24 major and minor keys in the united KEY-ALL database by total time (TT)

Key	A	A#	B	C	C#	D
TT	1 h 59 min	1 h 10 min	1 h 12 min	3 h 8 min	1 h 53 min	2 h 50 min
Key	F#m	Gm	G#m	Am	A#m	Bm
TT	24 min	1 h 18 min	15 min	1 h 47 min	24 min	46 min
Key	D#	E	F	F#	G	G#
TT	2 h 43 min	1 h 44 min	2 h 40 min	1 h 6 min	2 h 13 min	1 h 31 min
Key	Cm	C#m	Dm	D#m	Em	Fm
TT	1 h 45 min	43 min	1 h 3 min	17 min	1 h 13 min	1 h 26 min

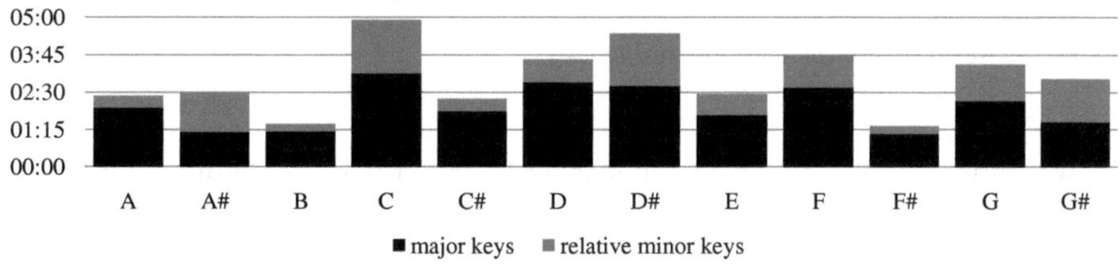

Fig. 11.12 Distribution of keys in the united KEY-ALL database in hours and minutes. Shown are the major keys and on *top* of these their according relative minor keys such as A minor in the case of C major for the 24 keys task. If one looks at the overall bar independent of the shading, one thus sees the distribution of the instances for the 12-key task [28]

by their relative major key, i.e., an A minor label, for example, turns into a C major label. Tests for parameter optimisation are based only on this more robust 12-key-task. In fact, knowledge of the used semitones suffices in the majority of applications. For increased realism, music is MP3 encoded after reading it from the CDs with constant parametrisation at 44.1 kHz, 16 bit, and 128 kbit/s fixed bit rate. These settings resemble typical minimum quality conditions. Then, the music is decoded back to waveform and down-mixed to monophonic by stereo channel addition. During this addition, it is ensured that the no distortion arises by first dampening the two channels evenly. In this way, the data can be processed easily, but the encoding / decoding ensures that the results are representative even in the case of lossy encoding.

11.4.2 Parameter Tuning

Figure 11.13 shows the chain of audio processing for key determination including preprocessing and extraction of the CHROMA features as is considered in the ongoing. A 12 dimensional CHROMA vector is obtained by spectral transformation (35 ms Hanning window and 50 % overlap) of the music, dB(A)-weighting, compensation of detuning, and mapping to pitch classes by semitone-interval based spectral band-pass filters [120, 121] and summation of the different octaves per note. Then follows the creation of derived music theoretic and perception-based features.

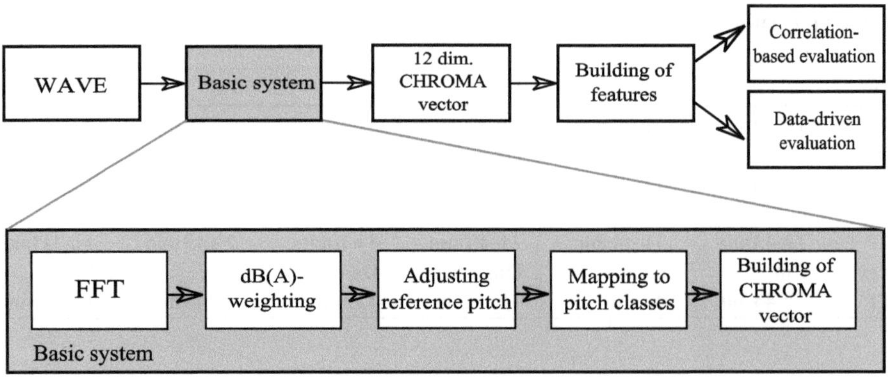

Fig. 11.13 Key detection chain of processing [28]

Two steps along this chain are non-standard: First, dB(A)-weighting according to the norm IEC/DIN 651 adapts to human perception and compensates low and very high frequencies amplification typically applied during mastering of CDs.

In order to compensate potential 'tape speed' variations, a method as follows was proposed.

First, the amount of 'detuning' (i.e., deviation) between the musical piece and the reference pitch classes is estimated. To this end, the 'prominent' frequency is determined as the one with the highest long-term energy. These typically range between 2 kHz to 3 kHz. In this range, one runs the risk to capture higher harmonics rather than fundamental frequencies. As these tend to show inaccuracies, the range was limited to 130 Hz to 1 kHz. The lower range end resembles C3, and is later shown to be well suited for musical key determination (cf. Table 11.11 in Sect. 11.4.5). The prominent frequency could be the root of the piece, but is not at all necessarily so. Next, the 'reference' frequency nearest to the measured prominent one is found, and a divergence factor D to the nearest properly tuned reference frequency is calculated:

$$D = \frac{nearest\ reference\ frequency}{measured\ prominent\ frequency}. \tag{11.25}$$

Prominent frequencies in the given range are usually found between 250 and 350 Hz. The value of the divergence thus maximally varies by $\pm 1.5\,\%$ and on average by $\pm 0.4\,\%$. The semi-tone-band filters for frequency-to-pitch class mapping (cf. Sect. 6.2.2.2) are then scaled by D as follows, where f_i represents the mid-frequency of each filter band:

$$f_i = D \cdot f_0 \cdot 2^{\frac{i}{12}},\ i = 0, 1, 2, \ldots, 88; f_0 = 27.5\,Hz \tag{11.26}$$

Table 11.8 in Sect. 11.4.5 shows the effect of this adjustment.

Table 11.8 WA for tape speed variation compensation (w/) or its omission (w/o)

WA [%]	w/o	w/
MTV	70.5	70.5
CHANSON	69.8	72.5
CLASSIC	82.0	82.0
JAZZ	58.5	59.8
KEY-ALL	75.2	76.2

SVM, ten-fold SCV, Gaussian filter, whole piece, range C3–C8, 12 keys

Table 11.9 WA for different semitone filter functions

WA [%]	Rectangle	Triangle	Triangle^2	Gaussian
MTV	71.5	71.5	72.0	70.5
CHANSON	71.1	67.8	69.1	72.5
CLASSIC	84.3	79.8	77.5	82.0
JAZZ	58.5	57.3	59.8	59.8
KEY-ALL	76.2	73.7	75.4	76.2

SVM, ten-fold SCV, whole piece, range C3–C8, 12 keys. Triangle^2 indicates the squared triangle function

For the band-pass filters, a rectangular, a triangular, a squared triangular, and a Gaussian filter are considered. From these, the Gaussian filter is preferred, based on the results that will be shown in Sect. 11.4.5, Table 11.9. However, it seems also intuitive that it leads to good results, as it prefers contributions of frequencies closer to the mid-frequencies as compared to, e.g., a rectangular filter. The standard deviation is selected as $\sigma = 0.125$, and the Gaussian filter $g_i(f)$ with the mean frequency f_i thus resembles:

$$g_i(f) = \frac{1}{0.125 \cdot \sqrt{2\pi}} \cdot \exp\left(-\frac{(\frac{f-f_i}{f_i-f_{i-1}})^2}{2 \cdot 0.125^2}\right) \tag{11.27}$$

Another aspect is the optimal length of the (macro) window of analysis [101]. As different alternatives, the first 30 s, 60 s, 90 s, 120 s, and complete length of a piece are considered with respect to the accuracy. This is depicted in Table 11.10 in Sect. 11.4.5.

Finally, the optimal frequency range for key extraction is analysed with different ranges covering four to seven octaves. The result is shown in Table 11.11 in Sect. 11.4.5.

11.4.3 Correlation-Based Analysis

Given the acoustic features, the key K that maximises the correlation with key templates is identified, where $\underline{\kappa}$ represents the input feature vector (11.29), and $t_{cor}(C)$

Table 11.10 WA for different 'gating' lengths from the beginning of a musical piece

WA [%]	First 30 s	First 60 s	First 90 s	First 120 s	Whole piece
MTV	59.5	70.0	71.5	69.0	70.5
CHANSON	72.5	75.8	75.2	71.1	72.5
CLASSIC	62.9	76.4	75.3	74.4	82.0
JAZZ	37.8	50.0	51.2	48.8	59.8
KEY-ALL	66.2	74.2	75.8	75.2	75.6

SVM, ten-fold SCV, range C3–C8, 12 keys

Table 11.11 WA of different frequency ranges. SVM, ten-fold SCV, 12 keys

WA [%]	C2–C6	C2–C7	C2–C8	C2–C9	C3–C7	C3–C8	C3–C9
MTV	58.0	63.5	67.5	67.0	64.0	70.5	69.5
CHANSON	57.0	61.7	71.8	71.8	59.7	72.5	71.1
CLASSIC	70.8	79.8	80.9	82.0	73.0	82.0	80.9
JAZZ	45.1	48.8	61.0	57.3	51.2	59.8	56.1
KEY-ALL	66.0	69.4	75.6	75.0	70.7	76.2	75.0

the corresponding correlation template vector (11.30). An example in the key of C major (as before) is shown—symbols are explained in Eq.11.32) where maj / min / cad / dom abbreviate major / minor / dominant / cadence as before. For the key K holds:

$$K = \arg\max_{k} \underline{\kappa}^T \cdot \underline{t}_{cor}(k) \tag{11.28}$$

$$\underline{\kappa} \in \left\{\underline{x}, \underline{s}, \underline{s}_{dom}, \underline{s}_{cad}, \underline{c}, \underline{c}_{dom}, \underline{c}_{cad}, \underline{p}_{maj}, \underline{p}_{maj,dom}, \underline{p}_{maj,cad}, \underline{p}_{min}, \underline{p}_{min,dom}, \underline{p}_{min,cad}\right\} \tag{11.29}$$

$$\underline{t}_{cor}(C) = \left[0,\ 0,\ 0,\ \underline{1},\ 0,\ 0,\ 0,\ 0,\ 0,\ 0,\ 0,\ 0\right]^T \tag{11.30}$$

The elements of $t_{cor}(k)$ weight the semitones of the feature vector, and the indices i correspond to the scale with ($i = 1 \widehat{=} A, i = 2 \widehat{=} A\#, ..., i = 12 \widehat{=} G\#$), and k corresponds to the root for which the correlation result is calculated. The root element's value is set to 1, and all remaining ones to zero as seen in Eq. (11.30). The key with the highest correlation value is then decided for. Thus, the maximum component of the input feature vector is searched for. Results per proposed feature type are given in Sect. 11.4.5.

A visualisation of the principle of the derived dominant and cadence features as were introduced in Sect. 6.2.2 is found in Fig. 11.14 exemplified by Charles Trénet—"*La mer*" in the key of C major of the CHANSON dataset: Shown is the distribution of correlation results for the basic feature *'scale'*, the derived features *'scale dominant'*, and *'scale cadence'* in the circle of fifths. As can be seen, with increasing relation to the original key, the correlation results monotonously increase. Figure 11.14a

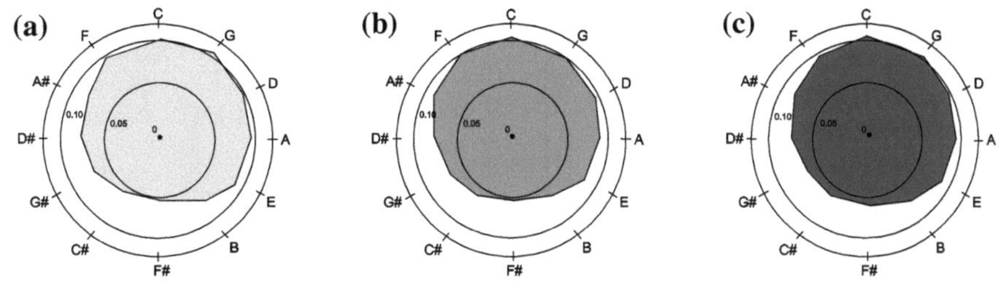

Fig. 11.14 Exemplary results for correlation, basic, and derived features in the circle of fifths [28]. **a** Correlation results for scale attribute, **b** Correlation results for scale dom attribute, **c** Correlation results for scale and attribute

shows the correlation results for scale features with a minimum for F# major and a maximum for G major. Thus, the key a fifth above the correct key C would be assumed. This can be avoided by addition of the dominant to enlarge the search mask to the two highest neighbouring values. In Fig. 11.14b the maximum value for the feature *'scale dominant'* is likewise shifted from G major to C major leading to the correct key assumption. Finally, in Fig. 11.14c the feature *'scale cadence'* is visualised: In the example the addition of the fifth above and below help to cope with the light variations of notes interfering in the feature *'scale'*.

In the case of distinction between 24 keys those feature types able to distinguish musical modes are concatenated to a 24-dimensional vector $\underline{\kappa}$ for correlation. These are PTR major and minor features and dominant and cadence features. The key is determined accordingly by retrieving the semitone k that maximises correlation in analogy to Eq. (11.29), yet. However, the 24-dimensional feature vector $\underline{\kappa}$ with

$$\underline{\kappa} \in \left\{ \left[\underline{p}_{maj}^T, \underline{p}_{min}^T \right]^T, \left[\underline{p}_{maj,dom}^T, \underline{p}_{min,dom}^T \right]^T, \left[\underline{p}_{maj,cad}^T, \underline{p}_{min,cad}^T \right]^T \right\} \tag{11.31}$$

is used with an according 24-dimensional correlation template vector $\underline{t}_{cor}(k)$ created using the previous $\underline{t}_{cor}(k)$ by appending 12 zero-entries at the end or beginning for major or minor keys, respectively. Thus, in the example in Eq. (11.30), 12 zero-entries would be appended at the end.

11.4.4 Data-Driven Analysis

Besides the knowledge-based method, a data-driven one based on SVMs with polynomial Kernel, SMO, and a one-versus-one multi-class discrimination strategy [122] is now described. This approach allows to combine all feature types in a super-vector $\underline{v}$. Given the 13 feature types with 12 features, each, its dimension resembles 156. The vector is shown in Eq. (11.32):

$$\underline{v} = \begin{bmatrix} \underline{x} \\ \underline{s} \\ \underline{s}_{dom} \\ \underline{s}_{cad} \\ \underline{c} \\ \underline{c}_{dom} \\ \underline{c}_{cad} \\ \underline{p}_{maj} \\ \underline{p}_{maj,dom} \\ \underline{p}_{maj,cad} \\ \underline{p}_{min} \\ \underline{p}_{min,dom} \\ \underline{p}_{min,cad} \end{bmatrix} \equiv \begin{bmatrix} CHROMA \\ scale \\ scale\ dom \\ scale\ cad \\ chords \\ chords\ dom \\ chords\ cad \\ PTR\ maj \\ PTR\ maj\ dom \\ PTR\ maj\ cad \\ PTR\ min \\ PTR\ min\ dom \\ PTR\ min\ cad \end{bmatrix} \tag{11.32}$$

In a data-driven approach, a higher number of non-correlated, yet information carrying features can lead to better results. However, focussing on the more informative ones seems reasonable. To this end, feature space optimisation by exhaustive group-wise elimination by WA is considered. As opposed to knowledge-based evaluation, data-driven evaluation requires a partitioning into training and test instances—ten-fold SCV is chosen to this end in the ongoing. This leads to mean values over all data instances. A comparison with correlation analysis results is therefore possible.

11.4.5 Performance

In the optimisation of parameters and methods, let us follow the chain of processing. As a first result, Table 11.8 shows key determination with and without tape speed variation compensation. The CLASSIC and MTV sets seem unaffected. However, the CHANSON and JAZZ ones containing several recordings first stored on analogue media show good improvement in WA.

Next is the influence of different filter functions for band-pass filtering of semitones. This is seen in Table 11.9, where the WA differs most for the CLASSIC set by a maximum difference of 6.8 % WA. For the other genres, the variation reaches 2.5 % at maximum. Least effective is the triangle filter. Between the remaining candidates, the Gaussian filter is preferred in the further results. The reason is its best results on the KEY-ALL set and the highest number of best WA across the other sets.

We will now take a look at the optimal time window from the beginning of a piece in Table 11.10. Lowest WA is observed for the first 30 s over all genres. This contradicts studies which recommend to focus on the beginning of a piece [91, 95] under the assumption that the 'home key' is present at the beginning and the key is easily determined thanks to a gradual addition of more and more instruments in typical introductions of a piece. Good choices are the longer alternatives of the first 90 s or the whole piece. Interestingly, the first 90 s lead to higher WA for the

CHANSON and MTV sets. This may be owed to transpositions of the chorus towards the end of a piece typical in these genres. In Classical and Jazz music, repeated changes to other keys such as the relative major/minor counter key are apparently better evened out looking at the overall piece. Given that on average over all genres the whole piece is the best choice, this variant is used here.

As a final parameter let us have a look at the effect of different frequency ranges for feature calculation in Table 11.11. In three of four genres, the range from C3–C8 is the best choice and thus used for the evaluations. At the higher end of the scale, the note C8 is two octaves above a human soprano singer's highest pitched note. Thus it appears that respecting higher harmonics seems reasonable in key determination. However, inclusion of the next octave up to C9 apparently degrades the results. An explanation for this behaviour can be seen in the weakness of higher harmonics in comparison to relatively stronger noise components from percussive sounds. At the other end of the scale, the optimum found coincides with a human tenor singer's range. It thus ignores lower bass components. This is different in the results for JAZZ where a benefit arises from an extension to C2. In fact, virtuoso bass solos are popular in this genre.

11.4.5.1 Evaluation of Feature Types and Performance

In addition to the WA of correctly classified keys, sub-dominant and dominant confusions for 12 keys are given and further the relative minor and relative major key confusions for 24 keys are added to the 'correctly' classified keys in the following. This adheres to the validation protocol introduced by the MIREX challenge in 2005.

We first look at 12 keys: Data-driven results (cf. Table 11.12) base on SVM in ten-fold SCV. This includes results per feature group and such for an 'optimised space' by supervised feature selection (cf. Sect. 11.4.4).

As single feature group, CHROMA features lead to the best result. There, single feature values 'clearly' represent the frequency characteristics of a musical piece. Within the derived feature types, these are partly 'blurred'. Further, WAs of derived features are generally lower. However, the additional features lead to better results when uniting all features—2.5 % WA absolute more than CHROMA—and also when selecting the best from the union of features—a further plus of 0.8 % WA and by that the overall maximum of 77.3 % WA. This difference is significant at the common level of 0.05 in a one-sided z-test.

Table 11.13 compares knowledge-based and data-driven key determination genre by genre. The optimal setting is chosen, each, for the two approaches, namely the 'scale cadence' features for correlation and the 'optimised space' for SVMs.

The correlation approach is superior in three out of four cases for the correct key. This changes, however, as more data for model-learning is available in the KEY-ALL case.

Switching to 24 keys, Table 11.14 first shows results for the data-driven approach with SVM in optimal parametrisation. Interestingly, no improvement is reached by space optimisation in this case. Given the double amount of classes available, the

Table 11.12 WA per feature type for data-driven SVMs: correct key (Key) and percentage of confusion with (sub-)dominant (Sub/Dom)

WA [%]	Key	Sub	Dom	Sum
All groups	76.2	7.3	9.8	93.3
Optimised space	77.3	7.1	9.8	94.2
CHROMA	73.7	7.5	11.2	92.4
Scale	59.6	13.3	18.1	91.0
Scale dom	55.2	15.0	19.6	89.8
Scale + cad	51.5	17.1	20.8	89.4
Chords	68.8	9.6	12.9	91.3
Chords dom	65.0	10.2	15.6	90.8
Chords + cad	54.4	15.6	20.4	90.4
PTR maj	68.5	8.8	13.8	91.0
PTR maj dom	63.7	9.0	16.5	89.2
PTR maj cad	56.7	14.4	19.0	90.1
PTR min	73.5	7.7	11.3	92.5
PTR min dom	72.5	8.5	11.5	92.5
PTR min cad	62.5	11.3	17.1	90.9

Database KEY-ALL, ten-fold SCV, 12 keys

Table 11.13 WA for correlation ('scale cadence' features) versus data-driven SVMs ('optimised space', ten-fold SCV) per genre

WA [%]	Key	Sub	Dom	Sum
Correlation				
MTV	74.5	9.0	8.5	92.0
CHANSON	70.5	12.8	9.5	92.8
CLASSIC	86.5	6.7	3.4	96.6
JAZZ	68.3	1.2	18.3	78.8
KEY-ALL	72.3	7.5	12.7	92.5
Data-driven SVMs				
MTV	73.0	8.0	10.0	91.0
CHANSON	72.5	8.1	10.1	90.7
CLASSIC	82.0	6.7	5.6	94.3
JAZZ	59.8	14.6	17.1	91.5
KEY-ALL	77.3	7.1	9.8	94.5

Correct key (Key) and percentage of confusion with (sub-)dominant (Sub/Dom), 12 keys

WA drops by roughly 15 % absolute to 62.1 % at maximum.Considering that pieces in major keys make up 71.5 % of the data, leaving only 28.5 % for minor keys, may explain the majority of confusions being in favour of relative major keys and almost none the other way round. This could be overcome by balancing. In this respect, interestingly, balancing by cyclic key-shift did not improve results [28].

In Table 11.15 a comparison is made for 24 keys as previously between the two assignment approaches in optimal configuration, each: 'PTR maj/min dominant' features for correlation and 'all' features for SVMs. SVMs prevail over correlation for

Table 11.14 WA per feature type for data-driven SVMs (no gain was reached by optimisation of the feature space)

WA [%]	Key	Sub	Dom	Sum	Min	Maj	Sum
All	62.1	4.0	6.7	72.8	2.9	9.2	84.9
CHROMA	59.6	4.6	7.9	72.1	1.2	11.0	84.3
Scale	43.3	10.4	13.1	66.8	0.0	10.0	76.8
Scale dom	42.1	11.3	12.7	66.1	0.0	9.6	75.7
Scale cad	40.0	11.9	14.0	65.9	0.0	8.8	74.7
Chords	49.4	6.5	11.2	67.1	0.0	13.3	80.4
Chords dom	46.2	8.7	12.1	67.0	0.0	11.0	78.0
Chords cad	41.9	11.2	13.5	66.6	0.0	9.6	76.2
PTR maj	48.8	6.3	11.9	67.0	0.0	13.5	80.5
PTR maj dom	46.3	8.3	11.9	66.5	0.0	10.8	77.3
PTR maj cad	42.5	9.2	12.5	64.2	0.0	9.4	73.6
PTR min	54.8	4.4	8.5	67.7	0.0	14.6	82.3
PTR min dom	53.3	4.4	10.0	67.7	0.0	14.2	81.9
PTR min cad	45.4	8.7	12.7	66.8	0.0	10.2	77.0

correct key (Key) and percentage of confusion with (sub-)dominant (Sub/Dom), and relative minor and major (Min/Maj). Database KEY-ALL, ten-fold SCV, 24 keys

Table 11.15 WA for correlation ('PTR maj/min dom' features) versus data-driven SVMs ('all' features, ten-fold SCV), subdivided per genre

WA [%]	Key	Sub	Dom	Sum	Min	Maj	Sum
Correlation							
MTV	46.5	5.0	11.5	64.0	14.0	0.5	77.5
CHANSON	66.4	8.0	10.8	85.2	0.0	2.0	87.2
CLASSIC	76.4	0.0	4.5	80.9	5.6	1.1	87.6
JAZZ	34.1	6.1	13.4	53.7	17.1	2.4	73.2
KEY-ALL	55.4	5.2	10.4	71.0	9.0	1.3	81.3
Data-driven SVMs							
MTV	52.0	3.0	9.5	64.5	7.0	9.0	80.5
CHANSON	67.8	8.7	12.8	89.3	0.0	0.0	89.3
CLASSIC	68.5	10.1	4.5	83.1	3.4	6.7	93.2
JAZZ	46.3	8.5	7.3	62.1	9.8	6.1	78.0
KEY-ALL	62.1	4.0	6.7	72.8	2.9	9.2	84.9

WA for the correct key (Key) and percentage of confusion with (sub-)dominant (Sub/Dom), and relative minor and major (Min/Maj), 24 keys

the correct key in all cases but the CLASSIC set. The JAZZ set benefits most from this trend. Interestingly, WA is more balanced across genres for the data-learnt method.

11.4.6 Summary

Within this section performance of musical key determination on originally recorded and MP3 encoded popular, Chanson, Classical, and Jazz music was demonstrated.

The main approaches by template correlation-based and data-driven modelling were opposed and evaluated on novel feature types. The data-driven model prevailed at a maximum of 77.3 % WA for 12 keys for the whole dataset. For correct recognition of six out of seven scale semitones, 94.2 % WA were reached. For individual datasets, the correlation approach partly showed better results, but SVMs were superior given sufficient data due to the ability to better cope with diversity: Perceptual studies of tonal hierarchies show genre and task dependency according to [123]. In the case of 24 keys the difference between these two approaches was amplified from 5.0 and 6.7 % absolute difference in WA. 62.1 % was the maximum WA for the correct key and 84.9 % WA for six out of seven notes.

As for parametrisation, an optimum has been found for adapting reference pitch classes to compensate for tape speed variation, using Gaussian filters for semitone filtering, analysing the whole piece, and using the frequency band from C3 to C8 or 130.8 to 4 186 Hz, respectively, for feature computation. The proposed feature types based on music theory and human perception were able to improve both approaches for key assignment.

Future design of features for key determination could consider non-CHROMA types such as bags of chords. In addition, further music theoretic or cognition inspired approaches, e.g., inspired by [124] could be targeted. For the acoustic features, the time-frequency representation could be improved, e.g., by wavelets [71, 125] or multi-resolution FFT. If one targets the mode instead of the 'absolute' key [126], hierarchical schemes could be established. Non-tonal music audio could be modelled as an additional class to cope with arbitrary music input [127]. Also, alternative minor scales apart from the considered natural relative minor scale can be added. In [128], PTR is given for harmonic and melodic minor scales which could be implemented directly in the presented approach.

Extending to pieces with changing key can be achieved based on local analysis [129]. Chunking for such local analysis could be based on beat and on-beat detection [6, 23] as presented in the previous two sections. Further, temporal context can then be integrated by the use of LSTM networks [23]. Further, the novel features could be used in related tonal analysis tasks [10], use key analysis to improve music structure analysis [30, 130], or exploit synergies by parallel key and progression analysis [131] or similar mutually dependent information [99]. Finally, the results demonstrate the complexity of key determination, and confidence measures and key hierarchies can be useful considerations for application in real-life systems.

11.5 Chords

A more fine-granular description beyond the musical key is provided by the chord progression in music. In the following, the method as presented in [10] and [29] is explained and benchmark results are presented.

To classify a chord, only the pitch classes, i.e., the note names without octave number, of the notes involved are relevant. A variety of different chord types exists and is characterised by the size of intervals between notes of the chords.

The automatic recognition and transcription of musical chords and their progression has manifold application potential:

In spontaneous improvisation sessions of musicians such as 'jams', the progression can be analysed and stored as a lead sheet, or media players can automatically identify and show the current chord in a musical piece for play-along—by humans or even the computer. Knowledge of the chord structure can also be used as meta-information in MIR tasks. A good example is genre recognition, as certain genres prefer typical progression patterns (e.g., Jazz: second, fifth, tonic successions or Blues: tonic, fourth, fifth as dominant sept chord successions).

Another example is musical mood recognition (e.g., ratio of major and minor chords—this will be shown in Sect. 11.7). Obviously, also key recognition can benefit from this information—and vice versa, which is why a simultaneous key and chord analysis seems promising. Structure analysis, e.g., for chorus retrieval [30] (cf. Sect. 11.6) can also be based on the chord progression, as it often differs between different parts of a musical piece such as verse, bridge, and chorus. Moreover, DJs can be provided with automatic on-line synthesis of chord matching notes as very low sub-basses or arpeggios, and with tools that allow to blend music with matched chord structure.

Finally, music similarity analysis, e.g., for plagiarism retrieval can be based on chord information. As an example, the chord progression of Johann Pachelbel's "*Canon in D*" ("*Canon per 3 Violini e Basso*"), is found in multiple contemporary popular songs, such as "*Go West*" first by the Village People, later covered by the Petshop Boys, or Ralph McTell's "*Streets of London*", The Farm's "*All Together Now*", Green Day's "*Basket Case*", Mattafix's "*Big City Life*" or Juanes's "*Volverte a Ver*".

To save labour-some manual labelling, an automatic beat-synchronous and data-driven approach is introduced here. The approach bases on the findings for tempo determination and key determination described in the previous sections. Early automatic chord recognition was based on pitch class profiles [100] (cf. Sect. 6.2.21). Later, HMMs were proven highly suited, e.g., in [104, 117]. Obviously, context modelling can improve the recognition rate [132], as chords tend to follow chords with certain properties such as neighbourhood in the circle of fifths (cf. Sect. 11.4). Exploiting these bases, results on realistic data are shown including a progression LM trained on a large corpus of 16 k songs to show reachable results on a database of mixed original recordings.

11.5.1 ChoRD Database

The *Chord Recognition Database*, respectively *ChoRD* database was introduced in [29].

Table 11.16 Distribution of the keys and chords in the ChoRD database

Root	# Key	# Major	# Minor	# Other
A	7	511	459	57
A#	8	567	171	86
B	7	480	213	61
C	16	854	278	105
C#	5	312	315	61
D	3	557	349	94
D#	8	533	141	61
E	12	643	362	21
F	13	728	272	52
F#	4	407	209	44
G	12	719	287	103
G#	5	353	196	41
Sum	100	6 664	3 252	786

To provide sufficient data for machine learning and testing, a total of 100 musical pieces—a representative variation of typically aired pop and rock music—was annotated with the tempo in BPM, the key, and each chord based on original scores as ground truth reference by three experienced musicians. The set contains 64 different artists, and on average, 1.6 pieces per artist. 18 artists are found more than once in the set: five songs are contained of each of Delta Goodrem, James Blunt, and Robbie Williams, followed by four songs, each, of Celine Dion, Coldplay, and Enya, three songs, each, of Bon Jovi, Bryan Adams, Cher, and finally two songs, each, of All Saints, Backstreet Boys, Britney Spears, Keane, Phil Collins, Roxette, and The Corrs.

All pieces have constant tempo. The complete list of songs is available for download.[4] The original recordings were compressed to 128 kbit/s MP3, and the total playtime is 6 h 58 min 12 s. 10702 bars are contained overall. The seven chord types annotated are major, minor, suspended second or fourth, augmented, diminished, and 'power chords'—i.e., the typical combinations of root and fifth with second, third, fourth or no further interval. Rather than seven chord types times 12 notes only $6 \cdot 12 + 4 = 76$ final chord classes were obtained, as only four different augmented chords exist. Due to sparseness of certain types, cover classes were used as follows: 36 major / minor / other chords (where other chords are augmented, diminished, power, and suspended), and 24 major / minor chords. The total of chord instances was kept constant throughout mapping by mapping according to the root and musical function in the context such as "Cno3" ("C" as power chord, i.e., without a third) onto "C" if the piece is in the key of C major or onto "Cm" if the piece is in the key of A major. Table 11.16 shows the frequencies of keys and chords within the ChoRD database by root note for the classes major, minor, and others.

[4] http://www.openaudio.eu/chord.txt

11.5.2 Methodology

Processing of the audio starts with conversion from MP3 to a monophonic, 44.1 kHz, 16 bit wave. MP3 compression was at first carried out to ensure a typical use-case scenario with higher realism in the sense that the algorithm can work on music delivered in a lossy compressed format. Then, the tempo, metre, and downbeat position, i.e., the position of the first beat of a measure[5] are determined by the comb-filter based approach described in Sect. 11.3. A musical piece is then partitioned into consecutive bars according to its tempo.

As features serve 12-dimensional CENS vectors per bar (cf. Sect. 6.2.2.3). The extraction chain of processing includes spectral transformation, dB(A) weighting for modelling of human frequency-dependent loudness perception according to norm IEC/DIN 651, compensation of tape speed variation and the mapping to pitch classes. The dB(A) weighting and pitch tuning are non-standard but reasonable steps and executed as described in Sect. 11.4.2. During tape speed variation compensation, the prominent frequency of a long-term analysis in the range 130 Hz–1 kHz is computed as was exemplified for key determination in Sect. 6.2.2.3. Then, the semi-tone filter-bank is shifted to the nearest reference frequency of the prominent one.

To model the neighbouring context of a chord instead of recognition of single chords on their own, a chord language model can be used in addition. In order to train the LM in the exemplary system, all chord lead sheets retrieved automatically from the On-Line Guitar Archive[6] were used after removal of doubles. Such sheets are usually made by users. Owing to this fact, they may be erroneous, simplified, e.g., by intention for easier playability, or transposed into easily playable keys on guitar such as G major which usually does not demand for the more complicated "barré" fingering patterns of chords. For the statistical chord language model, however, one is primarily interested in typical chord successions. The lead sheets often contain the chord succession only once. Thus, up-sampling by the following rule was used: Based on the mean of 60–100 bars for a typical rock or pop piece, the chord succession was repeated for pieces with less than 30 bars until 60–100 bars were reached.

Further, rule-based parsing such as clustering of different spelling variants of the same chords, elimination of bass-notes or of such being neither tonic, second, third, fourth, or fifth as well as the mapping rules for 'other' chords as explained above was used on the language model level as well. 19025 songs, and a total of 1573803 chords were used for the final model. In Table 11.17 the top-ranked uni- and bi-grams are shown by frequency of occurrence in the chord language model.

On the acoustic layer, a cross-correlation with a hard template serves as reference for an approach that is not based on data-learnt models. This happens in full analogy to the method of key determination described in Sect. 11.4.3. In the template, a "1" is used per note contained in the target chord, and a "0" marks out-of-chord notes.

[5] The term downbeat stems from orchestral conducting: The lowest point on the baton signals the downbeat.

[6] http://www.olga.net

Table 11.17 Top-ranked chord uni- and bigrams in the LM by frequency of occurrence

Rank	1-gram	#	2-gram	#
1	G	244 820	D-G	57 500
2	D	227 549	G-G	55 106
3	A	198 958	C-G	54 702
4	C	188 194	G-C	54 040
5	E	130 896	A-D	46 162
6	F	87 741	D-A	43 534
7	B	72 360	G-G	41 090
8	Am	58 929	A-A	40 161
9	Em	57 537	D-D	39 710
10	A#	32 583	E-A	36 659

Table 11.18 WA for the ChoRD corpus, LOSO evaluation

WA[%]	Correlation	SVM	HMM	HMM + LM
24 major / minor	39.41	40.24	58.57	60.13
36 major / minor / other	28.37	36.71	45.39	48.84

'Other' chords cover augmented, diminished, power, and sustained chords

As alternative data-driven processing methods, we compare SVMs to HMMs with and without the language model. A linear kernel, pairwise multi-class discrimination, and SMO learning proved as best choice for SVMs. For HMMs, one continuous model with one emitting state per beat was used. The models were trained with 20 Baum-Welch iterations [133]. A single Gaussian mixture component was the best choice. To enable Viterbi search for decoding, a 'word-loop' context free grammar modelled the chord sequence in the case where no data-driven language model was used. On the other hand, when the language model is enabled (HMM + LM), Laplace smoothed class-based Katz back-off-bigrams with a cutoff of one were found as best configuration.

11.5.3 Performance

A song-independent cyclic 'leave-one-song-out' (LOSO) training and testing was chosen as evaluation strategy under realistic conditions. Table 11.18 depicts observed WA for the different data-free and data-learnt chord determination strategies.

One notes that with increasing data inclusion on the AM and LM level and context modelling, the WA is increased. By that, HMM exceed SVM as they allow for contextual modelling. The mapping to and by that reduction to major and minor chords leads to higher WA despite still handling 'any input', if this appropriate in the context of the application.

11.5.4 Discussion

A method for fully automatic labelling of music chords was shown in this section. The feature extraction stage processes the audio beat-synchronous, and compensates for tape/playback speed variations. A chord progression modelling by a statistical chord language model is performed. The method was shown to be superior to 'open-domain' knowledge-driven cross-correlation and analysis of isolated bars. 60 % WA were reached on a mix of original MP3 compressed songs from diverse artists and genres. The difficulty of the task varied across genres: Songs, such as Enya—"*Silver Inches*" were recognised without mistake, while Prince—"*Purple Rain*" had the highest error rate with only every fourth chord determined correctly.

As additional advantage of the beat-synchrony the output can directly be turned into a lead-sheet. Future efforts should aim at the investigation of benefits arising from the use of source separation and stereophonic beam forming for the enhancement of the accompanying instruments over the noise and drum parts.

From an architectural point, BLSTM networks that allow for the modelling of knowledge of the whole song for every chord decision could be employed. To improve the reference by correlation, one could also consider perception-based and music theoretic variations of the templates, as were shown in Sect. 6.2.2.

On an even more fine-granular resolution, single note events can be targeted—the automatic transcription of music. Results and new approaches to this end were presented recently in [37], for example.

11.6 Structure

Apart from rhythm, metre, note events or chord changes, the structure of a musical piece can be of interest, such as the positions in time of the chorus section. In this section, we will highlight the findings presented in [30] and [31].

'Music thumbnails', i.e., the most mnemonic part of a musical piece, are precious in many applications such as 'teaser' creation in (on-line) music stores and radio stations. Teasers are a short part of the song that is very characteristic of it. More applications are the provision of samples to DJs or provision of samples for efficient browsing and arranging of large music archives [134] or on mobile devices with limited display space. Further, query by example systems (e.g., [41]) can exploit pre-extracted thumbnails for similarity matching. Today, such thumbnails are usually generated manually as sufficiently robust methods are still lacking.

Generally, highly repeated—preferably vocal—parts such as the chorus sections tend to be the most mnemonic part of musical pieces [135]. Thus, the following approach aims at localising the chorus by combining methods and approaches presented so far in the literature. The approaches towards localisation of the audio 'thumbnail' or general structure analysis in music can roughly be divided by the feature nature they are based upon and by the kind of further approach, such as calculation and analysis of self-similarity matrices (cf. Sect. 11.3) or segmentation with a subsequent clustering or classification.

The authors in [135] propose a modulated complex transformation logarithmised and reduced by oriented PCA for clustering of similar sequences. The clusters are then classified into different parts of a piece by scaled Renyi entropy and spectral flatness. In [136] MFCCs are used for clustering by modified KL distance. Another approach in the same article [136] uses ergodic HMMs for structure analysis. Ergodic HMMs are also used in [137]. These have three states. As features serve the spectral envelope with MFCCs, LPCCs and discrete cepstrum coefficients. A chunking can take place by a clustering algorithm to initialise the ergodic HMMs. GMMs initialised by a clustering step are applied in [138]. In [139], the music signal is chunked by an event detection function prior to dynamic time warping (DTW). Music visualisation by a self-similarity matrix for structure analysis was first based on MFCCs using the scalar product [140], and later using a normalised scalar product [141]. The authors in [142] use dynamic features which maximise the trans-information for computation of such a self-similarity matrix.

In [143] an unsupervised Bayesian clustering model is used. Its parameters are estimated by a modified EM algorithm. The authors in [120] perform a beat synchronous segmentation using a beat-tracker. Then, a self-similarity matrix is established based on CHROMA features. By uniform moving average filtering, a time-lag matrix is computed. Its maximum element is determined within limitations of the minimum lag and the maximum occurrence of a section. An extension is presented in [144]. It permits modulated repetitions and an adapted measure to determine the chorus sections. The authors of [145] suggest features based on harmonic information for the creation of self-similarity matrices.

From the above a number of findings can be distilled: In pre-processing, beat-synchrony seems advisable given robust beat detection. As for features, one should model the musical properties of the signal such as by PCPs or more specifically CHROMA, as these tend to be better suited than MFCCs or similar types. Further, temporal information should be modelled as by CENS features or similar [145]. As for the model, self-similarity matrices seem best suited. Given reliable beat-synchrony, dynamic modelling is not needed or might even downgrade results. In the remainder of this section, we consider a solution following these guide-lines and incorporating simple image processing methods for the processing of a self-similarity matrix. We will also need to define evaluation measures which are not settled for this task. Exemplary results will be given on a full day of MP3 compressed recorded music from multiple styles.

11.6.1 Methodology

As in the last section on chord progression analysis, CENS features (cf. Sect. 6.2.2.3) are used for the acoustic representation. These will be denoted from now on as

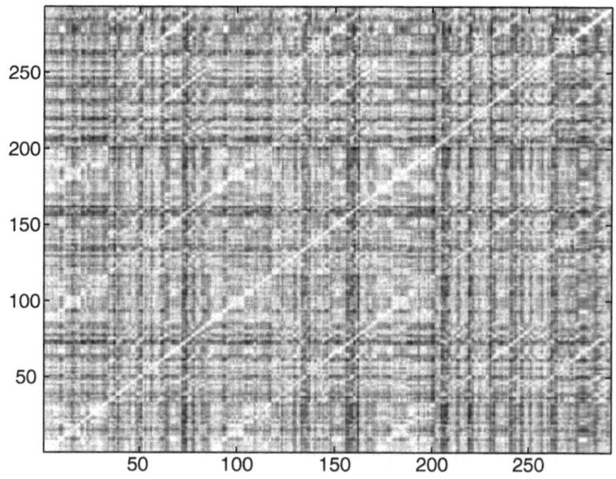

Fig. 11.15 Self-similarity matrix for Adriano Celentano—"*Azzurro*". Bright 45° diagonals indicate a high self-similarity [31]

$\underline{x} = (x_1, \ldots, x_{12})$. Further, beat-tracking as was described in Sect. 11.3 is used for beat-grid alignment.

An $N \times N$ self-similarity matrix $\underline{S}$ is calculated (cf. Sect. 11.3) based on the cosine distance [141] as follows:

$$\underline{S}(i,j) = \frac{\langle \underline{x}(i), \underline{x}(j) \rangle}{\|\underline{x}(i)\| \cdot \|\underline{x}(j)\|}. \tag{11.33}$$

If this matrix is visualised as heat-map, one now searches 'bright' diagonal segments parallel to the main diagonal at a 45 degree angle (cf. Fig. 11.15) in matrix $\underline{S}$ which indicate highly self-similar segments in a musical piece. To locate these, an edge filter can be used as given by

$$F_{Diag}(i,j) = \begin{cases} 1 & \text{for} \quad i = j \\ c & \text{for} \quad 0 < |i-j| \leq b \\ 0 & \text{for} \quad |i-j| > b \end{cases} \tag{11.34}$$

with $1 \leq i,j \leq 20$, $b = 5$ and $c = -\frac{2}{17}$. Then, a normalisation follows and a threshold δ is subtracted from the filtered 'image'. This results in the matrix $\widehat{\underline{S}}$ and is carried out for reduction of noise introduced by the edge filter. The threshold δ can be chosen as the highest value exceeded by at least $10 \cdot N$ values in the filtered 'image'. Next, a binary matrix $\underline{S}_b$ is computed by

$$\underline{S}_b(i,j) = \begin{cases} 1 & \text{for} \quad \widehat{\underline{S}}(i,j) > 0 \\ 0 & \text{for} \quad \widehat{\underline{S}}(i,j) \leq 0 \end{cases}. \tag{11.35}$$

Now, regions of interest (ROIs) are found. Start and end of potential chorus sections are determined as follows: Let $d(i, j)$ be the temporal derivative along a diagonal segment $\underline{S}_b(i+1, j+1) - \underline{S}_b(i, j)$. Then, segment bounds are estimated by setting start points at i and at j if $d(i, j) > 0$ and corresponding end points at i and j if $d(i, j) < 0$.

To improve these initial values, let us define a counter c_k per segment k and a threshold δ_{sim} corresponding to the highest value exceeded by at least $0.1 \cdot N^2$ entries of the matrix $\underline{S}$. Starting from the middle (m_x, m_y) of each segment, c_k is incremented if $\underline{S}(m_x, m_y)$ falls below δ_{sim}, and decremented if it exceeds δ_{sim} up to a minimal counter value of $c_k = 0$ [30]. In the case that c_k is smaller than a threshold C, the next value $(m_x, m_y) := (m_x - 1, m_y - 1)$ is processed. Otherwise, one aborts and the current start $(m_x + C, m_y + C)$ is stored. This is repeated for the other direction with $(m_x, m_y) := (m_x + 1, m_y + 1)$, again proceeding from the middle of each segment k, but aiming at the corrected end $(m_x - C, m_y - C)$. $C = 4$ has proven to be the optimal choice in the experiments described here. Figure 11.16 exemplifies this approach.

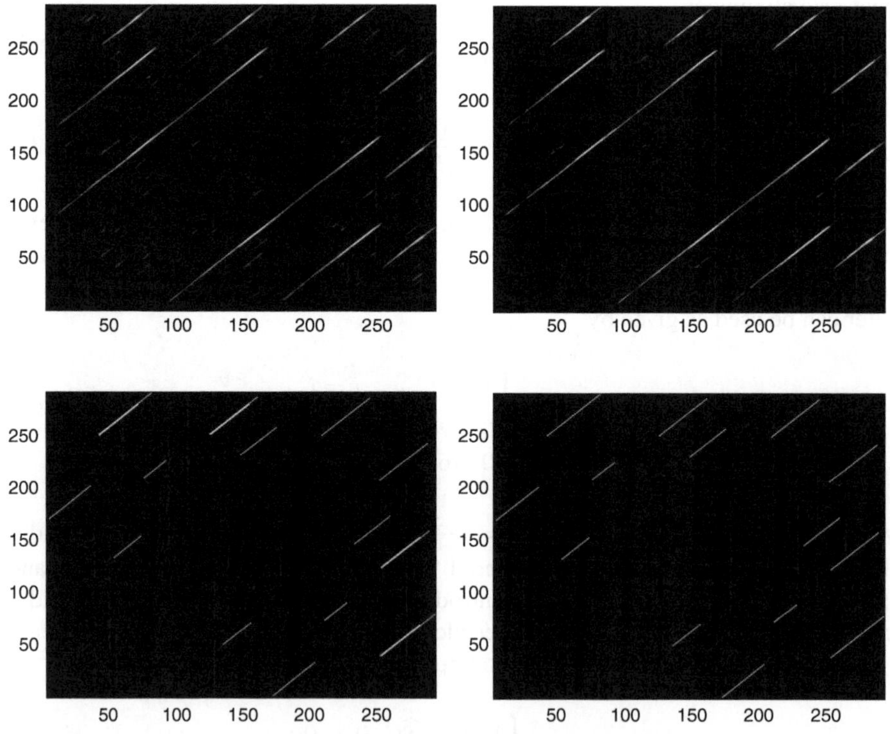

Fig. 11.16 Self-similarity matrix after the different steps of processing: First, edge filtering (*top left*), then, dynamic thresholding (*top right*). From the resulting matrix $\widehat{\underline{S}}$, respectively its binary representation $\underline{S}_b$, ROIs are determined by length and characteristics of each segment (*bottom left*). Last is the combination of adjacent segment (*bottom right*) [31]

Imposing lower and upper limits for the segment length l helps to reduce the number of ROIs. A 'good' dynamic lower bound was observed to be

$$l \cdot m_{sim} > 8.7\,\mathrm{s}, \tag{11.36}$$

where m_{sim} denotes the mean similarity of the segment in the self-similarity matrix $\underline{S}$. The value is a good trade-off to avoid detection of short re-occurring sections that are not relevant in the search of the chorus, such as (drum) fills, etc. Further, a static upper bound of 29.1 s helped to avoid the choice of longer non-chorus sections—usually the verse or verse plus chorus. Finally, adjacent segments are combined given that they do not provide further information.

The music 'thumbnail' is then the remaining segment with highest mean similarity m_{sim}. The rationale behind is the assumption that the chorus sections are the most similar sequences within the ROIs. Three best such segments are kept in the ongoing as a musical piece might be better characterised by more than one and the chorus might not be the first best hit. Optionally, these can be aligned to the automatically located beats—ideally to the on-beat, unless it is 'too far' from the assumed start of the thumbnail.

11.6.2 Performance

A set of 360 pieces of different genres with a total playtime of 23 h 47 min was used for computation of benchmark results. 250 pieces are divided in five genres with 50 songs, each, within Electronic Dance, Pop, Rock, German Folk, and Oldies. 110 pieces add to the Rock music and 'Oldies'.

For performance assessment, the start points of automatically found chorus thumbnails are compared with the gold standard of manual annotation. A tolerance of deviations with the time T_{max} is introduced when comparing these two. Another relaxation is the allowance of the correct position to be within the 'Top N', i.e., within the N best assumed positions. Table 11.19 contains the results for $T_{max} = 1, 2, 3\,\mathrm{s}$. The majority of 'wrong' thumbnails in the sense of the manual annotation resemble other characteristic parts of a piece—in particular the chorus section with higher temporal deviation than T_{max}, a chorus variant which is not the typical chorus repetition, the verse, or the bridge. In the literature, quantitative benchmarks by deviation of

Table 11.19 Correctly determined chorus thumbnails within the top one, two or three best candidates, and different tolerance of time deviation

T_{max}[s]	Top 1 [%]	Top 2 [%]	Top 3 [%]
1	22.6	37.8	45.8
2	48.6	67.2	73.3
3	60.6	76.1	81.4

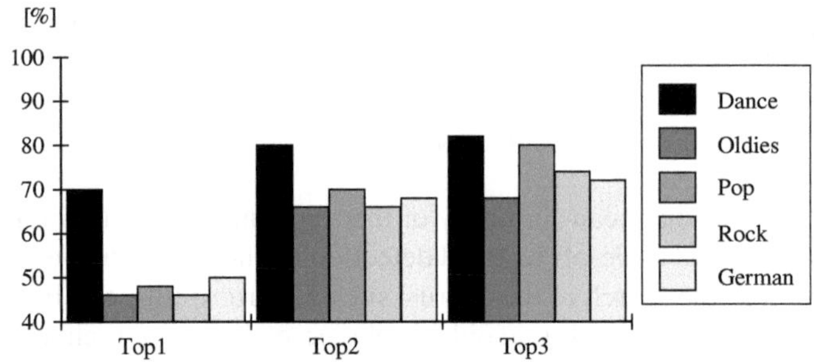

Fig. 11.17 Correctly located chorus thumbnails by genre for the maximum deviation $T_{max} = 2\,\text{s}$ [31]

automatically generated thumbnails from the actual chorus sections are lacking—evaluations are mostly of perceptive nature with individual listener ratings.

Figure 11.17 shows results by genre. Visibly, the task is best solved for electronic dance music. This can be explained by the high similarity present in this genre given the 'perfect' electronic tones and computer aided sequencing. We could further speculate that the structure is less complex and less variations exist.

11.6.3 Summary

Within this section automatic generation of music thumbnails was shown. The approach mainly based on a self-similarity matrix established on chromagram-type features in combination with basic methods of image processing to locate diagonals. In addition, beat positions were used as information. Best results were observed for electronic dance music: There, the chorus location was determined correctly in 70 % of the pieces when allowing for a maximum deviation of 2 s. Averaged over all considered genres, this value dropped to 48.6 %.

Future efforts could incorporate analysis of key changes [115] (cf. Sect. 11.4), chord patterns [29, 105] (cf. Sect. 11.5), or by classifying vocal and non-vocal sequences [9] (cf. Sect. 11.8). Obviously, machine learning could also be introduced as well as alternative matching techniques such as in [146].

11.7 Mood

So far, we dealt with measurable characteristics of music. In the following section, we take a look at music mood classification (cf. [32])—similar to the analysis of emotion in speech (cf. Sect. 10.4.2). While we will be looking at mood classes in a discrete way, a natural extension is to model continuous dimensions, as was later shown in [33].

Rather than choosing music by artist or album one sometimes wishes for music that 'fits the occasion' or one's mood, such as when jogging, relaxing, or perhaps having dinner for two. Thus, tags such as 'activating', 'calming' or 'romantic' would be of help in music retrieval [147, 148]. Manual annotation by individual users seems rather labour intensive, but some services exist that provide such tags such as Allmusic,[7] often based on several users' ratings. Regrettably, this information is not always reliable, as the tags are often only attached to artists rather than to single tracks. This leads to the desire of automated mood classification of music. In this section, we will thus have a look at audio features suited for this particular task, and benchmark results reachable with state-of-the-art approaches under real-world conditions—without pre-selection of instances, e.g., by limiting analysis to those with majority agreement of annotators.

Features for mood recognition can be extracted from the raw audio stream, but also from metadata. Those derived from the audio can be added by mid-level ones basing on pre-classification. This means that, apart from the LLDs and functionals as introduced in Sect. 11.6, knowledge from other classification tasks such as the ones introduced for music processing in this chapter can be used as mid-level feature information describing concepts such as rhythm or tonal structure. Metadata on the other hand includes all types of textual information available on a music track such as title, artist, genre, year of release or lyrics.

In the literature so far some commonalities are visible: In [149] a 30 element feature vector containing timbre, pitch, and rhythm information is used. The work in [150] employs timbre features by spectrum centroid, bandwidth, roll off, and spectral flux, and seven octave-interval sub-bands' minimum, maximum, and average amplitude plus RMS energy. For rhythm information the lowest sub-band was used. Edge detection with a Canny estimator led to a rhythm curve. In this curve peaks are assumed to indicate bass instruments' onsets, and their strength as indication for the degree of rhythm presence. Further, analysis by ACF serves as measure for rhythm steadiness, and the common divisor of the correlation peaks for the tempo. In [151] an extension is presented for rhythm analysis by addition of all sub-band onset curves. The authors of [152] also use rhythm and timbre features: Two tempo candidates in BPM are based on peaks in a beat histogram ACF. From this histogram amplitude ratios and sum of its ranges are added. Timbre is based on 13 MFCCs [153] and spectral centroid, flux, and roll off. Mean and standard deviation of the features over all frames were also included. In [154]—a MIREX 2008[8] audio mood classification task contribution—MFCC, CHROMA, and spectral crest and flatness describe whether the signal spectrum contains peaks, e.g., in case of sinusoidal signals or it is flat indicating noise.

The learning algorithms vary strongly for this task, just as the mood taxonomies do (cf. Sect. 5.3.2). In fact, the diverse mood models certainly influence the selection of the learning algorithm. As an example, in [150, 151] a four-class dimensional model is handled by GMMs as basis for a hierarchical classification system (HCS):

---

[7] Allmusic (http://www.allmusic.com)

[8] MIREX 2008 (http://www.music-ir.org/mirex/2008)

A binary arousal classification by rhythm and timbre features is first. Then, valence is classified by different features.

In the following results, popular music of the NTWICM database is analysed. Thereby feature extraction is based on the entire duration of musical pieces. Real-world conditions will be emphasised by using only meta information as available on-line, and processing music that has been coded by lossy MP3 compression.

11.7.1 Methodology

Let us first detail out features by type. These comprise spectral, rhythmic, and tonal audio low level descriptors and mid-level features such as pre-classified chords, and features based on information retrieved from public databases.

11.7.1.1 Audio Features

The music is decoded and converted to mono. Then, a FFT is computed [155], and a number of selected functionals is applied: centre of gravity, standard deviation, skewness, kurtosis, and band energies and energy densities for seven octave based frequency intervals (0–400 Hz, ..., and 6.4–12.8 kHz).

The rhythm features used in this section base on those presented in Sect. 11.3. Tempo estimation is performed in two steps: First, the Tatum tempo is estimated. To this end, the pre-processed music is fed into 57 comb filters, and the filter outputs form the unnormalised Tatum vector $\underline{T}'$. Then, the 82 features as in Sect. 11.3 are computed. They are augmented by the following five derived features:

- The main tempo θ_B is computed based on the metre vector $\underline{M}$. In principle, the tempo resonating best with the musical piece is decided for.
- The tracker tempo θ_{BT} is the main tempo refined by beat-tracking. Ideally, θ_B and θ_{BT} should vary not at all or only slightly owing to rhythm inaccuracies.
- The base metre M_b and the final metre M_f are the estimates on duple or triple metre of the musical piece.
- The tatum maximum T_{max} equals the maximal entry of $\underline{T}'$.
- The tatum mean T_{mean} equals the mean value of $\underline{T}'$.

Overall, this leads to 87 additional features based on Tatum and metre vector elements.

11.7.1.2 Chords

Each musical chord (cf. Sect. 11.5) type can be associated with moods induced in or perceived by human listeners owing to its characteristic sound. Examples for frequent chord types are shown in Table 11.20 following [156].

Table 11.20 Chord types and emotions associated with these [156]

Chord type	Example	Associated emotions
Major	B	Happiness, cheerfulness, confidence, satisfaction, brightness
Minor	Bm	Sadness, darkness, sullenness, apprehension, melancholy, depression, mystery
Suspended fourth	Bsus4	Delightful tension
Major seventh	B^7	Funkiness, moderate edginess, soulfulness
Minor seventh	Bm^7	Mellowness, moodiness, jazziness
Major Major seventh	B^{maj7}	Romance, softness, jazziness, serenity, exhilaration, tranquillity
Ninth	B^9	Openness, optimism
Diminished	Bdim	Fear, shock, spookiness, suspense
Seventh, Minor ninth	$B^{7/9\flat}$	Creepiness, ominousness, fear, darkness
Added ninth	B^{add9}	Steeliness, austerity

Table 11.21 Recognised chord types

Chord type	Example
Major	$D^\sharp$
Minor	Em
Major seventh	C^7
Minor seventh	Am^7
Major Major seventh	$F^{\sharp maj7}$
Minor Major seventh	$C^\sharp m^{maj7}$
Augmented	A^+
Diminished	Fdim
Diminished seventh	$Edim^7$

For chord determination in the original music file, a fully automatic algorithm [157] is used as was explained in Sect. 11.5. It basically compares the chromagram with predefined chord templates (cf. Sect. 11.4), and outputs the chord type (e.g., major, minor, diminished) and the chord base tone (e.g., C, F, G♯).

As chord features, 'bag-of-chords' is used with the frequency of occurrence of a chord normalised to the total number of chords in a musical piece. Overall, 22 numeric features are obtained, of which the last simply is the number of recognised chords (cf. Table 11.21 for those recognised).

11.7.1.3 Metadata

Rich meta-information for all music in the NTWICM database is hard to obtain given its large size (cf. Sect. 5.3.2). Thus, it is limited to the artist, title, and year of release which is available for each song. The year of release is used 'as is' as a numeric feature. As for artist and title, by standard word delimiters text strings are chunked to words. Then, the Porter stemming algorithm [158] is used and binary BoW features are generated (cf. Sect. 6.3). A minimum term frequency helps to keep the number of generated features in reasonable limits. As for the NTWICM database, generated artist features appear database-specific—just as one might expect. In addition, they are not too helpful in an artist independent evaluation. As opposed to this, title-based features tend to—intuitively—partly contain more direct relation to mood including words such as "feel", "love", or "sweet". Overall, the meta-data features comprise one numeric year feature, and 152 binary numeric word occurrence BoW-type features.

11.7.1.4 Lyrics

From the recognition of emotion in spoken language, it is known that the spoken content also bears information on the affect besides the acoustics of the speaker [159, 160]. Thus, the 'sung language' is considered as feature source, yet, based on lyrics as retrieved from the Internet. Two different strategies of lyrics-to-feature conversion are followed as were introduced in Sect. 6.3.

First, a knowledge-based method is based on the use of *ConceptNet* [161, 162], as was detailed in Sect. 6.3, and as was shown comparably effective for valence prediction in film reviews in Sect. 10.4.1 [162]. The subset of concepts is classified into one of the 'big six' emotional categories [163] (anger, disgust, happiness, fear, sadness, and surprise). Now the emotional affect of unclassified concepts that are extracted from the song's lyrics can be calculated by finding and weighting paths which lead to those classified concepts. To give an illustrative example, the output for Cutting Crew—"*(I Just) Died In Your Arms*" is: ('sad', 0.579), ('happy', 0.246), ('fearful', 0.134), ('angry', 0.000), ('disgusted', 0.000), and ('surprised', 0.000).

This probability information is used as features—one per one of the six emotions. In addition, six features contain the first, second, etc., ranked emotion. Further variants would exist: In [164], arousal and valence probabilities are used directly, but the vocabulary was more limited than here. Also, the other alternatives which exploit further on-line knowledge sources – as were described in Sect. 10.4.1—could be used.

In addition, a data-learnt method as for the meta-information is considered, as was also observed well suited for sentiment detection in Sect. 10.4.1. Again, the raw text is chunked into words and punctuation is deleted. Then, Porter stemming follows (cf. Sect. 6.3.1), and one BoW feature is generated per word stem outside the stop-word list with a minimum term-frequency of ten per class in the database. A binary representation for the BoW features is chosen, but normalised to the number of terms in the current piece's lyrics to model the prevalence of the current term in relation to the text length. It lies in the nature of BoW that the word order is ignored. However, BoNG or BoCNG (cf. Sect. 6.3.2) would require considerably more learning material than the roughly 3 k instances which NTWICM offers, in relation to the 'explosion' of the feature space dimensionality. In the case of the 100 k instance Metacritic database, however, this was possible (cf. Sect. 10.4.1). Despite the typically higher number of terms in a song's lyrics in comparison to the key-statement of a film critic no improvement was thus observed when opting for BoNG or BoCNG.

Table 11.22 summarises all feature subsets as were presented above. A subset 'No-Lyrics' will be used in the ongoing as well, to compare the influence of lyrics processing in comparison to the information that comes directly from the audio and artist, title, and year information of a song. It has to be noted at this point that 25 % (675) of the pieces in the NTWICM database have no lyrics included as these were not contained in the two used on-line lyric databases—they were left as they are (empty BoW vector), owing to the philosophy to stick with realism as given for a working system in a typical use-case.

11.7.2 Performance

Given the imbalance of instances across classes in the NTWICM database (cf. Sect. 5.3.2 balancing is reasonable to avoid a bias towards class throughout classifier learning. This was realised by random up-sampling up to perfect balance with the default random seed in Weka [122] (cf. Sect. 7.5.1). This required a target size of 200 %.

As classifier serve SMO-trained SVMs with pairwise multi-class discrimination, linear kernel, and a complexity c of 1.0 at first. The complexity was optimised on the development set of NTWICM on the A3 and V3 tasks (cf. Sect. 5.3.2) and $c \in \{0.5, 1.0, 1.5, 2.0, 2.5\}$. Higher order polynomial kernels did not lead to an improvement in terms of UA and WA. For reference, performance by RFs will also be shown.

All results are provided by UA and WA.

First, a feature selection from the 691 overall features was carried out to reveal promising features and reduce the complexity for the classifier. For easily interpretable feature analysis results, the groups as shown in Table 11.22 are evaluated individually by classification with the target classifier. An even more compact, but less interpretable representation in the feature space is then additionally reached by SFFS—also with the target classifier 'in the loop'. The gold standard throughout feature selection was given by the rounded median. Table 11.23 summarises the results of these computations, and Figs. 11.18 and 11.19 visualise the confusions made by the classifier per feature type.

Table 11.22 Feature subsets used

Name	Description	#
Spectral	For spectral features	24
Rhythmic	For rhythmic features	87
Chords	Recognised chord features	22
Meta-Info	Date, artist, and title related	153
Lyrics-CN	ConceptNet's mood on lyrics	12
Lyrics-BoW	Word occurrences in the lyrics	393
All	Unision of the above	691
No-Lyrics	All without Lyrics-BoW and Lyrics-CN	286

Table 11.23 UA and WA for classification on AllInst test data against different attribute subsets for the A3 and V3 tasks, SVM

Type	Arousal		Valence	
[%]	UA	WA	UA	WA
Spectral	49.0	47.6	48.8	47.5
Rhythmic	54.0	52.4	57.7	56.4
Chords	50.0	47.0	49.2	47.6
Meta-Info	37.4	36.1	39.3	35.5
Lyrics-CN	33.5	28.9	35.9	38.4
Lyrics-BoW	39.4	36.8	37.8	40.5
All	50.5	50.0	50.9	51.3
No-Lyrics	54.1	53.3	58.8	58.5

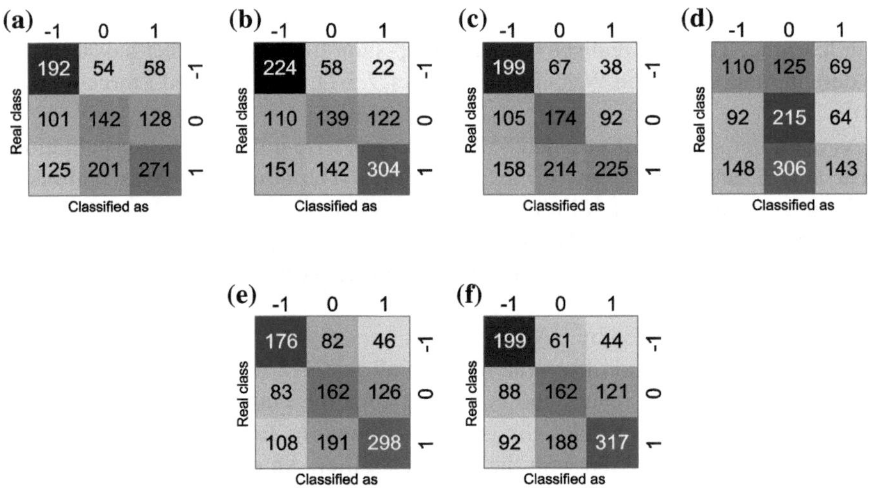

Fig. 11.18 Arousal confusions in the A3 classification task for selected feature subsets. Classifier SVM, dataset AllInst of NTWICM [28]. **a** Spectral, **b** Rhythmic, **c** Chords, **d** Lyrics-BoW, **e** All, **f** No-Lyrics

As can be seen, the task is demanding, and there are pronounced differences across individual feature groups: The lyrics features hardly surpass chance level, but the rhythm features, almost reach the best performance of all features except for lyrics features. Given this best result for the No-Lyrics set, it will be used in the ongoing. All features combined being inferior to this reduced set can be seen as indication of a too high dimensionality of the feature space. Further, the good results for the chord-based features show the suitability of the 'mid-level' features that base on decisions. The differences between arousal and valence are less pronounced within a type. In the confusion matrices for the No-Lyrics and Rhythmic feature sets confusions luckily occur mostly between neighbouring classes, i.e., negative or positive is mostly confused with neutral.

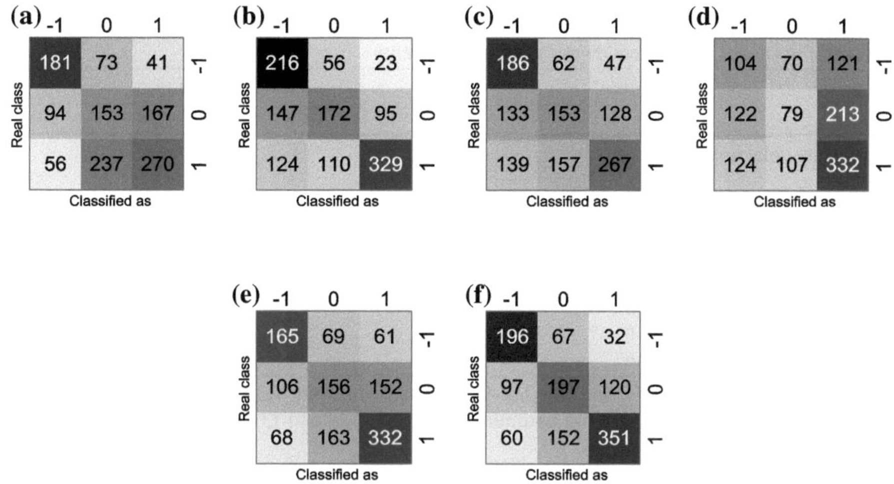

Fig. 11.19 Valence confusions in the V3 classification task for selected feature subsets. Classifier SVM, dataset AllInst of NTWICM [32]. **a** Spectral, **b** Rhythmic, **c** Chords, **d** Lyrics-BoW, **e** All, **f** No-Lyrics

Table 11.24 UA and WA for the different raters (A–D) by on the A3 and V3 tasks

Rater	Arousal		Valence	
[%]	UA	WA	UA	WA
A	43.4	43.6	58.5	57.6
B	63.8	60.0	48.5	48.1
C	53.0	52.0	55.3	53.5
D	47.8	46.9	54.2	56.3

Feature set No-Lyrics, set AllInst of NTWICM, SVM

Let us next consider the differences across raters with respect to the UA and WA in Table 11.24 on the A3 and V3 task and set AllInst. There, the training and testing was carried out exclusively on the ratings of one rater, each. As can be seen, either the learnt classifier has varying difficulties to model raters, or raters' mood models are more or less consistent. Interestingly, ratings of the professional DJ (rater A, cf. Sect. 5.3.2) lead to the best result for valence. In the case of arousal, the deltas in UA and WA are even more pronounced.

For an impression on the effect of exclusion of instances with lower agreement—as is common practice in most other work—Table 11.25 shows the effect of limiting test instances to the ones with a minimum agreement of two or three out of the four raters. For training, however, all instances are used. According to one's intuition, the UA and WA increases up to 8 % with increasing limitation to such prototypical cases, in particular in the case of arousal. The table further shows the effect of the SFFS feature selection to increase performance. In fact, results are improved by this step except for prototypical arousal.

Table 11.26 contains additional results for the Random Forests (RFs) classifier (based on decision trees) as an example of sub-sampling the feature space and boot-

Table 11.25 UA / WA for training with the training and development instances of AllInst, but testing on instances in different degrees of prototypicality

Type	Arousal		Valence	
[%]	UA	WA	UA	WA
w/o feature selection				
AllInst	54.1	53.3	58.8	58.5
Min2/4	56.7	54.8	61.1	60.1
Min3/4	64.9	60.9	65.5	61.4
with feature selection				
AllInst	56.2	55.2	61.2	61.0
Min2/4	59.6	57.2	64.1	63.0
Min3/4	64.8	60.9	68.6	64.1

AllInst of NTWICM, Min2/4, and Min3/4. No-Lyrics feature set, A3 and V3 tasks, selection by SFFS (out of the 286 features 131 as optimum for the A3 task, and 132 for the V3 task)

Table 11.26 UA / WA for SVM versus RF classification on AllInst of NTWICM with the No-Lyrics feature set for the A3 and V3 tasks

Type	Arousal		Valence	
[%]	UA	WA	UA	WA
SVM	54.1	53.3	58.8	58.5
RF	56.2	58.7	58.3	61.0

Feature set No-Lyrics

strapping of the data. 250 trees were found as optimum within the search range of 100 to 250 on the development set.

One observes RFs to outperform SVMs without feature selection on the task. Given RFs' minor transparency owing to the random injection in the feature and data selection process, SVMs had been preferred in the previous experiments.

11.7.3 Summary

In this section, automatic music mood classification was discussed. Features were based on musical features, meta-information, and lyrics. The mood model had three degrees of arousal and valence.

As mood was annotated per song, ambiguous cases that contain different polarities of arousal or valence might have been handled as neutral—in the future an extra class may help change this behaviour. Alternatively, shorter segments of music could be analysed to allow for a change of mood within a musical piece as in [151] for classical music. This can be combined with automatic music structure analysis [41, 120] as was presented in the last section.

As for features, rhythmic, chord-based, and spectral ones were found best-suited. Lyrics information, however, could not lead to further improvements, which may be partly owing to the database—in [8] a different trend was found on another database. A gain was also not achieved by meta-information such as artist and title. More information could, however, be added such as usage statistics [165]. Also, other forms of representation of the lyrics can be considered, potentially integrating other variants of on-line knowledge source integration.

The requirement to process all music was handled by establishing a gold standard based on the (rounded) median to deal with cases of complete rater disagreement.

In addition, the effect of prototypicality was investigated by limitation to the test-cases with clear rater agreement in different levels. UA and WA were raised from roughly 60 % to around 70 % with this limitation in the three-class tasks of arousal and valence classification. Confusions were mostly made between neighbouring classes, thus increasing applicability. However, further improvements are needed for real-life usage. In this respect the high differences between performance depending on the individual raters have to be named indicating the subjective character of music mood. Future efforts could consider other feature combination methods, such as individual feature streams. Further, other dimensions could be added, such as 'dominance', which is often used in speech emotion analysis [166]. These dimensions can also be handled by regression approaches (cf. [162, 167]). To that end, more labeller tracks should ideally be added to approach genuine numeric continuity across the dimensions. First results for a regression approach with the four raters on NTWICM are reported in [33].

11.8 Singer Traits: Age, Gender, Height, Race

Extending the assessment of speaker traits to sung speech, and bridging from assessing mood in music, one can also aim at the assessment of singers' traits. This was first shown in [34], then refined in [35], and later extended for more traits in [36].

Such singer trait classification, that is, automatically recognising meta data such as age and gender of the performing vocalist(s) in recorded music, is currently still an under-researched topic in MIR in contrast to the increasing efforts devoted to that area in paralinguistic speech processing. Applications in music processing can be found in categorisation and query of large databases with potentially unknown artists—that is, artists for whom not enough reliable training data is available for building singer identification models as, e.g., in [168]. Robustly extracting a variety of meta information can then allow the artist to be identified in a large collection of artist meta data. In addition, exploiting gender information can be useful for building and adapting models for other MIR tasks such as automatic lyrics transcription [169]. In comparison to speaker trait determination as was shown in Sect. 10.4.3, recognition of *singer* traits can be expected to be an even more challenging task due to high variability of the singer's pitch, instrumental accompaniment, and simultaneous presence of multiple vocalists.

Little, if any, research dealt with the recognition of singer traits other than gender in music. Apart from gender, three further tasks are thus investigated in the following:

age, height, and race.[9] To this end, we will also show how to improve the extraction of the leading voice beyond the harmonic enhancement, i.e., filtering of the drum accompaniment, as was shown in Sect. 11.1.

11.8.1 UltraStar Singer Traits Database

To test such automatic singer-independent classification, the UltraStar database, as was first introduced in [34], was enriched with according detailed annotation of singer traits, particularly continuous age and gender. The database contains 581 songs corresponding to over 37 h total play time commonly used for the 'UltraStar' karaoke game. The focus on highly popular artists was needed for the establishment of solid ground truth as information on these can be retrieved with sufficient certainty. To ensure transparent partitioning and singer independence, the first letter of the name of the performer is used for assignment to training, development, and test sets. The UltraStar meta-data provides ground truth tempo and lyrics aligned to beats. The singer(s) identity was annotated at beat level wherever possible. In the case of more than one singer per song the 'singer diarisation'—i.e., the alignment of singer identity to the music—was manually determined with the help of the corresponding official music video for precise results. Subsequent to this step, gender, height, birth year, and race of the 516 distinct singers was collected and repeatedly verified from on-line textual (IMDB,[10] and Wikipedia[11]) and audiovisual (YouTube[12]) knowledge sources. The two male raters (24 and 28 years old) were experts for popular music.

In fact, a considerable amount of the contained songs has two or more singers present simultaneously. To ensure realistic 'non-preselected' analysis, the following scheme was derived in such a case: In case of the nominal traits gender and race, beats were marked as 'unknown' except if all simultaneously present singers share the same attribute value. In case of the continuous-valued traits age and height, the mean over present singers was used. In the same way, musical pieces were treated where an exact singer diarisation could not be reached. Finally, beats were also marked as 'unknown' if an attribute was missing for at least one of the present singers.

Figure 11.20a,b visualise the obtained distribution of gender and race among the 516 singers. In Fig. 11.20c,d the continuous-valued age and height are shown with

---

[9] The annotation scheme is inspired by the TIMIT corpus as was used in Sect. 10.4.3. As such, the term 'race' is adopted from the corpus' meta-information—though modern biology often neither classifies the homo sapiens sapiens by race nor sub-categories for collective differentiation in both physical and behavioural traits. Opposing current molecular biologic and population genetic research's view that a systematic categorisation may be insufficient to describe the enormous diversity and fluent differences between geographic population, it can be argued that, when aiming at an end-user information retrieval system, a categorisation into illustrative, archetypal categories can be useful.

[10] http://www.imdb.com

[11] http://www.wikipedia.org

[12] http://www.youtube.com

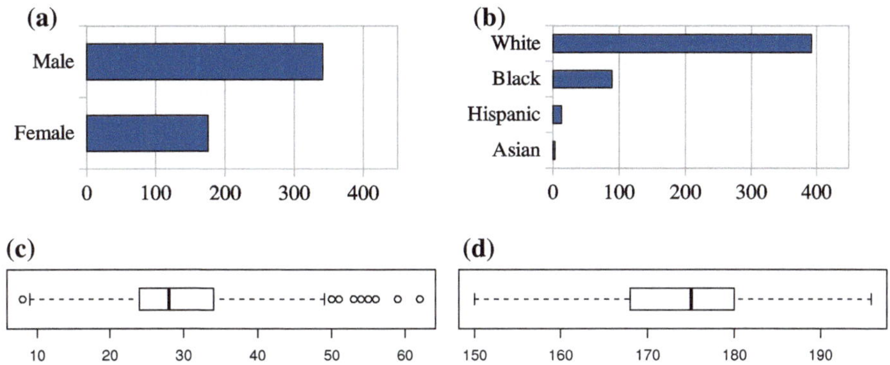

Fig. 11.20 UltraStar Singer Trait Database's distribution of traits among its 516 contained singers [36]. **a** Gender, **b** Race, **c** Age [years], **d** Height [cm]

boxes ranging from the first to the third quartile and values exceeding this range by more than a factor of 1.5 shown as outliers by circles. The fact that singer age is a function of a musical piece's recording date was taken into account.

For automatic assessment, the tasks were constrained to binary and ternary classification tasks on frame (beat) level as well as on song level. This decision needed to be made owing to the challenging real-world conditions given when assessing singer traits in polyphonic music. Such binary classification provides a simple categorisation per singer trait, and ternary classification is carried out to perform simultaneous singing activity detection on frame level in order to provide full realism. Height and age were discretised to 'small' (s, $<$175 cm) and 'tall' (t, $\geq$175 cm), respectively 'young' (y, $<$30 years) and 'old' (o, $\geq$ 30 years). From the annotated race classes the sparse classes 'Asian', 'Black', and 'Hispanic' were clustered as opposed to 'White' singers.

The number of beats for task evaluation are shown in Table 11.27. The annotation is available for reproduction of results.[13]

11.8.2 Methodology

Given the challenging condition of person trait recognition under singing in polyphonic music, finding the optimal preprocessing by suited singer separation becomes a focus issue. To this end, harmonic enhancement as was shown in Sect. 11.1 basing on openBliSSART (cf. Sect. 11.8 is used as a first means. This will now be followed by targeted extraction of the leading voice as in [170]. Different sets of NMF components shall be used in different parts of a song for higher flexibility of the algorithm. A song is therefore chunked into frame-synchronous non-overlapping chunks of 881 664 samples ($\approx$ 20 s at 44.1 kHz sample rate) as in [35]. Then, the leading voice

[13] http://www.openaudio.eu/UltraStar_Singers.arff

Table 11.27 Number of beats per trait, class and partition in the UltraStar singer trait database

# Beats	Train	Devel	Test	Sum
No voice (0)	90076	75741	48948	214765
Gender				
Female (f)	32308	23071	9739	65118
Male (m)	55505	49497	37686	142688
?	86	253	771	1110
Race				
White (w)	67525	62003	40479	170007
b/h/a	16378	9465	7136	32979
?	3996	1353	581	5930
Age				
Young (y)	48510	42056	25682	116248
Old (o)	34074	24596	18712	77382
?	5315	6169	3802	15286
Height				
Small (s)	29638	24946	8562	63146
Tall (t)	30177	30146	23452	83775
?	28084	17729	16182	61995
Sum	177975	148562	97144	423681

'b/h/a': black / hispanic / asian. 'Unknown' (?) includes simultaneous performance of artists of different gender/race, and those with unknown ground truth

separation approach as described in [170, 171] is additionally applied: Starting from the STFT of the audio signal at frame n, denoted $[\underline{S}]_{:,n}$, the spectrum is expressed as the sum of two independent components as $[\underline{S}]_{:,n} = [\underline{V}]_{:,n} + [\underline{M}]_{:,n}$, where $[\underline{V}]_{:,n}$ is the STFT of the leading voice, and $[\underline{M}]_{:,n}$ is the one of the background musical signal parts. $[\underline{V}]_{:,n}$ and $[\underline{M}]_{:,n}$ are assumed to be centre proper complex Gaussian variables[14]:

$$[\underline{V}]_{:,n} \sim \mathcal{N}_c(0, \mathrm{diag}(\sigma^2_{[\underline{V}]_{:,n}})), \tag{11.37}$$

$$[\underline{M}]_{:,n} \sim \mathcal{N}_c(0, \mathrm{diag}(\sigma^2_{[\underline{M}]_{:,n}})), \tag{11.38}$$

where $\sigma^2_{[\underline{V}]_{:,n}}$ or respectively $\sigma^2_{[\underline{M}]_{:,n}}$ is the power spectral density (PSD) of the leading voice or respectively of the background music at frame n. Assuming independence between the two components, the STFT of the observed signal then is also a proper Gaussian vector:

$$[\underline{S}]_{:,n} \sim \mathcal{N}_c(0, \mathrm{diag}(\sigma^2_{[\underline{V}]_{:,n}} + \sigma^2_{[\underline{M}]_{:,n}})). \tag{11.39}$$

Then, $\sigma^2_{[\underline{V}]_{:,n}}$ and $\sigma^2_{[\underline{M}]_{:,n}}$ are estimated per signal frame n. In the present use-case, the approach is completely unsupervised, i.e., no learning takes place. Rather, it relies on

[14] A complex random variable whose real and imaginary parts are independent and follow a real Gaussian distribution, with mean equal to 0 and identical variance or co-variance matrix in case of a multi-variate distribution.

the typical constraints of the voice's signal share assuming it follows a source filter production model as was introduced in Sect. 6.2.1.4 where the source is a periodic signal as given by periodic glottal pulses. For the background music no constraints or assumptions are made owing to the high potential variability. Model parameters are estimated iteratively based on NMF techniques in two steps: First, an initial estimate is made of the parameters, each. Then, a constrained re-estimation stage refines the leading melody estimation. In this step also sudden 'octave jumps', i.e., doubling of the frequency, are avoided. After determination of the final PSDs $\sigma^2_{[\underline{V}]_{:,n}}$ and $\sigma^2_{[\underline{M}]_{:,n}}$, the separated singing voice signal is obtained by frame-wise Wiener filtering:

$$\widehat{[\underline{V}]}_{:,n} = \frac{\sigma^2_{[\underline{V}]_{:,n}}}{\sigma^2_{[\underline{V}]_{:,n}} + \sigma^2_{[\underline{M}]_{:,n}}} [\underline{S}]_{:,n}. \tag{11.40}$$

To foster reproducibility of results also in this stage, it was decided for an open-source implementation[15] of the algorithm. Default parameters were chosen. Chunking is identical as in [35].

After applying the leading voice extraction algorithm to popular music, parts of the drum track may remain in the voice part. Thus, 'harmonic enhancement' by drum-beat separation (cf. Sect. 11.1) can be sequentially added either after leading voice separation (LV-HE), or the other way round (HE-LV). To use different NMF parametrisations for the two algorithms, time domain signals are re-synthesised in between the separate separation stages.

Features for singer trait classification were extracted per beat [34] again using openSMILE [172] as was shown in Sect. 6.5. The LLD set first consists of the short-time energy, mean-crossing rate, ZCR, voicing probability, and HNR—all known to be well suited to indicate vocal presence. Additional LLDs focus on speaker trait [168]: F_0, and MFCCs 0–12; respective first-order delta regression coefficients of these. Overall, this results in a set of 46 LLDs.

Sequence classification is considered using BLSTM RNNs which have been observed superior to individual beat-wise static classification by SVM or Hidden Naive Bayes for vocalist gender recognition [35]. As in this study, the BLSTM networks for the results described here use one hidden layer with 80 LSTM memory cells, each, for forward and backward processing. The size of the input layer equals the 46 LLDs, and the number of outputs equals the (varying) number of classes—two or three. The softmax function is used for the output activations, ensuring a restriction to the interval [0; 1] and unity sum to represent the posterior class probabilities. Songs were presented frame-wise in correct temporal order to the input layer. For a final decision, each frame was assigned to the class with the highest output probability.

[15] http://www.durrieu.ch/phd/software.html

Table 11.28 Beat-wise BLSTM-RNN classification on the UltraStar test set on 2- and 3-class tasks

[%]		–		HE		LV		LV-HE		HE-LV	
Task	Classes	UA	WA	UA	WA	UA	WA	UA	WA	UA	WA
Voice	0/1	74.55	74.50	73.82	73.84	75.77	75.81	75.40	75.41	75.09	75.11
Gender	0/m/f	63.75	68.54	65.65	68.91	69.29	71.31	67.90	70.41	68.52	70.44
	m/f	86.67	91.09	88.45	91.91	86.93	91.12	89.61	93.60	87.76	92.50
Race	0/w/b+h+a	48.17	63.84	47.46	63.02	49.37	65.46	49.23	63.63	48.40	63.77
	w/b+h+a	60.44	65.82	63.30	76.98	55.05	76.18	62.57	78.67	62.78	75.16
Age	0/y/o	51.02	57.61	50.00	57.14	53.50	59.85	51.26	58.86	50.01	57.72
	y/o	55.30	55.60	57.55	56.56	53.93	53.63	55.97	54.89	54.69	54.17
Height	0/s/t	53.94	66.79	52.35	66.57	58.15	69.30	57.67	68.41	58.91	69.53
	s/t	64.70	72.73	62.31	70.67	66.54	73.00	69.65	77.49	72.07	78.26

Preprocessing: harmonic enhancement by drum-beat separation (HE), leading voice extraction (LV), and sequential combination of these two

11.8.3 Performance

Supervised training of the networks followed a random initialisation of the network weights with a Gaussian distribution with zero mean and 0.1 as standard deviation. For improved generalisation, the order of the input sequences was randomised, and Gaussian noise with zero mean and 0.3 as standard deviation was added to the input activations. Resilient propagation was used for iterative update of the network's weights during training. Once no improvement over 20 epochs had been observed on the validation set, the training was stopped. To cope with the race task's high imbalance, a fixed number of 20 epochs was run to avoid overfitting to the validation set, and the standard deviation of the Gaussian noise added to the input activations was increased to $\sigma = 0.9$.

The general imbalance of instances across classes and tasks on the beat level (cf. Table 11.27) renders UA the major performance measure of interest. Singer presence detection reaches over 75 % UA with the use of leading voice extraction. On the 2-class gender recognition task, the combination of source separation algorithms leads to the best result with drum-beat separation as last step at 89.61 % UA. For height recognition, the combination of the pre-processing steps—albeit in inverse order—also leads to optimal results and 72.07 % UA are reached, increasing UA by more than 7 % absolute compared to no preprocessing. On the 3-class task, the best UA is 69.29 % UA when using exclusively the isolation of the singing voice.With the same pre-processing, 2-class recognition of race and age is solved best at 63.30 % and 57.55 % UA. Age recognition falls behind results on spoken language (cf. Sect. 10.4.3), but the result is significantly above the chance level of 50 % UA according to a z-test ($p < 0.001$).

To evaluate semantic singer trait tagging of entire songs, accuracies of a majority vote on the beat level compared to the most frequent ground truth class on the beat level are shown in Table 11.29. Obviously, such a gold standard is 'more of heuristic nature' given phenomena such as mixed gender duets. On the song level, gender

Table 11.29 Song-wise BLSTM-RNN predictions on the UltraStar test set by beat-wise majority vote among 3-class tasks (ignoring beats classified as 0) or 2-class tasks

[%]	Vote on	–		HE		LV		LV-HE		HE-LV	
Task		UA	WA	UA	WA	UA	WA	UA	WA	UA	WA
Gender	0/m/f	80.9	87.0	81.7	85.6	87.7	90.9	91.3	92.4	87.7	90.9
	m/f	86.9	90.1	89.0	90.9	87.7	90.9	89.6	93.9	89.6	93.9
Race	0/w/b+h+a	49.8	78.8	53.5	79.7	51.0	78.2	54.0	75.2	48.9	72.2
	w/b+h+a	52.8	59.8	62.6	75.9	54.7	73.7	64.4	78.9	61.7	74.4
Age	0/y/o	55.2	54.5	54.6	54.1	56.0	54.1	56.9	57.4	50.9	51.6
	y/o	54.5	54.5	57.0	55.7	52.2	51.6	53.4	52.5	58.9	58.2

Singer height is not included on this level due to sparseness: only 88 songs have a known ground truth. Preprocessing steps are equivalent to the results shown in Table 11.28

recognition reaches 91.3 % UA, race recognition 64.4 % UA and age recognition 58.9 % UA. Interestingly, for gender, voting on all beats is more robust than voting exclusively over beats with voice presence. This might be explained by the fact that BLSTM RNNs model bi-directional context and thus consider neighbouring frames in their decisions, ie the predictions for parts without vocals are influenced by the features of the vocal parts. Across the tasks and settings, the combination of the two pre-processing steps, first leading voice extraction, then drum-beat separation, gives best results.

11.8.4 Summary

Fully automatic assessment of paralinguistic traits (age, height and race) was demonstrated in this section based on vocals in original pop-music rather than on more or less clean speech as was shown in Sect. 10.4.3. Gender recognition was observed to give 'application-ready' results even on the beat level for unseen test data. Race and height classification show general feasibility, even in a such highly realistic setting. An interdependency of race and musical genre might be given—yet, taking the fact into account that source separation generally improved performance can be seen as indication that the networks at least partly are capable of race recognition.

The quite good results for height classification certainly stem from the correlation with gender. Age recognition results were lower than those reported on speech in Sect. 10.4.3, where four classes of age were discriminated rather than two here—at around similar performance. The challenge besides music 'disturbance' and singing voice may be owing to the considered type of 'chart' music, where many singers are at a similar age. Using only males for training and testing of age classification in an additional test-run, however, led to 61.63 % UA—female singers were too sparse in the set.

Next efforts could analyse the influence of longer units of analysis than the beat level, such as the supra-segmental functionals as were used for paralinguistic analysis in speech (cf. Sect. 10.4.3). In that case, however, feature variation owing to the singing—especially for pitch—will require suited methods of adaptation or transformation. Further, extending the database to reach higher musical variation, e.g., by Jazz or non-Western music would be of interest. Finally, multi-task learning could help to exploit singer trait interdependencies in learning, given the observations for height assessment in speech as was described in Sect. 10.4.3.

References

1. Casey, M., Slaney, M.: Fast recognition of remixed music audio. In: IEEE Proceedings International Conference on Audio Speech and Signal Processing (ICASSP), vol. IV, pp. 1425–1428 (2007)
2. Schuller, B., Zobl, M., Rigoll, G., Lang, M.: A hybrid music retrieval system using belief networks to integrate queries and contextual knowledge. In: IEEE Proceedings 4th IEEE International Conference on Multimedia and Expo, ICME 2003, vol. I, pp. 57–60. Baltimore (2003)
3. Schuller, B., Rigoll, G., Lang, M.: Multimodal music retrieval for large databases. In: IEEE Proceedings 5th IEEE International Conference on Multimedia and Expo, ICME 2004, vol. 2, pp. 755–758. Taipei, Taiwan (2004)
4. Downie, J.: Music information retrieval. Ann. Rev. Inform. Sci. Technol. **37**, 295–340 (2003)
5. Scheirer, E.D.: Tempo and beat analysis of acoustic musical signals. Acoust. Soc. Am. **103**(1), 588–601 (1998)
6. Schuller, B., Eyben, F., Rigoll, G.: Tango or waltz?—putting ballroom dance style into tempo detection. EURASIP J. Audio, Speech, Music Process. Spec. Issue Intell. Audio, Speech, Music Process. Appl. (Article ID 846135), 12 (2008)
7. Tzanetakis, G., Cook, P.: Musical genre classification of audio signals. IEEE Trans. Speech Audio Process. **10**(5), 293–302 (2002)
8. Schuller, B., Hage, C., Schuller, D., Rigoll, G.: "mister d.j., cheer me up!": Musical and textual features for automatic mood classification. J. New Music Res. **39**(1), 13–34 (2010)
9. Berenzweig, A., Ellis, D.: Locating Singing Voice Segments Within Musical Signals. In: Proceedings of International Workshop on Applications of Signal Processing to Audio and Acoustics (WASPAA), pp. 119–123. Mohonk, New York (2001)
10. Schuller, B., Hörnler, B., Arsić, D., Rigoll, G.: Audio chord labeling by musiological modeling and beat-synchronization. In: IEEE Proceedings 10th IEEE International Conference on Multimedia and Expo, ICME 2009, pp. 526–529. New York (2009)
11. Bellmann, H.: About the determination of key of a musical excerpt. In: Computer Music Modeling and Retrieval, vol. 3902 LNCS, pp. 76–91. Springer, Berlin (2006)
12. Dannenberg, R., Goto, M.: Music structure analysis from acoustic signals. In: Havelock, D., Kuwano, S., Vorländer, M. (eds.) Handbook of Signal Processing in Acoustics, vol. 1, pp. 305–331. Springer (2009)
13. Dixon, S., Pampalk, E., Widmer, G.: Classification of dance music by periodicity patterns. In: Proceedings of the 4th International Conference on Music, Information Retrieval, pp. 159–165 (2003)
14. Foote, J., Uchihashi, S.: The beat spectrum: a new approach to rhythm analysis. In: Proceedings of International Conference on Multimedia and Expo (ICME), IEEE, Tokyo (2001)
15. Hu, N., Dannenberg, R.B., Tzanetakis, G.: Polyphonic audio matching and alignment for music retrieval. In: Proceedings of the IEEE Workshop on Applications of Signal Processing to Audio and Acoustics, pp. 185–188 (2003)
16. Klapuri, A.P., Eronen, A.J., Astola, J.T.: Analysis of the meter of acoustic musical signals. IEEE Trans. Speech Audio Process. **14**(1), 342–355 (2006)

17. Müller, M., Ellis, D., Klapuri, A., Richard, G.: Signal processing for music analysis. IEEE J. Sel. Top. Sig. Process. **5**(6), 1088–1110 (2011)
18. Orio, N.: Music retrieval: a tutorial and review. Found. Trends Inf. Retrieval **1**, 1–90 (2006)
19. Uhle, C., Rohden, J., Cremer, M., Herre, J.: Low complexity musical meter estimation from polyphonic music. In: Proceedings of the AES 25th international conference, pp. 63–68. London, UK (2004)
20. Weninger, F., Schuller, B., Liem, C., Kurth, F., Hanjalic, A.: Music information retrieval: an inspirational guide to transfer from related disciplines. In: Müller, M., Goto, M. (eds.) Multimodal Music Processing, vol. 11041, pp. 195–215. Seminar of Dagstuhl Follow-UpsSchloss Dagstuhl, Germany (2012)
21. Schuller, B., Lehmann, A., Weninger, F., Eyben, F., Rigoll, G.: Blind enhancement of the rhythmic and harmonic sections by nmf: Does it help? In: Proceedings International Conference on Acoustics Including the 35th German Annual Conference on Acoustics, NAG/DAGA 2009, pp. 361–364, Acoustical Society of the Netherlands, DEGA, Rotterdam, The Netherlands (2009)
22. Böck, S., Eyben, F., Schuller, B.: Onset detection with bidirectional long short-term memory neural networks. In: Proceedings Annual Meeting of the MIREX 2010 Community as Aart of the 11th International Conference on Music Information Retrieval, ISMIR. p. 2. Utrecht, Netherlands (2010)
23. Eyben, F., Böck, S., Schuller, B., Graves, A.: Universal onset detection with bidirectional long-short term memory neural networks. In: Proceedings 11th International Society for Music Information Retrieval Conference, ISMIR 2010, pp. 589–594. Utrecht, The Netherlands (2010)
24. Schuller, B., Eyben, F., Rigoll, G.: Fast and robust meter and tempo recognition for the automatic discrimination of ballroom dance styles. In: IEEE Proceedings 32nd IEEE International Conference on Acoustics, Speech, and Signal Processing, ICASSP 2007, vol. I, pp. 217–220. Honolulu, HY (2007)
25. Eyben, F., Schuller, B., Reiter, S., Rigoll, G.: Wearable assistance for the ballroom-dance hobbyist—holistic rhythm analysis and dance-style classification. In: IEEE Proceedings 8th IEEE International Conference on Multimedia and Expo, ICME 2007, pp. 92–95. Beijing, China (2007)
26. Eyben, F., Schuller, B.: Tempo estimation from tatum and meter vectors. In: Proceedings Annual Meeting of the MIREX 2010 Community as Part of the 11th International Conference on Music Information Retrieval, ISMIR, p. 1. Utrecht, Netherlands (2010)
27. Böck, S., Eyben, F., Schuller, B.: Tempo detection with bidirectional long short-term memory neural networks. In: Proceedings Annual Meeting of the MIREX 2010 community as Part of the 11th International Conference on Music Information Retrieval, ISMIR, p. 3. Utrecht, Netherlands (2010)
28. Schuller, B., Gollan, B.: Music theoretic and perception-based features for audio key determination. J. New Music Res. **41**(2), 175–193 (2012)
29. Schuller, B., Eyben, F., Rigoll, G.: Beat-synchronous data-driven automatic chord labeling. In: Proceedings 34. Jahrestagung für Akustik, DAGA, DEGA, pp. 555–556. Dresden, Germany (2008)
30. Schuller, B., Dibiasi, F., Eyben, F., Rigoll, G.: One day in half an hour: music thumbnailing incorporating harmony- and rhythm structure. In: Proceedings 6th Workshop on Adaptive Multimedia Retrieval, AMR 2008, p. 10. Berlin, Germany (2008)
31. Schuller, B., Dibiasi, F., Eyben, F., Rigoll, G.: Music thumbnailing incorporating harmony- and rhythm structure. In: Detyniecki, M., Leiner, U., Nürnberger, A. (eds.) Adaptive Multimedia Retrieval: 6th International Workshop, AMR 2008, June 26–27, Berlin, Germany (2008). Revised Selected Papers, vol. 5811/2010, Lecture Notes in Computer Science (LNCS), pp. 78–88. Springer, Berlin (2010)
32. Schuller, B., Dorfner, J., Rigoll, G.: Determination of non-prototypical valence and arousal in popular music: Features and performances. EURASIP J. Audio, Speech, Music Process. Spec. Issue Scalable Audio-Content Anal. (Article ID 735854), 19 (2010)

33. Schuller, B., Weninger, F., Dorfner, J.: Multi-modal non-prototypical music mood analysis in continuous space: Reliability and performances. In: Proceedings 12th International Society for Music Information Retrieval Conference, ISMIR 2011, pp. 759–764. Miami (2011)
34. Schuller, B., Kozielski, C., Weninger, F., Eyben, F., Rigoll, G.: Vocalist gender recognition in recorded popular music. In: Proceedings 11th International Society for Music Information Retrieval Conference, ISMIR 2010, pp. 613–618. Utrecht, The Netherlands (2010)
35. Weninger, F., Durrieu, J.-L., Eyben, F., Richard, G., Schuller, B.: Combining monaural source separation with long short-term memory for increased robustness in vocalist gender recognition. In: IEEE Proceedings 36th IEEE International Conference on Acoustics, Speech, and Signal Processing, ICASSP 2011, pp. 2196–2199. Prague, Czech Republic (2011)
36. Weninger, F., Wöllmer, M., Schuller, B.: Automatic assessment of singer traits in popular music: Gender, age, height and race. In: Proceedings 12th International Society for Music Information Retrieval Conference, ISMIR 2011, pp. 37–42. Miami (2011)
37. Grosche, P., Schuller, B., Müller, M., Rigoll, G.: Automatic transcription of recorded music. Acta Acustica united with Acustica. **98**(2), 199–215(17) (2012)
38. Schuller, B., Rigoll, G.: Self-learning acoustic feature generation and selection for the discrimination of musical signals. In: Proceedings 32. Jahrestagung für Akustik, DAGA 2006, pp. 285–286. Braunschweig, Germany (2006)
39. Schuller, B., Wallhoff, F., Arsić, D., Rigoll, G.: Musical signal type discrimination based on large open feature sets. In: IEEE Proceedings 7th IEEE International Conference on Multimedia and Expo, ICME 2006, pp. 1089–1092. Toronto, Canada (2006)
40. Schuller, B., Schmitt, B.J.B., Arsić, D., Reiter, S., Lang, M., Rigoll, G.: Feature selection and stacking for robust discrimination of speech, monophonic singing, and polyphonic music. In: Proceedings 6th IEEE International Conference on Multimedia and Expo, ICME 2005, pp. 840–843. Amsterdam, The Netherlands (2005)
41. Schuller, B., Rigoll, G., Lang, M.: Hmm-based music retrieval using stereophonic feature information and framelength adaptation. In: Proceedings 4th IEEE International Conference on Multimedia and Expo, ICME 2003, vol. II, pp. 713–716. Baltimore (2003)
42. Schuller, B., Rigoll, G., Lang, M.: Matching monophonic audio clips to polyphonic recordings. In: Proceedings 31. Jahrestagung für Akustik, DAGA, 2005, DEGA, pp. 299–300, Munich, Germany (2005)
43. Weninger, F., Amir, N., Amir, O., Ronen, I., Eyben, F., Schuller, B.: Robust feature extraction for automatic recognition of vibrato singing in recorded polyphonic music. In: Proceedings 37th IEEE International Conference on Acoustics, Speech, and Signal Processing, ICASSP 2012, pp. 85–88. Kyoto, Japan (2012)
44. Helén, M., Virtanen, T.: Separation of drums from polyphonic music using non-negative matrix factorization and support vector machine. In Proceedings of EUSIPCO, Antalya, Turkey (2005)
45. Paulus, J., Virtanen, T.: Drum transcription with non-negative spectrogram factorisation. In: Proceedings of EUSIPCO, p. 4, EURASIP, Antalya, Turkey (2005)
46. Moreau, A., Flexer, A.: Drum transcription in polyphonic music using non-negative matrix factorisation. In: Proceedings of 8th International Conference on Music Information Retrieval (ISMIR), September 23–27, pp. 353–354, Vienna, Austria (2007)
47. Uhle, C., Dittmar, C., Sporer, T.: Extraction of drum tracks from polyphonic music using independent subspace analysis. In: Proceedings of 4th International Symposium on Independent Component Analysis and Blind Signal Separation (ICA), April 2003, pp. 843–848, Nara, Japan (2003)
48. Smaragdis, P., Brown, J.C.: Non-negative matrix factorization for polyphonic music transcription. In: IEEE Proceedings of WASPAA, pp. 177–180 (2003)
49. Virtanen, T., Ryynänen, M.: A. Mesaros. Combining pitch-based inference and non-negative spectrogram factorization in separating vocals from polyphonic music. In: ISCA Tutorial and Research Workshop on Statistical And Perceptual Audition, SAPA 2008, pp. 17–22, ISCA, Brisbane (2008)

50. Sethares, W.A.: Local consonance and the relationship between timbre and scale. J. Acoust. Soc. Am. **94**(3), 1218–1228 (1993)
51. Gouyon, F., Klapuri, A.P., Dixon, S., Alonso, M., Tzanetakis, G., Uhle, C., Cano, P.: An experimental comparison of audio tempo induction algorithms. IEEE Trans. Audio, Speech, Lang. Process. **14**(5), 1832–1844 (2006)
52. Dixon, S.: Onset detection revisited. In: Proceedings of DAFx-06, pp. 133–137, Montreal, Canada (2006)
53. Zhou, R., Reiss, J.: Music onset detection combining energy-based and pitch-based approaches. In: Proceedings of MIREX as part of the 8th International Conference on Music Information Retrieval (ISMIR). Sept 23–27. P. 4, Vienna, Austria (2007)
54. Röbel, A.: Onset detection by means of transient peak classification in harmonic bands. In: Proceedings of MIREX as part of the 10th International Conference on Music Information Retrieval (ISMIR), P. 2, Kobe, Japan (2009)
55. Klapuri, A.: Sound onset detection by applying psychoacoustic knowledge. In Proceedings of ICASSP, vol. 6, pp. 3089–3092 (1999)
56. Bello, J., Daudet, L., Abdallah, S., Duxbury, C., Davies, M., Sandler, M.: A tutorial on onset detection in music signals. IEEE Trans. Speech Audio Process. **13**(5), 1035–1047 (2005)
57. Duxbury, C., Bello, J.P., Davies, M.: M. Sandler. Complex domain onset detection for musical signals. In: Proceedings of Digital Audio Effects Workshop (DAFx-03) pp. 1–4, London, UK (2003)
58. Collins, N.: Using a pitch detector for onset detection. In Proceedings of ISMIR, pp. 100–106 (2005)
59. Basseville, M., Nikiforov, I.V.: Detection of Abrupt Changes: Theory and Application. Prentice-Hall, Englewood Cliffs (1993)
60. Lacoste, A., Eck, D.: Onset detection with artificial neural networks. In: Proceedings of MIREX as part of the 6th International Conference on Music Information Retrieval (ISMIR), P. 4, London, UK (2005)
61. Graves, A.: Supervised sequence labelling with recurrent neural networks. Ph.D. Thesis, Technische Universität München (2008)
62. Collins, N.: A comparison of sound onset detection algorithms with emphasis on psychoacoustically motivated detection functions. In: Proceedings of AES Convention 118, pp. 28–31 (2005)
63. Handel, S.: Listening: An Introduction to the Perception of Auditory Events. MIT Press, Cambridge (1989)
64. Böck, S.: Onset Detector 2011. In: Proceedings Annual Meeting of the MIREX 2011 Community as Part of the 12th International Conference on Music Information Retrieval. p. 2. ISMIR, ISMIR (2011)
65. Böck, S., Arzt, A., Krebs, F., Schedl, M.: Online real-time onset detection with recurrent neural networks. In Proceedings of the 15th International Conference on Digital Audio Effects (DAFx-12), p. 4. New York, UK (2012)
66. Klapuri, A.P.: Musical meter estimation and music transcription. In: Proceedings of Cambridge Music Processing Colloquium, Cambridge University, UK (2003)
67. Gouyon, F., Herrera, P.: Determination of the meter of musical audio signals: seeking recurrences in beat segment descriptors. In: AES 114th Convention, Amsterdam, The Netherlands (2003)
68. Gouyon, F., Dixon, S., Pampalk, E., Widmer, G.: Evaluating rhythmic descriptors for musical genre classification. In: Proceedings of the AES 25th International Conference, pp. 196–204. London, UK (2004)
69. Grosche, P., Müller, M., Kurth, F.: Cyclic tempogram—a mid-level tempo representation for music signals. In: Proceedings of ICASSP, pp. 5522–5525. Dallas, TX (2010)
70. Kirovski, D., Attias, H.: Beat-ID: identifying music with beat analysis. In: Proceedings of the International Workshop on Multimedia Signal Processing, IEEE, pp. 190–193, St. Thomas, US Virgin Islands (2002)

71. Kurth, F., Gehrmann, T., Muller, M.: The cyclic beat spectrum: Tempo-related audio features for time-scale invariant audio identification. In: Proceedings of the 7th International Conference on Music Information Retrieval (ISMIR 2006), pp. 35–40, Victoria, Canada (2006)
72. Goto, M., Muraoka, Y.: A real-time beat tracking system for audio signals. In: Proceedings of the 1995 International Computer Music Conference, pp. 171–174 (1995)
73. Goto, M., Muraoka, Y.: Real-time rhythm tracking for drumless audio signals—chord change detection for musical decisions. In: Proceedings of the IJCAI-97 Workshop on Computational Auditory Scene, Analysis, pp. 135–144 (1997)
74. Goto, M.: An audio-based real-time beat tracking system for music with or without drumsounds. J. New Music Res. **30**(2), 159–171 (2001)
75. Seppänen, J.: Computational models of musical meter recognition. Master's thesis, Tampere University of Technology (2001)
76. Dixon, S.: Automatic extraction of tempo and beat from expressive performances. J. New Music Res. **30**, 39–58 (2001)
77. Hainsworth, S., Macleod, M.: Beat tracking with particle filtering algorithms. In: 2003 IEEE Workshop on Applications of Signal Processing to Audio and Acoustics, pp. 91–94 (2003)
78. Alonso, M., David, B., Richard, G.: Tempo and beat estimation of musical signals. In: Proceedings of the International Conference on Music Information Retrieval, pp. 158–163 (2004)
79. Sethares, W.A., Staley, T.W.: Meter and periodicity in musical performance. J. New Music Res. **22**(5), 1–11 (2001)
80. Brown, J.C.: Determination of meter of musical scores by autocorrelation. J. Acoust. Soc. Am. **94**(4), 1953–1957 (1993)
81. van Noorden, L., Moelants, D.: Resonance in the perception of musical pulse. J. New Music Res. **28**(1), 43–66 (1999)
82. Rabiner, L., Juang, B.-H.: Fundamentals of Speech Recognition. Prentice Hall, Englewood Cliffs (1993)
83. Fu, Z., Lu, G., Ting, K.M., Zhang, D.: A survey of audio-based music classification and annotation. IEEE Trans. Multimedia **13**(2), 303–319 (2011)
84. Ballroomdancers.com. Preview audio examples of ballroom dance music. https://secure.ballroomdancers.com/Music/style.asp, (2006)
85. Zwicker, E., Fastl, H.: Psychoacoustics—Facts and Models. 2nd edn. Springer, Berlin (1999)
86. Paulus, J., Klapuri, A.P.: Measuring the similarity of rhythmic patterns. In: Proceedings of the2002 International Conference on Music Information Retrieval (ISMIR 2002). France, Paris (2002)
87. Gouyon, F., Dixon, S.: Dance music classification: A tempo-based approach. In: Proceedings of Fitfth International Conference on Music Information Retrieval, ISMIR, p. 4, Barcelona, Spain (2004)
88. Daniel, A., Emiya, V., David, B.: Perceptual-based evaluation of the errors usually made when automatically transcribing music. In: Proceedings of 9th International Symposium on Music Information Retrieval (ISMIR), pp. 550–555. Philadelphia (2008)
89. Gomez, E.: Estimating the tonality of polyphonic audio files: cognitive versus machine learning modelling strategies. In: Proceedings of 5th International Conference on Music Information Retrieval, Barcelona, Spain (2004)
90. Gomez, E.: Key estimation from polyphonic audio. In: Proceedings of 1st Annual Music Information Retrieval Evaluation eXchange (MIREX'05), London, UK (2005)
91. Izmirli, O.: Template based keyfinding from audio. In: Proceedings of International Computer Music Conference (ICMC), pp. 211–214. Barcelona, Spain (2005)
92. Mardirossian, A., Chew, E.: Skefis a symbolic (midi) key-finding system. In: Proceedings of 6th International Symposium on Music Information Retrieval (ISMIR), no pagination, pp. 1–8. London, UK (2005)
93. Pauws, S.: Musical keyextraction from audio. In: Proceedings of 5th International Symposium on Music Information Retrieval (ISMIR), pp. 96–99. Barcelona, Spain (2004)
94. Chuan, C., Chew, E.: Fuzzy analysis in pitch class determination for polyphonic audio keyfinding. In: Proceedings of 6th International Symposium on Music Information Retrieval (ISMIR), pp. 296–303. London, UK (2005)

95. Peeters, G.: Chroma-based estimation of musical key from audio-signal analysis. In: Proceedings of 7th International Symposium on Music Information Retrieval (ISMIR). Victoria, Canada (2006)
96. Chuan, C.H., Chew, E.: Audio key finding: considerations in system design and case studies on chopins 24 preludes. EURASIP J. Adv. Sig. Process. **2007**(056561) (2006)
97. Noland, K., Sandler, M.: Key estimation using a hidden markov model. In: Proceedings of 7th International Symposium on Music Information Retrieval (ISMIR), pp. 121–126. Victoria, Canada (2006)
98. Mandel, M.I., Ellis, D.P.W.: Song-level features and support vector machines for music classification. In: Proceedings 6th International Conference on Music Information Retrieval (ISMIR), pp. 594–599. London, UK (2005)
99. Mauch, M., Dixon, S.: Simultaneous estimation of chords and musical context from audio. IEEE Trans. Audio, Speech Lang. Process. **18**(6), 1280–1289 (2010)
100. Fujishima, T.: Realtime chord recognition of musical sound: a system using common lisp music. In: Proceedings of International Computer Music Conference, pp. 464–467. Bejing, China (1999)
101. Gomez, E.: Tonal description of polyphonic audio for music content processing. INFORMS J. Comput. **18**(3), 294–304 (2006)
102. Temperly, D.: An algorithm for harmonic analysis. Music Percept. **15**, 31–68 (1997)
103. Madsen, S.T., Widmer, G.: Key-finding with interval profiles. In: Proceedings International Computer Music Conference (ICMC), p. 4, Copenhagen, Denmark (2007)
104. Lee, K., Slaney, M.: A unified system for chord transcription and key extraction using hidden markov models. In: Proceedings of 8th International Symposium on Music Information Retrieval (ISMIR). Vienna, Austria (2007)
105. Lee, K., Slaney, M.: Acoustic chord transcription and key extraction from audio using key-dependent hmms trained on synthesized audio. IEEE Trans. Audio, Speech, Lang. Process. **16**, 291–301 (2008)
106. Purwins, H., Blankertz, B., Dornhege, G., Obermayer, K.: Scale degree from audio investigated with machine learning techniques. In: Proceedings Audio Engineering Society 116th Convention (2004)
107. Purwins, H.: Profiles of Pitch Classes Circularity of Relative Pitch and Key—Experiments, Models, Computational Music Analysis, and Perspectives. Ph.D. thesis, Technische Universität, Berlin (2005)
108. Cremer, M., Derboven, C.: A system for harmonic analysis of polyphonic music. In: Proceedings 25th International AES Conference. London, UK (2004)
109. Sun, J., Li, H., Li, L.: Key detection through pitch class distribution model and ANN. In: 2009 16th International Conference on Digital Signal Processing. Santorini, Hellas (2009)
110. Cabral, G., Briot, J.-P., Pachet, F.: Impact of distance in pitch class profile computation. In: Proceedings of 10th Brazilian Symposium on Computer Music (SBCM2005), pp. 135–144. Belo Horizonte, Brazil (2005)
111. Izmirli, O.: Audio key finding using low-dimensional spaces. In: Proceedings of 7th International Conference on Music Information Retrieval (ISMIR). Victoria, Canada (2006)
112. Izmirli, O.: An algorithm for audio key finding. In: Proceedings of Music Information Retrieval Evaluation Exchange (MIREX2005), as Part of the 6th International Symposium on Music Information Retrieval (ISMIR). London, UK (2006)
113. Zhu, Y.: An audio key finding algorithm. In: Proceedings of the 1st Annual Music Information Retrieval Evaluatione Xchange(MIREX'05). London, UK (2005)
114. Zhu, Y.: Music key detection for musical audio. In: Proceedings of 11th International Multimedia Modelling Conference (MMM'05). Melbourne, Australia (2005)
115. Chew, E.: An algorithm for determining key boundaries. In: Proceedings 2nd International Conference on Music and Artificial Intelligence (ICMAI). Edinburgh, Scotland (2002)
116. Shenoy, A., Mohapatra, R., Wang, Y.: Key determination of acoustic musical signals. In: Proceedings of International Conference on Multimedia and Expo(ICME). Singapore (2004)

117. Sheh, A., Ellis, D.: Chord segmentation and recognition using emtrained hidden markov models. In: Proceedings of ISMIR 2003, pp. 183–189. Baltimore, Maryland (2003)
118. Chai, W., Vercoe, B.: Detection of key change in classical piano music. In: Proceedings 6th International Conference on Music Information Retrieval (ISMIR), pp. 468–474. London, UK (2005)
119. Casey, M., Veltkamp, R., Goto, M., Leman, M., Rhodes, C., Slaney, M.: Content-based music information retrieval: current directions and future challenges. Proc. IEEE **96**(4), 668–696 (2008)
120. Bartsch, M.A., Wakefield, G.H.: To catch a Chorus: using chroma-based representations for audio thumbnailing. In: IEEE Workshop on Applications of Signal Processing to Audio and Acoustics 2001, pp. 15–18, New Paltz, New York (2001)
121. Wakefield, G.: Mathematical representation of joint time chroma distributions. In: Proceedings of SPIE, vol. 3807, pp. 637–645. Denver, Colorado (1999)
122. Hall, M., Frank, E., Holmes, G., Pfahringer, B., Reutemann, P., Witten, I.H.: The WEKA data mining software: an update. SIGKDD Explor. **11**(1), 10–18 (2009)
123. Krumhansl, C.: Tonal hierarchies and rare intervals in music cognition. Music Percept. Interdisc. J. **7**(3), 309–324 (1990)
124. Gatzsche, D., Gatzsche, G., Mehnert, M., Brandenburg, K.: A symmetry based approach for musical tonality analysis. In: Proceedings of 8th International Society for Music Information Retrieval Conference (ISMIR), no pagination, Vienna, Austria (2007)
125. Bello, J.P., Daudet, L., Sandler, M.B.: Automatic piano transcription using frequency and time-domain information. IEEE Trans. Audio, Speech Lang. Process. **14**(6), 2242–2251 (2006)
126. Duan, Z., Lu, L., Zhang, C.: Audio tonality mode classification without tonic annotations. In: Proceedings of 8th IEEE International Conference on Multimedia and Expo (ICME), pp. 1361–1364. Hannover, Germany (2008)
127. Izmirli, O.: Tonal-atonal classification of music audio using diffusion maps. In: Proceedings of 10th International Society for Music Information Retrieval Conference (ISMIR 2009), pp. 687–691. Kobe, Japan (2009)
128. Vuvan, D., Prince, J., Schmuckler, M.: Probing the minor tonal hierarchy. Music Percept. Interdisc. J. **28**(5), 461–472 (2011)
129. Papadopoulos, H., Peeters, G.: Local key estimation based on harmonic and metric structures. In: Proceedings of 12th International Conference on Digital Audio Effects (DAFx-09), pp. 1–8. Como, Italy (2009)
130. Goto, M., Muraoka, Y.: An audio-based real-time beat tracking system and its applications. In: Proceedings International Computer Music Confernce, pp. 17–20. ICMA, San Francisco (1998)
131. Rocher, T., Robine, M., Hanna, P., Oudre, L.: Concurrent estimation of chords and keys from audio. In: Proceedings of 11th International Society for Music Information Retrieval Conference (ISMIR), pp. 141–146. Utrecht, The Netherlands (2010)
132. Bello, J.B., Pickens, J.: A robust mid-level representation for harmonic content in music signals. Proc. ISMIR **2005**, 304–311 (2005)
133. Young, S., Evermann, G., Gales, M., Hain, T., Kershaw, D., Liu, X., Moore, G., Odell, J., Ollason, D., Povey, D., Valtchev, V., Woodland, P.: The HTK book (v3.4). Cambridge University Press, Cambridge (2006)
134. Stober, S., Nürnberger, A.: Towards user-adaptive structuring and organization of music collections. In: Proceedings of 6th Workshop on Adaptive Multimedia Retrieval (AMR 2008), Berlin, Germany (2008)
135. Burges, C.J.C., Plastina, D., Platt, J.C., Renshaw, E., Malvar, H.S.: Duplicate detection and audio thumbnails with audio fingerprinting. Technical Report MSR-TR-2004-19, Microsoft Research (MSR), March (2004)
136. Logan, B., Chu, S.: Music summarization using key phrases. Proc. ICASSP **2**, 749–752 (2000)
137. Aucouturier, J.-J., Pachet, F., Sandler, M.: The way it sounds: timbre models for analysis and retrieval of music signals. IEEE Trans. Multimedia **7**(6), 1028–1035 (2005)

138. Aucouturier, J.-J., Sandler, M.: Segmentation of musical signals using hidden markov models. In: Proceedings of the 110th AES Convention, AES (Audio Engineering Society), Amsterdam, The Netherlands (2001)
139. Jehan, T.: Hierarchical multi-class self similarities. In: Proceedings of IEEE Workshop on Applications of Signal Processing to Audio and Acoustics, pp. 311–314 (2005)
140. Foote, J.: Visualizing music and audio using self-similarity. In: Proceedings of 7th ACM International Conference on Multimedia (Part 1), pp. 77–80 (1999)
141. Cooper, M., Foote, J.: Automatic music summarization via similarity analysis. In: Proceedings of 3rd ISMIR, pp. 81–5 (2002)
142. Peeters, G., Burthe, A.L., Rodet, X.: Toward automatic music audio summary generation from signal analysis. In: Proceedings of 3rd ISMIR, pp. 94–100 (2002)
143. Abdallah, S.A., Noland, K., Sandler, M., Casey, M., Rhodes, C.: Theory and evaluation of a bayesian music structure extractor. In: Proceedings of 6th ISMIR, pp. 420–425 (2005)
144. Goto, M.: A chorus section detection method for musical audio signals and its application to a music listening station. IEEE Trans. Audio, Speech, Lang. Process. **14**(5), 1783–1794 (2006)
145. Müller, M., Kurth, F.: Enhancing similarity matrices for music audio analysis. Proc. ICASSP **5**, 9–12 (2006)
146. D'Aguanno, A., Vercellesi, G.: Automatic synchronization between audio and partial music score representation. In: Proceedings of 6th Workshop on Adaptive Multimedia Retrieval (AMR 2008). Berlin, Germany (2008)
147. Tolos, M., Tato, R., Kemp, T.: Mood-based navigation through large collections of musical data. In: Proceedings of 2nd CCNC 2005, pp. 71–75. Las Vegas, NV (2005)
148. Feng, Y., Zhuang, Y., Pan, Y.: Popular music retrieval by detecting mood. In: Proceedings 26th International SIGIR Conference on Research and Development in Information Retrieval, pp. 375–376. Toronto, ACM, Canada (2003)
149. Li, T., Ogihara, M.: Detecting emotion in music. In: Proceedigns of ISMIR, pp. 239–240. Baltimore (2003)
150. Liu, D.: Automatic mood detection from acoustic music data. In: Proceedings International Conference on Music Information Retrieval, pp. 13–17 (2003)
151. Lu, L., Liu, D., Zhang, H.: Automatic mood detection and tracking of music audio signals. IEEE Trans. Audio, Speech, Lang. Process. **14**(1), 5–18 (2006)
152. Trohidis, K., Tsoumakas, G., Kalliris, G., Vlahavas, I.: Multi-label classification of music into emotions. In: Proceedings 9th International Conference on Music Information Retrieval (ISMIR), pp. 325–330. Philadelphia (2008)
153. Logan, B.: Mel frequency cepstral coefficients for music modeling. In: Proceedings of ISMIR. Plymouth, USA (2000)
154. Peeters, G.: A generic training and classification system for MIREX08 classification tasks: Audio music mood, audio genre, audio artist and audio tag. In: Proceedings of MIREX as part of the 9th International Conference on Music Information Retrieval (ISMIR), ISMIR, Philadelphia, PY (2008)
155. Boersma, P.: Praat, a system for doing phonetics by computer. Glot Int. **5**, 341–345 (2001)
156. Chase, W.: How Music REALLY Works!. 2nd edn. Roedy Black Publishing, Vancouver, Canada (2006)
157. Harte, C.A., Sandler, M.: Automatic chord identification using a quantised chromagram. 118th Convention of the AES, May (2005)
158. Porter, M.F.: An algorithm for suffix stripping. Program **3**(14), 130–137 (1980)
159. Chuang, Z.-J., Wu, C.-H.: Emotion recognition using acoustic features and textual content. In: Proceedings of ICME, pp. 53–56. Taipei, Taiwan (2004)
160. Schuller, B., Batliner, A., Steidl, S., Seppi, D.: Emotion recognition from speech: putting asr in the loop. In: Proceedings 34th IEEE International Conference on Acoustics, Speech, and Signal Processing, ICASSP 2009, pp. 4585–4588. Taipei, Taiwan (2009)
161. Liu, H., Singh, P.: ConceptNet—a practical commonsense reasoning tool-kit. BT Technol. J. **22**(4), 211–226 (2004)

162. Schuller, B., Schenk, J., Rigoll, G., Knaup, T.: The godfather versus "chaos": comparing linguistic analysis based on online knowledge sources and bags-of-n-grams for movie review valence estimation. In: IAPR, IEEE Proceedings 10th International Conference on Document Analysis and Recognition, ICDAR 2009, pp. 858–862, Barcelona, Spain (2009)
163. Ekman, P., Sorenson, E., Friesen, W.: Pan-cultural elements in facial displays of emotions. Science **164**, 86–88 (1969)
164. Bradley, M.M., Lang, P.J.: Affective norms for english words (anew): Stimuli, instruction manual, and affective ratings. Technical Report C-1, Center for Research in Psychophysiology, University of Florida, Gainesville, Florida (1999)
165. Hu, X., Downie, J.S.: Exploring mood metadata: relationships with genre, artist and usage metadata. In: Proceedings 8th International Conference on Music Information Retrieval (ISMIR), Vienna, Austria (2007)
166. Wöllmer, M., Eyben, F., Reiter, S., Schuller, B., Cox, C., Douglas-Cowie, E., Cowie, R.: Abandoning emotion classes—towards continuous emotion recognition with modelling of long-range dependencies. In: Proceedings INTERSPEECH 2008, 9th Annual Conference of the International Speech Communication Association, incorporating 12th Australasian International Conference on Speech Science and Technology, SST 2008, ISCA/ASSTA, ISCA, pp. 597–600. Brisbane, Australia (2008)
167. Schuller, B., Müller, R., Eyben, F., Gast, J., Hörnler, B., Wöllmer, M., Rigoll, G., Höthker, A., Konosu, H.: Being bored? recognising natural interest by extensive audiovisual integration for real-life application. Image Vis. Comput. Spec. Issue Vis. Multimodal Anal. Hum. Spontaneous Behav. **27**(12), 1760–1774 (2009)
168. Mesaros, A., Virtanen, T., Klapuri, A.: Singer identification in polyphonic music using vocal separation and pattern recognition methods. In: Proceedings of ISMIR, pp. 375–378 (2007)
169. Mesaros, A., Virtanen, T.: Automatic recognition of lyrics in singing. EURASIP J. Audio, Speech, Music Process. Article ID 546047 (2009)
170. Durrieu, J.-L., Richard, G., David, B., Févotte, C.: Source/filter model for unsupervised main melody extraction from polyphonic audio signals. IEEE Trans. Audio, Speech, Lang. Process. **18**(3), 564–575 (2010)
171. Durrieu, J.-L., Richard, G., David, B.: An iterative approach to monaural musical mixture de-soloing. In: Proceedings of ICASSP, pp. 105–108, Taipei, Taiwan (2009)
172. Eyben, F., Wöllmer, M., Schuller, B.: Opensmile—the munich versatile and fast open-source audio feature extractor. In: Proceedings of the 9th ACM International Conference on Multimedia, MM 2010, pp. 1459–1462. ACM, Florence, Italy (2010)

Chapter 12
Applications in Intelligent Sound Analysis

If you develop an ear for sounds that are musical it is like developing an ego. You begin to refuse sounds that are not musical and that way cut yourself off from a good deal of experience.

—John Cage

Apart from the more specific types of sound considered so far—speech and music—general sound can also carry relevant information. This is, however, a considerably less researched field up to-date. Most prominent in this area are the tasks of acoustic event detection (AED) and classification (AEC) [1] that can be subsumed under the area of computational auditory scene analysis (CASA) [2]. For these tasks international evaluation campaigns exist that have mostly seen HMM and SVM approaches with various acoustic features [1]. Fields of application include media retrieval [3] including affective content analysis [4] or human-machine and human-robot interaction [5], animal vocalisation recognition [6], and monitoring of industrial processes [7]. Mostly, closed-set recognition is addressed, i.e., training and testing classes are the same. Recently, however, also open-set recognition is faced, the so-called novelty detection [8, 9].

As before, examples of application have been chosen for illustration of obtainable performances and methods employed. Three applications have been chosen to cover a good variety of the above named use cases: Firstly, recognition of animal vocalisation [10], then, acoustic event classification including unsupervised learning to exploit the availability of sheer infinite amounts of sound on the Internet [11], and finally prediction of the emotion evoked in human listeners of sound [12] in analogy to the sections on speech and music.

B. W. Schuller, *Intelligent Audio Analysis*, Signals and Communication Technology, DOI: 10.1007/978-3-642-36806-6_12,

12.1 Animal Vocalisations

As a first example of application in more general sound analysis, we will consider animal voices instead of human voices. The following application investigates the recognition of animal vocalisations 'in the wild' [10].

In the field of bioacoustics, a multiplicity of approaches exists for classifying animal sounds, for example to monitor populations of certain species, such as whales [13] or birds [14]. More recently, increasing efforts are invested in digitisation of sound archives. Similarly as in the case of MIR, this demands for efficient indexing and retrieval methods. For example, in [15], an effective indexing algorithm for animals with curve-like harmonic vocalisations, such as various species of birds, was presented and evaluated on bird songs contained in the Animal Sound Archive ("Tierstimmenarchiv") of the Humboldt-University of Berlin [16]. This data set will be referred to as 'HU-ASA database' in the ongoing. In the past, SVM-based static classification using segment-wise functionals [17] (e.g., mean and standard deviation) was proposed for animal sounds classification [18]. Alternatively, dynamic classification, e.g., by HMMs [19] or by suited neural networks [6] is reported successful in the literature. Hence, we will consider SVMs, HMMs with different topologies, and LSTM recurrent neural networks on the HU-ASA database in the ongoing.

12.1.1 HU-ASA Database

The evaluation database builds on the large HU-ASA database of animal vocalisations. It is annotated with the species and additional meta-data such as recording conditions and the type of vocalisation for each audio file. 1418 audio files are available in MP3 encoding. These were obtained from the on-line archive.[1] Per species, the audio files with (biological) class were automatically annotated (e.g., *Aves*, *Mammalia*), order (e.g., *Passeriformes*, *Primates*), and family (e.g., *Felidae*, *Canidae*) according to the Linnaean rank-based biological classification as retrieved from Wikipedia.[2] The majority of the HU-ASA's instances consist of bird (*Aves*) and mammal (*Mammalia*) sounds, as shown in Table 12.1. The class 'Others' include *Sauropsida*, *Hexapoda*, and recordings without automatic annotation, where according information was missing in Wikipedia. The total audio duration is 20423 s (5 h 40 min 23 s). *Amphibia*, *Insecta*, and *Reptilia* were not included in the described experiments given their sparseness (cf. Table 12.1).

Two tasks of practical interest were derived from the biological classification, as shown in Table 12.2. The first (2-class) task aims at classification of songbirds (*Passeriformes*) versus non-songbirds (*Non-Passeriformes*). Non-songbirds include by number of instances the orders *Anseriformes*, *Charadriiformes*, *Galliformes*, *Psitacciformes*, *Gruiformes*, and 24 other orders—often with sparse instances.

[1] http://www.tierstimmenarchiv.de, accessed mid 2010.

[2] http://www.wikipedia.org

Table 12.1 Number of instances, as well as min(imum), mean, max(imum), and total recording length (Σ) of the audio files by the biological class of the species in the HU-ASA database

(Biological) Class	# Instances	Duration [s]			
		Min	Mean	Max	Sum
Aves	868	2.4	14.8	64.7	12 210
Mammalia	487	1.0	14.7	37.7	6 954
Amphibia	27	1.8	19.6	65.9	529
Reptilia	7	11.2	22.5	39.6	157
Insecta	19	2.3	16.0	30.1	287
Other	10				133
Sum	1 418				20 423

Table 12.2 Distribution of instances in the 2-class (*Passeriformes / Non-Passeriformes*) and 5-class tasks as defined on the HU-ASA database

Class	# Instances
Passeriformes	282
Non-Passeriformes	586
Sum	868
Primates	90
Canidae	43
Felidae	62
Sum	1 063

The more complex 5-class task adds mammals (*Mammalia*) of the families *Felidae* and *Canidae*, as well as the instances of the biological order *Primates* (cf. Table 12.2). A particular challenge arises from the real-world nature of the database: vocalisations of the same species often vary considerably, depending on the situation and stance (i.e., aggression or warning calls), and age of the animals, from young to full-grown. The recordings are further corrupted by background noises—even of other animal species.

12.1.2 Methodology

Static classification by SVMs bases on linear kernel SVM. For dynamic classification, two topologies of HMMs and LSTM RNNs are compared. A typical HMM topology in audio (and general sequence) classification is a linear (left-right) layout: With N as the number of states in total, state transitions are allowed from state $i = 1, \ldots, N-1$ to states i and $i + 1$. However, animal vocalisations are often highly repetitive, motivating the usage of a *cyclic* topology. In such a layout a transition from state N to the first state is added. In the following experiments the number of states is fixed to $N = 8$ basing on a series of evaluations.

As for neural networks, e.g., a feedforward MLP was used for classifying animal vocalisations in [6]. To enhance the neural network paradigm by extended memory capabilities, LSTM networks are considered here with one hidden layer of 100 LSTM memory cells. The size of the input and output layers was equal to the number of features and classes to discriminate. Softmax functions were applied to the output activations, and the resulting values were normalised to the sum one to provide posterior class probabilities.

MFCCs 1–12 along with energy and their first (δ) and second order ($\delta\delta$) regression coefficients were chosen as features for frame-level classification due to their suitability across a multiplicity of Intelligent Audio Analysis tasks [17–19]. In [19], these features were found superior to the MPEG-7 spectral projection features as used in [15] for sound classification with HMMs. The overall 39-dimensional feature set will be denoted by 'MFCC'.

For static classification of larger audio chunks, functionals are applied. In [17], mean and standard deviation were proposed. The functionals considered in the ongoing also include extremes and higher-order moments [20]. Additional LLDs for include HNR, pitch and ZCR by using openSMILE's (cf. Sect. 6.5, [21]) INTERSPEECH 2009 Emotion Challenge set [20], as described in Table A.1. This choice could allow to discriminate between animals with voiced and unvoiced sounds. The functionals of the 32 LLD will be denoted by 'IS09-func'. For better comparability of classifier paradigms less dependent of the acoustic features used, the functionals listed in Table A.1 were also computed only from the MFCCs 1–12 along with energy; this feature set will be called 'MFCC-func'. The IS09-func and MFCC-func feature sets consist of 384 and 312 features, respectively.

12.1.3 Performance

Ten-fold SCV is used for evaluation with partitioning by the Weka toolkit [22] with the default random seed of 0 for easy reproducibility. 10 % of the data were used for evaluation, and 10 % for validation whenever needed, e.g., for neural network training. HMMs were trained by the EM algorithm: Gaussian mixtures were consecutively added and re-estimated after six initial iterations until 16 Gaussian mixtures were reached for each state. For network training, supervised learning with early stopping and MVN was used. The network weights were initialised randomly from a Gaussian distribution ($\mu = 0, \sigma = 0.1$). Then, each training sequence was presented frame by frame. For improved generalisation ability, the order of the input sequences was determined randomly, and Gaussian noise ($\mu = 0, \sigma = 0.3$) was added to the input activations. The network weights were iteratively updated by resilient propagation. Further, the performance (in terms of WA) on the validation set was evaluated after each training epoch. Training was stopped in case of no improvement over 20 epochs or after a total of 100 training epochs. Then, the network with the best performance on the validation set was selected as the final network. SVMs were trained using SMO and a complexity constant of 0.1 on MVN processed features.

Table 12.3 Results of the 2-class and 5-class tasks of the HU-ASA database with various classifiers and feature sets

Classifier	[%]	2-class		5-class	
	Features	UA	WA	UA	WA
SVM	IS09-func	69.0	72.0	46.4	57.2
SVM	MFCC-func	73.9	75.6	42.2	56.0
Left-right HMM	MFCC	79.0	79.8	47.3	63.4
cyclic HMM	MFCC	79.0	79.6	49.5	64.0
LSTM	MFCC	80.0	81.3	41.1	62.3

The training set was up-sampled for each fold for the LSTM-RNN and SVM classifiers. This was done by copying training instances of minority classes until a near-uniform class distribution is achieved. This step was not necessary in the case of HMMs, as each class is learnt by an individual model, and classification is performed with HMMs and the maximum likelihood criterion, i.e., class priors, were not used in the decision rule. For classification with the LSTM RNN each sequence in the test set was presented frame by frame to the input layer, and each frame was assigned to the class with the highest probability as indicated by the output layer. Then, a majority vote over the frame-level decisions was made to label the sequence.

Table 12.3 depicts results by UA and WA for the 2-class and 5-class tasks of the HU-ASA database, as defined in Table 12.2. Always deciding for the majority class leads to WA and UA of 55.1 % and 20.0 % (5-class task), and 67.5 % and 50.0 % (2-class task).

In SVM classification on the 2-class task, the MFCC-func feature set outperforms the IS09-func set in terms of WA by 3.6 % absolute, being significant at the 5 % level (one-tailed z-test). However, the IS09-func feature leads to a significantly higher UA (4.4 % absolute improvement) for the 5-class task. Both types of HMMs outperform static classification by SVM. Further, the cyclic HMM is superior to the left-right HMM justifying the made assumption of partly quasi-periodic vocalisations. Yet, this observation is not significant on the 5 % level. To explain this, the estimated 'cycle probability' $a_{N,1}$ of the HMMs is shown for each class, on average across the ten folds, in Table 12.4. There, the cycle probabilities are around 28 % in the models for songbirds (*Passeriformes*) and primates, but below 10 % for *Felidae*.

The additional LLDs from Table A.1 as input features for the HMMs could not improve the above results. The impact of a varying number of Gaussian mixtures

Table 12.4 Cycle probabilities $a_{N,1}$ after training of the cyclic HMMs for comparison among each other given for each class in the 5-class task, averaged over ten folds

Class	$a_{N,1}$ [%]
Passeriformes	28.1
Non-Passeriformes	17.2
Canidae	14.2
Felidae	9.9
Primates	28.0

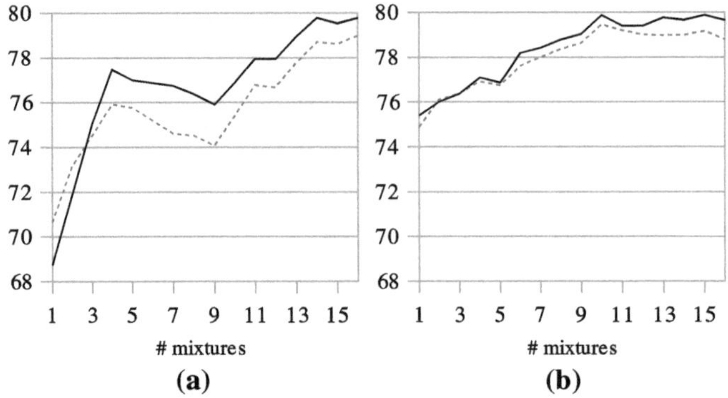

Fig. 12.1 UA and WA on the HU-ASA database by 8-state HMMs with left-right and cyclic topologies, depending on the number of mixtures per state. Solid line: WA, dashed line: UA [10] **a** left-right HMM, 2-class task, **b** cyclic HMM, 2-class task

for the HMMs is selectively shown in Fig. 12.1 for the 2-class task. Interestingly, the cyclic HMM performs better than the left-right HMM for a small numbers of mixtures. Further, the UA on the 5-class task seems to be largely unaffected by the number of mixtures. This is surprising given that, ML classification partially compensates for the unequal class distribution. LSTM RNNs outperform—not significantly ($p > 5\,\%$)—the HMMs on the 2-class task. Yet, they have the lowest UA for the 5-class task. Additional variation of the network layout may change this behaviour. However, the lower performance for the 5-class is likely partly owing to the sparseness of the non-bird classes as LSTM RNN have a comparably high demand of training data.

12.1.4 Summary

In this section, an evaluation framework was shown for a challenging real-world database of animal vocalisations. The performances of static and dynamic classifiers, including LSTM networks, were compared. Dynamic classification provided higher accuracy. In the comparison of 'standard' MFCC features with an enhanced feature set containing pitch and voicing information no clear preference could be determined. Further evaluations in this direction are needed to reveal the relevance of different LLD and functional types for the classification of animal vocalisations.

From a classifier point of view, a hierarchical classification framework, e.g., by combining the songbird/non-songbird classifier with a bird song recogniser could be attempted.

12.2 Acoustic Events

In the next application of sound analysis, baseline results for the recognition of sound events are given. At the same time, this shall serve as an example of the usage of unlabelled data—sound event archives exist in masses on the Internet and can be exploited in semi-supervised learning even if no labels are given [11].

Recently, there is increasing interest in sound event classification in the field of acoustic signal analysis. This comes, apart from interest for application in multimedia search based on sound, as it is one of the key components to acoustically analyse environments, e.g., in surveillance [23, 24], monitoring of people in need of care, or detecting, and classifying sources of interest in real time [25]. There is also a benefit for humanoid and general robots [26] if they are able to better understand their acoustic environment. Finally, speech and music enhancement may be improved given a reliable identification of disturbing sound events. So far, most of research efforts in this direction base on rather prototypical and small databases with less than or around 1 000 instances (e.g., as in [24, 27–32]), or a few thousands of instances [26, 32, 33].

In this section, we will focus on sound events classification in a large scale database, covering sound classes that reach from nature (such as animals) over human beings (i.e., people) to artificial sounds (i.e., office, musical instruments, noise makers, and vehicles) as was introduced in Sect. 5.3.3.

Semi-supervised learning will be used to have the machine by itself label additional data instances as "there is no data like more data" and human labelling can easily become tedious and is expensive. Given a sufficiently robust automatic sound event classification system, unlabelled data can be classified and used in an iterative re-training process. Unlabelled sound data is practicably available in 'infinite' amounts: Recordings of real-life audio can be easily collected and typically contain various kinds and huge numbers of sound events [34]. Further, audio data can be added from the Internet. The semi-supervised adaptation of AMs and LMs in ASR [35, 36] and affective speech analysis [37] demonstrates that addition of unlabelled training data can lead to improvements in accuracy of classification systems. However, typically at least twice or sometimes up to around ten times as much unlabelled data is needed as compared to labelled data. Thus, AEC is shown in this book as an example for semi-supervised learning to improve a sound event classifier.

12.2.1 Methodology

openSMILE's (cf. Sect. 6.5, [21]) 'AVEC 2011' set as shown in Table A.1 in the Annex is used for AEC. It consists of 1 941 features, composed of 25 energy and spectral related LLD x 42 functionals, 6 voicing related LLD x 32 functionals, 25 delta coefficients of the energy/spectral LLD x 23 functionals, 6 delta coefficients of the voicing related LLD x 19 functionals, and 10 voiced/unvoiced durational features.

As classifier, Random Forests as ensemble of decision trees are used. This choice is motivated by their good ability to cope with large feature spaces, as feature sub-spaces are randomly assigned to the trees in the forest. A good configuration proved to be 30 trees, and 150 randomly assigned features for each tree. For further reproducibility besides using an open-source feature extractor and the FindSounds database (cf. Sect. 5.3.3) that can be retrieved from the Internet, the classifier implementation provided by the Weka toolkit [22] is chosen again.

12.2.2 Performance

Considering the imbalance of instances among the classes, UA will be the measure of primary interest. Further, WA is partly provided in addition, as well as recall, precision, and F_1-measure. The experiments base on random partitioning of the FindSounds database into three stratified folds to provide two training and one completely disjoint testing set. The first fold (F1, 5 646 instances) is always used with its original manually assigned labels for training. The second fold (F2, 5 646 instances) is used either without its original labels ($F2_U$) or with these labels (F2) to be able to compare to using this fold in a semi-supervised or supervised manner for training. The third and last fold (5 645 instances) is always used for testing. Random partitioning is carried out with Weka's default random seed.

Table 12.5 shows the occurred confusions for seven categories of sound event classification using the original labels training on fold one and two and testing on the third fold. This is the 'best case' given the entirely supervised learning with utmost data and serves as upper benchmark. Most confusions can be explained well by common sense, such as those of sounds from people with sounds of animals or sounds from vehicles with sounds of noise makers.

Table 12.5 'Best case' confusions when automatically classifying seven sound categories on the FindSounds database with original labels for both training folds F1 and F2 (cf. line 'supervised F1 + F2' in Table 12.6)

Truth [#]	Classified as						
	People	Animals	Nature	Vehicles	Noisemakers	Office	Instruments
People	564	153	11	26	17	25	50
Animals	126	717	7	35	23	20	18
Nature	18	35	157	42	44	10	6
Vehicles	37	37	26	476	86	15	45
Noisemakers	22	43	36	77	372	72	48
Office	29	37	1	16	111	364	31
Instruments	32	33	6	31	47	16	1 395
Confusions	264	338	87	227	328	158	198

Table 12.6 Recall for seven sound categories and UA/WA with un-/supervised learning on the FindSounds database

	[%]	UA	WA	People	Animals	Nature	Vehicles	Noisemakers	Office	Instruments
supervised	F1	61.1	67.0	61.7	68.2	39.7	60.2	52.7	57.9	87.2
semi-supervised	$2{\cdot}F1 + F2^1_U$	63.1	68.5	61.7	72.5	47.4	61.8	51.9	58.4	87.9
supervised	F1 + F2	66.5	71.7	66.7	75.8	50.3	65.9	55.5	61.8	89.4

To establish a reference if the fold two data is not used at all, let us now consider exclusively fold one with its original labels for training (line "supervised F1" shown in Table 12.6) and fold three for testing. Then, for semi-supervised learning, fold one with the original manually assigned labels and fold two without the original labels, but labelled automatically by a system which was trained on fold one with application of diverse strategies (line "semi-supervised" in the same table). Testing is again carried out on fold three. Finally, the upper benchmark of using both folds with the original labels is shown in the table (line "supervised F1 + F2")—again with fold three for testing.

For semi-supervised learning, the confidence of the Random Forests—the percentage of trees agreeing on the class—is taken into account. Evaluated confidence levels are > 0.7 and > 0.8. This is needed to suppress data likely labelled wrong by the machine. Two additional strategies are investigated: up-sampling of the originally labelled data to emphasise more on definitely correctly labelled data and repeated iteration of the semi-supervised learning process. Table 12.7 shows the UA of up to three iterations of semi-supervised learning, i.e., repeated re-labelling of the unlabelled data in fold two using all fold one data and selected fold two data in training with labels from the last iteration, and only using instances with sufficient confidence level. Without up-sampling (1·F1), a gain is also obtained (62.0 % vs. 61.6 % UA for confidence level > 0.7, and 63.0 % vs. 62.1 % UA for confidence level > 0.8). However, one notices that the benefit of iteration is limited, as UA partly begins to decrease after the third iteration. A larger number of iterations did not lead to improvements (not shown in numbers). Finally, the up-sampling and iterating strategies are

Table 12.7 UA of iterative semi-supervised learning on the FindSounds database with minimum confidence values 0.7 and 0.8 combined with up-sampling or not up-sampling of originally labelled data

UA [%]	Confidence level			
	>0.7		>0.8	
	F1	2·F1	F1	2·F1
$F2^1_U$	61.6	63.1	62.1	62.5
$F2^2_U$	62.0	62.2	63.0	62.6
$F2^3_U$	62.0	61.7	62.6	63.2

2·F1: up-sampling (doubling up) fold 1 instances; $F2^1_U$, $F2^2_U$, $F2^3_U$: first, second, and third iteration of semi-supervised learning combined expecting synergies. Looking at line "2·F1" in Table 12.6, up-sampling improves over the baseline setting in four out of six cases. Table 12.6 also shows detailed results for the case up-sampling by copying (2·F1) and confidences higher than 0.7.

Looking again at UA values in Table 12.7, as one would expect, the best average result is obtained using the original labels and data of fold one and fold two for training (66.5 % UA). Then, semi-supervised learning significantly (one-sided z-test, $p < 0.05$) boosts the performance of sound event classification by an increase in UA of 2 % absolute over not using fold two data at all. This boost is almost half the one achieved by supervised training (5.4 %) with all data over only using fold one. The nature class being the most sparse one, benefited most from semi-supervised learning. This effectively demonstrates the potential gain of semi-supervised learning for exploitation of unlabelled audio data.

12.2.3 Summary

The potential of semi-supervised learning on a large scale AEC task was investigated. In the result, adding unlabelled data with high classifier confidence level to the human-labelled training data can enhance recognition performance. Up-sampling of originally labelled data and iterating the semi-supervised learning process both boosted classification accuracy in the experiments by emphasising on originally labelled data. Combining both strategies gradually increases the advantage of semi-supervised learning. As one would expect, accuracy of semi-supervised learning is below the gain that can be expected when adding labelled data of the same amount. Yet, given the considerable efforts and costs involved in human labelling of thousands of instances and the large amounts of sound event data publicly available makes consideration of semi-supervised learning a promising approach in future machine-based sound analysis.

Future efforts could continue to focus on agglomeration of huge amounts of unlabelled sound event data and its application in analysis of real-life sound streams—ideally in combination with blind audio source separation.

12.3 Emotion

Similarly to the analysis of speech and music, where we first looked at 'what' was being said or played before looking at the affective side of speech and music, one can also attempt to automatically predict the emotion a sound event is likely to evoke in a listener. This will be the last application example presented in this book. It was first introduced in [12].

In fact, literature on emotion recognition from the acoustic channel—be it the emotion a listener thinks is contained or that she or he feels when listening—, is dominated by studies dealing with speech [20, 38], and next follows music [39]. However, as shown in the last two sections, there is a rich variety of sounds besides speech and music in a real acoustic environment. These sounds certainly are also loaded with emotional connotation for a human listener. As an example, the shrill sound of a fire alarm would be less pleasant than the gentle sound of waves drilling the sand beach to the majority of listeners. In fact, listeners feed back emotion to any sound they are listening to in their daily life. This is independent of the kind of sound and its subjective or objective nature. Sound perception is thus linked with emotional response: New-borns' first attempts to overcome anxiety are centred on sound making [40]. Thus, for future intelligent systems it may be useful or relevant to understand emotion connotated with general sound. In 'sound information retrieval' emotional content may help in the design and dubbing of audio plays and films. For example, one might look for a furious door slam or a spooky door creek, etc. Research in this direction is utmost limited up to the present day: The only work besides the work by Schuller et al. is the very recent one presented in [41] basing on 120 clips of the BBC Sound Effects Library labelled in three affective dimensions. The approach uses mean and standard deviation per one second of 12 MFCC features as acoustic feature information. In this section, the focus is set on sound emotion recognition in realistic conditions.

A crucial problem is the lack of specialised sound databases for emotion research. There some freely accessible sound databases [42], but usually without emotional labelling. The Emotional FindSounds database, which was described in Sect. 5.3.3 solves these issues. In emotion recognition from speech, emphasis is usually put on the subject's expressed emotion rather than listeners' emotions evoked by sound. This is more mixed for music emotion recognition. In fact, knowledge upon the emotion elicited on the listener side may help identify human reaction ahead. In this section, 'sound emotions' refer to the listeners' induced emotions.

12.3.1 Methodology

The audio feature set used is the openSMILE toolkit's 'AVEC 2011' set with 1941 features as shown in Table A.1 in the Annex and as was used in the last section for AEC. For recognition, random subspace meta-learning is used again owing to its good generalisation properties—the sounds are highly varied and require this feature. The base classifier is a decision tree. Based on experience, trees are not pruned. A subspace size of 0.05 is chosen, which means that 97 features out of the 1941 are assigned by random to each tree in the forest. The forest is grown from 500 trees [12]. The labelling and the feature extractor including the configuration are available for reproduction.[3] This principle was kept by again deciding for Weka for the implementation of the trees.

[3] Available at http://www.openaudio.eu

Table 12.8 Automatic regression results by CC with different types of gold standard

CC		#trees		
		100	200	500
Arousal	EWE	0.611	0.608	0.606
	median	0.553	0.555	0.548
	mean	0.558	0.563	0.559
Valence	EWE	0.458	0.469	0.473
	median	0.446	0.449	0.454
	mean	0.467	0.484	0.485

EWE, median, and mean in ten-fold SCV. The number of trees is varied

12.3.2 Performance

A ten-fold SCV—again with reproducible partitioning by Weka's default random seed—is carried out on the emotionally tagged partition of the FindSounds database as introduced in Sect. 5.3.3. Table 12.8 shows the CCs for arousal and valence employing the Evaluator Weighted Estimator (EWE), median, and mean to establish a gold standard by merging the evaluation results of the four evaluators. In this table, numbers of trees in the forest are additionally varied. Visibly, the regression of sound emotion performs well with CCs of around 0.61 (arousal) and up to 0.49 (valence) when evaluating on the EWE. The tendency that arousal is the 'easier' task is well in line with experience from speech and music emotion analysis based on acoustics [20, 43]. CC as evaluated on EWE usually exceeds the other two methods of gold standard establishment—mean and median. Median is found on the other end of the scale probably due to its instability when evaluators show huge disagreement. In Table 12.9 the CC and its relation to sound category is highlighted for one exemplary configuration. There, arousal prediction is roughly stable across sound categories. As for valence, especially *Noisemakers* and *Nature* can be identified well above others in

Table 12.9 Automatic regression results by CC per sound category for EWE and 500 trees in ten-fold SCV

Class	CC	
	Arousal	Valence
	0.601	0.474
Animals	0.643	0.448
Musical instruments	0.516	0.217
Nature	0.688	0.589
Noisemaker	0.579	0.778
People	0.604	0.048
Sports	0.682	0.198
Tools	0.590	−0.057
Vehicles	0.579	0.279

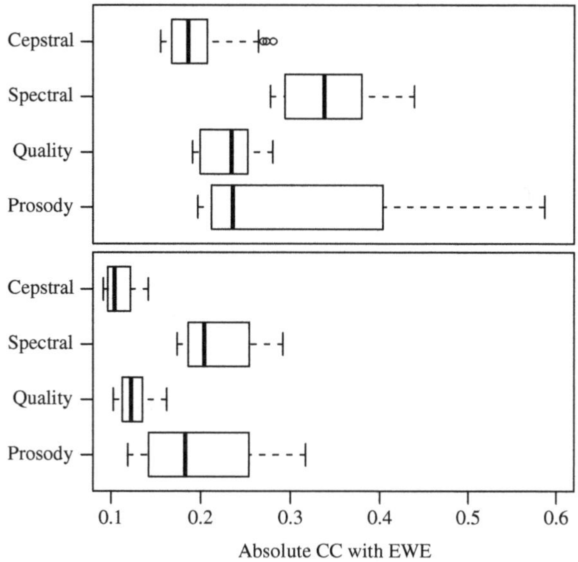

Fig. 12.2 Boxplots of the 30 highest absolute CCs of features with the EWE. Features are grouped in four cover classes for arousal (top) and valence (bottom). The 'Quality' group contains voicing probability, log HNR, jitter, and shimmer based features. 'Prosody' groups loudness, F0, and ZCR [12]

terms of CC. In comparison with the gold standard as was shown in Fig. 5.8, one may argue that the regressor is not only implicitly recognising the sound category. In fact, the values of valence for *Noisemakers* are rather widespread despite considerable differences in the mean valence.

As there exists practically no experience on feature relevance for this particular task, it seems worth to have a look at this issue. The 30 best features were ranked by their CC with the EWE as gold standard. The result is shown as boxplots per dimension for the groups cepstral, spectral, 'sound quality' in analogy to voice quality, and prosody in Fig. 12.2. Independent of arousal or valence, spectral features are the most relevant group. Interestingly, the best single feature is prosody-related for these two dimensions. From the full list of the 30 best features (not shown) the following is found: Arousal is highly correlated with loudness, and loudness features almost reach the CC with the EWE of the learnt regressor. The highest CC is observed for the root quadratic mean of loudness (0.587).

Next, valence is correlated with loudness as well, but not as strongly and negatively, which seems intuitive, as loud sounds are likely unpleasant. The highest absolute CC with the EWE can be reported for the third quartile of loudness (−0.316). Spectral flux also shows good (negative) CC, i.e., large spectral variations seem to be perceived as unpleasant: The CC of the inter quartile range 1–2 of spectral flux is −0.292. Finally, spectral harmonicity is negatively correlated: Apparently quasi-sinusoidal sounds are unpleasant. The CC of 50 % up-level time of harmonicity is −0.241.

12.3.3 Summary

Automatic recognition of emotion evoked by general sound events was shown and found in the rough range of typical dimensional speech and music emotion recognition when operating in high realism comparable to the results in Sect. 10.4.2 and Sect. 11.7. The sound events considered here were completely independent of each other and often of lower acoustic quality. Spectral features were shown to be most important as a group after individual prosodic features for this task.

Future efforts may aim at creation of larger sound emotion resources, e.g., by crowd sourcing or similar. Deeper analysis of feature relevance per sound category will also shed more light on optimal acoustic feature spaces. Finally, multi-task learning of the sound category and the evoked emotion seems a promising approach to improve both tasks as was suggested in speech and music processing before.

References

1. Temko, A., Nadeu, C., Macho, D., Malkin, R., Zieger, C., Omologo, M.: Acoustic event detection and classification. In: Waibel, A., Stiefelhagen, R. (eds.) Computers in the Human Interaction Loop, pp. 61–73. Springer, London (2009)
2. Wang, D., Brown, G.: Computational auditory scene analysis: Principles, algorithms, and applications. IEEE Press (2006)
3. Huang, Q., Cox, S.: Using high-level information to detect key audio events in a tennis game. In: Proceedings INTERSPEECH 2010, 11th Annual Conference of the International Speech Communication Association, pp. 1409–1412. Makuhari, Japan, Sept 2010. ISCA
4. Xu, M., Chia, L., Jin, J.: Affective content analysis in comedy and horror videos by audio emotional event detection. In: Proceedings 6th IEEE International Conference on Multimedia and Expo, ICME 2005,p. 4. Amsterdam, The Netherlands, IEEE, July 2005
5. Okuno, H., Ogata, T., Komatani, K., Nakadai, K.: Computational auditory scene analysis and its application to robot audition. In: Proceedings of the International Conference on Informatics Research for Development of Knowledge Society Infrastructure, pp. 73–80. IEEE (2004)
6. Gunasekaran, S., Revathy, K.: Content-based classification and retrieval of wild animal sounds using feature selection algorithm. In: Proceedings of International Conference on Machine Learning and Computing (ICMLC), pp. 272–275. IEEE Computer Society, Bangalore, India, Feb 2010
7. Wan, C., Mita, A.: An automatic pipeline monitoring system based on PCA and SVM. World Acad. Sci. Eng. Technol. **45**, 90–96 (2008)
8. Bach, J., Anemuller, J.: 11th Annual Conference of the International Speech Communication Association, pp. 2206–2209. ISCA, Makuhari, Japan, Sept 2010
9. Geiger, J.T., Lakhal, M.A., Schuller, B., Rigoll, G.: Learning new acoustic events in an hmm-based system using map adaptation. In: Proceedings INTERSPEECH 2011, 12th Annual Conference of the International Speech Communication Association, pp. 293–296. ISCA, Florence, Italy, Aug 2011
10. Weninger, F., Schuller, B.: Audio recognition in the wild: Static and dynamic classification on a real-world database of animal vocalizations. In: Proceedings of 36th IEEE International Conference on Acoustics, Speech, and Signal Processing, ICASSP 2011, pp. 337–340. IEEE, Prague, Czech Republic, May 2011

11. Zhang, Z., Schuller, B.: Semi-supervised learning helps in sound event classification. In: Proceedings of 37th IEEE International Conference on Acoustics, Speech, and Signal Processing, ICASSP 2012, pp. 333–336. IEEE, Kyoto, Japan, March 2012
12. Schuller, B., Hantke, S., Weninger, F., Han, W., Zhang, Z., Narayanan, S.: Automatic recognition of emotion evoked by general sound events. In: Proceedings of 37th IEEE International Conference on Acoustics, Speech, and Signal Processing, ICASSP 2012, pp. 341–344. IEEE, Kyoto, Japan, March 2012
13. Mellinger, D.K., Clark, C.W.: Recognizing transient low-frequency whale sounds by spectrogram correlation. J. Acoust. Soc. Am. **107**(6), 3518–3529 (2000)
14. Härmä, A.: Automatic recognition of bird species based on sinusoidal modeling of syllables. In: Proceedings of ICASSP, vol. 5, pp. 545–548. Hong Kong, April 2003
15. Bardeli, R.: Similarity search in animal sound databases. IEEE Trans. Multimedia **11**(1), 68–76 (2009)
16. Frommolt, K.-H., Bardeli, R., Kurth, F., Clausen, M.: The animal sound archive at the Humboldt-University of Berlin: current activities in conservation and improving access for bioacoustic research. Adv. Bioacoustics **2**, 139–144 (2006)
17. Guo, G., Li, S.Z.: Content-based audio classification and retrieval by support vector machines. IEEE Trans. Neural Networks **14**(1), 209–215 (2003)
18. Mitrovic, D., Zeppelzauer, M., Breiteneder, C.: Discrimination and retrieval of animal sounds. In: Proceedings of Multi-Media Modelling Conference, IEEE, Beijing, China, Jan 2006
19. Kim, H.-G., Burred, J.J., Sikora, T.: How efficient is MPEG-7 for general sound recognition?In: Proceedings of AES 25th International Conference, London, UK, June 2004
20. Schuller, B., Batliner, A., Steidl, S., Seppi, D.: Recognising realistic emotions and affect in speech: State of the art and lessons learnt from the first challenge. Speech Commun. **53**(9/10):1062–1087 (2011) (Special Issue Sensing Emotion and Affect-Facing Realism in Speech Processing)
21. Eyben, F., Wöllmer, M., Schuller, B.: Opensmile—the munich versatile and fast open-source audio feature extractor. In: Proceedings of the 9th ACM International Conference on Multimedia, MM 2010, pp. 1459–1462. ACM, Florence, Italy, October 2010
22. Hall, M., Frank, E., Holmes, G., Pfahringer, B., Reutemann, P., Witten, I.H.: The WEKA data mining software: An update. SIGKDD Explor. **11**(1), 10–18 (2009)
23. Temko, A., Malkin, R., Zieger, C., Macho, D., Nadeu, C.: Acoustic event detection and classification in smart-room environments: Evaluation of chil project systems. In: Proceedings of the IV Biennial Workshop on Speech Technology, pp. 1–6. Zaragoza, Spain (2006)
24. Clavel, C., Ehrette, T., Richard, G.: Events detection for an audio-based surveillance system. In: Proceedings of ICME, pp. 1306–1309. Amsterdam (2005)
25. Ferguson, B.G., Lo, K.W.: Acoustic cueing for surveillance and security applications.In: Proceedings of SPIE, Orlando, FL, USA (2006)
26. Kraft, F., Malkin, R., Schaaf, T., Waibel, A.: Temporal ICA for classification of acoustic events in a kitchen environment. In: Proceedings of INTERSPEECH, pp. 2689–2692. Lisbon, Portugal (2005)
27. Temko, A., Nadeu, C.: Classification of acoustic events using SVM-based clustering schemes. Pattern Recogn. **39**, 682–694 (2006)
28. Zieger, C., Omologo, M.: Acoustic event classification using a distributed microphone network with a GMM/SVM combined algorithm. In: Proceedings of INTERSPEECH, pp. 115–118. Brisbane, Australia (2008)
29. Heittola, T., Klapuri, A.: TUT acoustic event detection system 2007. In: Proceedings of Multimodal Technologies for Perception of Humans: International Evaluation Workshops CLEAR 2007 and RT 2007, pp. 364–370. Springer, Berlin, Heidelberg (2008)
30. Ntalampiras, S., Potamitis, I., Fakotakis, N.: Automatic recognition of urban environmental sound events. In: Proceedings of CIP2008, Eurasip, pp. 110–113 (2008)
31. Peng, Y., Lin, C., Sun, M., Tsai, K.: Healthcare audio event classification using hidden markov models and hierarchical hidden markov models. In: Proceedings of ICME, pp. 1218–1221. Piscataway, NJ, USA (2009)

32. Dat, T.H., Li, H.: Probabilistic distance svm with hellinger-exponential kernel for sound event classification. In: Proceedings of ICASSP, pp. 2272–2275. Prague, Czech Republic (2011)
33. Chu, S., Narayanan, S., Kuo, C.-C.J.: Environmental sound recognition with time-frequency audio features. Trans. Audio Speech Lang. Process. **17**(6), 1142–1158 (2009)
34. Mesaros, A., Heittola, T., Eronen, A., Virtanen, T.: Acoustic event detection in real life recordings. In: Proceedings of EUSIPCO, Aalborg, Denmark (2010)
35. Hakkani-Tur, D., Tur, G., Rahim, M., Riccardi, G.: Unsupervised and active learning in automatic speech recognition for call classification. In: Proceedings of ICASSP, pp. 429–432. Montreal, Canada, (2004)
36. Tur, G., Stolcke, A.: Unsupervised language model adaptation for meeting recognition. In: Proceedings of ICASSP, pp.173–176. Honolulu, Hawaii, USA (2007)
37. Zhang, Z., Weninger, F., Wöllmer, M., Schuller, B.: Unsupervised learning in cross-corpus acoustic emotion recognition. In: Proceedings of 12th Biannual IEEE Automatic Speech Recognition and Understanding Workshop, ASRU 2011, pp. 523–528. IEEE, Big Island, HY, Dec 2011
38. Gunes, H., Schuller, B., Pantic, M., Cowie, R.: Emotion representation, analysis and synthesis in continuous space: a survey. In: Proceedings of the International Workshop on Emotion Synthesis, representation, and Analysis in Continuous spacE, EmoSPACE 2011, held in Conjunction with the 9th IEEE International Conference on Automatic Face & Gesture Recognition and Workshops, FG 2011, pp. 827–834. IEEE, Santa Barbara, CA, March 2011
39. Kim, Y., Schmidt, E., Migneco, R., Morton, B., Richardson, P., Scott, J., Speck, J., Turnbull, D.: Music emotion recognition: a state of the art review. In: Proceedings of ISMIR, pp. 255–266. Utrecht, The Netherlands (2010)
40. Forrester, M.: Auditory perception and sound as event: theorising sound imagery in psychology. J. Sound, http://www.kent.ac.uk/arts/sound-journal/forrester001.html (2000)
41. Sundaram, S., Schleicher, R.: Towards evaluation of example-based audio retrieval system using affective dimensions. In: Proceedings of ICME, pp. 573–577. Singapore, Singapore (2010)
42. Gygi, B., Shafiro, V.: Development of the database for environmental sound research and application (DESRA): Design, functionality, and retrieval considerations. EURASIP J. Audio Speech Music Process. pp. 12 (2010). Article ID: 654914
43. Schuller, B., Dorfner, J., Rigoll, G.: Determination of non-prototypical valence and arousal in popular music: Features and performances. EURASIP J. Audio Speech Music Process. (Special Issue on Scalable Audio-Content Analysis, 2010) pp. 19 (2010). (Article ID 735854)

Part IV
Conclusion

This last part will first provide a discussion on Intelligent Audio Analysis in the light of the content of the book up to this point followed by best practice recommendations and remaining challenges. In a second chapter, a more general summary will be given. Future directions of research will conclude this book.

Chapter 13
Discussion

A scientist's aim in a discussion with his colleagues is not to persuade, but to clarify.

—Leo Szilard

Picking up on the aims of this book as were outlined in Sect. 2.2, these are discussed one by one as follows. This includes a statement on how the state-of-the-art in the field was advanced more recently. Basing on these, a distilled 'best practice' recommendation is given to the reader, before a critical discussion of missing aspects and remaining research steps.

13.1 Picking Up on the Goals

(I) A unified perspective on audio analysis tasks was provided in this book in the hope to bridge between the disciplines of speech and language processing—including sub-disciplines such as ASR, computational paralinguistics or opinion mining and sentiment analysis—music analysis, i.e., MIR, and general sound analysis. Even though these often co-exist side-by-side and transfer is often limited, the current book by intention provided a unified perspective to reach first and foster future synergies. Further, a broad overview was given on recent advancements in these fields by presentation of manifold exemplary applications and performances in realistic conditions. In particular, realistic conditions were mostly ignored in the field so far by a number of simplifications.

(II) Recent methods were presented and shown in application. These aim to improve robustness and reliability of today's Intelligent Audio Analysis systems. In particular, the enhancement and isolation of the signal of interest by suited methods of blind separation of sources was emphasised on [1–9]. This is sparsely found in the field at this time coupled with subsequent 'intelligent' processing. Instead, research usually focuses either on the separation or on intelligent processing of 'clean' and

DOI: 10.1007/978-3-642-36806-6_13,

idealised audio material. After this step, systematic feature brute-forcing of up to thousands of audio features was shown as highly efficient means in particular also in the case of handling of novel Intelligent Audio Analysis tasks. This was seen in many of the presented tasks [10–12]. At the same time, individually tailored feature types were shown, in particular such basing on NMF activations [4–6, 13–16] or music theory and human perception [17]. The idea is to demonstrate limitations of unification. Further, memory-enhanced learning algorithms such as by (B) LSTM RNNs [3, 16, 18–30] and their synergistic combination with DBNs [5, 31–42] were shown to prevail in many tasks. For example, two MIREX 2010 Challenges for music onset detection [22] and tempo determination [21] were won by this method. Then, suited GM topologies such as SLDM and SLDS were presented in their successful application for highly noise robust ASR [43]. In their combination, the overall efforts led to the best result in the CHiME 2011 Challenge for highly robust keyword spotting when using only a single microphone source [5]. Subsequent to the Challenge, the overall best result—beating also those approaches that exploit multiple microphone sources—could be reached based on combining the presented approaches towards source separation with NMF activation features, and a triple-stream topology of a DBN with BLSTM RNN feed [6]. To ease the ever-present bottleneck of data sparseness, a series of methods was further suggested and shown to be beneficial. These can be added by synthesis of training material [44], and the collaboration of machine and human for the labelling of data guided by the machine: The machine first by itself labels the data it is sufficiently confident it can assign the correct label itself in a semi-supervised learning step [45, 46]. Then, it asks for human's help if it cannot assign a label with sufficient confidence, but thinks the data may be interesting, for example because it covers a sparse class. This is the 'active learning' step. Finally, it decides that some instances might not be of interest as they are too similar to already seen data in an active learning step. Further, transfer learning methods can help to use data with 'similar' conditions.

(III) The very broad range of Intelligent Audio Analysis application was shown. These include the recognition of speaker states and traits such as age [47], height [48], interest [27, 49–54], intoxication [55, 56], and sleepinesss [12, 57], singer traits in polyphonic music [2, 3, 58] such as age, gender, height, and race, the recognition of ballroom dance style [59–61] in music, or the recognition of emotion evoked in the listener of sounds [62]—to name the most recent ones of the examples.

(IV) Benchmark results and standardised test-beds were shown for a broader range of audio analysis tasks. Especially in the field of paralinguistic speech analysis these were entirely lacking until very recently. Instead, comparability between research results in the field was considerably low. Apart from different evaluation strategies, the diversity of corpora is high, as many early studies report results on their individual and proprietary corpora. Additionally, there was practically no same feature set found twice: High diversity is not only found in the selection of LLD, but also in the perceptual adaptation, speaker adaptation, and—most of all—selection and implementation of functionals. This opposes the more or less settled and clearly defined feature types MFCC, RASTA or PLP that allow for higher comparability in speech recognition. A series of consecutive annual research challenges held at INTERSPEECH

2009–2012 changed this recently: the INTERSPEECH 2009 Emotion Challenge [63–65], the INTERSPEECH 2010 Paralinguistic Challenge [47, 66], the INTERSPEECH 2011 Speaker State Challenge [55], and the INTERSPEECH 2012 Speaker Trait Challenge [67]. Further, the first and second International Audio/Visual Emotion Challenge and Workshop (AVEC) in 2011 [68, 69] and 2012 [70] as satellites of the HUMAINE International Conference on Affective Computing and Intelligent Interaction (ACII) and ACM International Conference on Multimodal Interaction contained speech analysis tasks. The aim of this succession of challenges has been two-fold: First, the concept of a strict partitioning of data into train, development, and test sets, together with well-defined measures of performance was accomplished—this is known from established fields such as ASR—in the broad and divergent field of paralinguistics. Second, the research in these fields mostly lacks in two respects: small, preselected, prototypical, and often non-natural data sets and the named low comparability of results. All of these events featured very high participation of the research community—in the latest of these events 52 research teams participated. By using methods as presented in this book, best results on these challenge tasks could be obtained [30, 54, 56, 67, 71]. These and several further benchmark results were presented in this book constantly emphasising on reproducibility and accessibility of data and algorithms by the research community.

Standards were further provided by the openSMILE [72] and openEAR [73] toolkits as presented in this book (cf. Sect. 6.5). As open-source software, they are entirely transparent, and the standardised feature sets (cf. Annex for four of these) can provide a good starting point for many audio analysis tasks. For source separation, the openBliSSART toolkit [4, 74] as was shown in Chap. 8.

Another part of standardisation is found in the datasets introduced, which are mostly freely accessible to interested readers and by now found manifold usage,[1] including the following nine that cover a broad range of Intelligent Audio Analysis tasks: HU-ASA [75], TUM AVIC [53], Metacritic [76], BRD [61], NTWICM [77], Audio Key [17], FindSounds [45], Emotional FindSounds [62], UltraStar [2].

(V) Deficiencies in current approaches and future perspective in and for the field were shown in detail for all presented exemplary tasks in the respective sections and chapters. However, this book shall be concluded by a more general perspective on Intelligent Audio Analysis best practice, remaining challenges and a vision on the future of this field.

Finally, it should be noted that fusion with other modalities—in particular image and video processing—can lead to improvements for many of the tasks discussed such as non-linguistic vocalisation recognition [26, 78] or emotion recognition [79–83]. Further, successful transfer of the introduced methods such as feature brute-forcing and LSTM-modelling can be of interest, as was shown for 3D gesture recognition in [23] or for CAN-bus data analysis in the car in [28].

---

[1] http://www.openaudio.eu

13.2 Best Practice Recommendations

In the following, best practice recommendations as based on the presented content of the book are given. These again follow the chain of processing from data provision to its enhancement, feature extraction, classification or regression, and output encoding for optimal system embedding.

High realism: [84] In order to evaluate systems for Intelligent Audio Analysis in conditions close to real-life application, realistic data are needed [66]. However, progress in this direction is often slow in the field. This is likely owing to the high effort of collecting and annotating such data. Realism concerns in particular the choice of testing instances. To assess an Intelligent Audio Analysis's system performance in a realistic way, these may not be restricted to prototypical, straightforward cases. If a pre-selection is applied at all, e.g., to gain performance bounds, it needs to be based on objective and transparent criteria rather than on 'intuitive' expert-selection. While methods such as semi-supervised or synthesis of training material have been named in this book, they are less suited for the collection of test-instances. Even crowd-sourcing may—depending on the task—be more appropriate for the collection of training data if laymen are involved. Realism further touches pre-processing such as chunking according to acoustic or symbolic, e.g., linguistic criteria. In most real-life applications, chunking will be expected to work automatically and should be oriented on acoustic LLDs. An example is a audio activity based chunking, which easily becomes challenging in reverberant or noisy acoustic conditions. If additional meta-information or common knowledge is exploited in the analysis process, the information should be automatically retrieved from publicly available knowledge-sources, e.g., by web-based queries as was shown in this book, e.g., for chord lead sheets, lyrics and word information. If such information includes individual experts' knowledge on the test cases this may result in a considerable bias of accuracies to be expected for unseen material. Finally, real-life applications imply highest possible independence of training and test conditions in most cases. This can be established by partitioning into train, development and test sets [63]. Today, however, often random cross-validation without known random seed for partitioning is employed especially in case of small data sets, to ensure significance of results. Using an independent and stratified subdivision according to simple criteria (e.g., splitting according to instance IDs, by speaker or composer, etc.) is a transparent alternative to keep the statistical significance. Otherwise, the random seed should be provided together with the toolkit for reproduction of the partitions or a download for an archive containing the instance or file list may be provided.

Standardised, multi-faceted and machine-aided data collection: Publicly available audio data with rich annotation are still sparse [84]. Even with a recently increasing number of available databases ready for experimentation, these often come with different labelling schemes such as discrete versus continuous task representation. This can make cross-corpus evaluation and data agglomeration [85] partly difficult. 'Translation' schemes and standards are therefore needed to map from one task representation to another or for the task representation itself and should be followed if existent—at least in addition to individual solutions. Multi-corpus and cross-corpus evaluation, such as for age and gender and for emotion [86, 87], is crucial to assess generalisation of AMs and LMs. In fact, experiments in cross-corpus manner indicate overfitting to single corpora [87]. This trend can only partly be eased by corpus adaptation and normalisation. In addition, optimisation such as feature selection or parameter optimisation for the learning algorithm may exhibit low cross-data generalisation [88]. Still, unification of the labelling schemes as mentioned above introduces 'information loss'. A late fusion of multiple classifiers trained on single corpora with different labelling schemes may help overcome this in the future. In addition, the efficacy of semi-supervised learning to leverage unlabelled audio data for its computationally intelligent analysis has been repeatedly demonstrated [46, 89–91]. This may be turned into large-scale studies across multiple tasks using large amounts of data acquired from the web. Finally, a promising technique is synthesis of training data: In fact, it has been shown that generalisation properties of models in a cross-corpus setting can be improved through joint training with both human and synthetic speech [44] or human-played and MIDI-synthesised music [92]. These results are very promising since synthetic audio can be easily produced in large quantities, and a variety of combinations can be simulated. It is hoped that this will yield good generalisation of models and facilitate learning of multiple tasks and their interdependencies. In any case, acquisition of more and well-defined data for building robust and generalising models can thus be seen as major challenges for the future.

Source separation: The results cited in this book clearly demonstrated the gain obtained by source separation for the enhancement of the signal of interest in real-life audio streams. As particularly suited algorithm NMF and its derivatives was shown—e.g., in the openBliSSART implementation. This may be added by methods exploiting multiple sources such as ICA for stereophonic information or microphone array feed exploitation.

Feature brute-forcing: The features used in early Intelligent Audio Analysis research were often motivated by the fields of ASR and speaker recognition as these were among the earliest and the driving forces. As a consequence, usage of spectral or cepstral features such as MFCCs prevails to the present day [84]. In the meantime, manifold expert-crafted acoustic features, including perceptually motivated ones [55, 93, 94] or such basing on pre-classification [95] were introduced. These have often been successfully evaluated for diverse audio analysis tasks as was also shown in this book, along with the addition of more or less brute forced features. Furthermore, it has repeatedly been shown that enlarging the feature space can help boost accuracy [11, 55]. Such large spaces can be brute-forced by toolkits as openSMILE and serve as broad basis for subsequent space optimisation—in particular when approaching novel audio analysis tasks. A promising additional direction is the semi-supervised learning of features, e.g., through deep belief networks [86] or bottleneck topologies [27, 40].

Temporal evolution modelling: It has been shown in several chapters of this book touching all three fields speech, music, and sound that explicit storage of temporal context—in particular with learning the optimal amount of such context—

outperforms common approaches in the field that do not provide this option. The LSTM architectures shown are clearly suited in this respect, and whenever the application allows for modelling of temporal context dependencies, these or future alternatives should be considered—be it on their own or in combination with other machine learning algorithms to provide them with dynamic warping abilities such as the shown tandem DBN-BLSTM architectures.

Coupling of tasks: [84] A number of interdependencies is already visible in the tasks that were considered in this book. By addition of further or novel tasks, this dependency is likely to be amplified. For example, in speaker analysis, long term traits are coupled to some degree, e.g., height with age, gender, and race as were shown interdependent in the determination of age, gender, and height in this book. Other examples include emotional manifestation being dependent on personality [96], and gender-dependencies of non-linguistic vocalisations such as laughter [97]. In this book an example was also shown in the music domain for the interdependence of ballroom dance style, metre, and tempo. It seems obvious that the further introduced sound analysis tasks of event and evoked emotion are also interdependent.

Such knowledge can be integrated by keeping separate models depending on the other tasks, adaptation or normalisation, or considering additional information on related 'side tasks' [98, 99] as in the above named examples shown in this book. An alternative to such explicit modelling of dependencies is to automatically learn them from training data. For example, the rather simple strategy of using pairs of age and gender classes as learning target instead of each attribute individually was shown to be beneficial in this book and [47]. In the future, enhanced modelling of multiple correlated target variables should be commonly targeted in multi-task learning. The input features are then shared among tasks, such as the internal activations in the hidden layer of a neural network [100]. A challenge may then arise from the different representation of task variables by various data types (continuous, ordinal, nominal), which may additionally use different time scales (e.g., dance style is mostly constant in a musical piece, but tempo may vary). Considering such suited methods for multi-scale fusion and multi-task learning [101], future Intelligent Audio Analysis should not focus on tasks in isolation, but aim at a 'more holistic' analysis of tasks.

Standardisation: Arguably, the more mature and closer to real-life application the field of Intelligent Audio Analysis gets, the greater is the need for standardisation [84]. Similarly as before, standardisation efforts can be categorised along the signal processing chain. They include definition of the task modelling such as given in the MPEG-4 standard for emotion in audio or the MIREX tasks and ID3 tag categories for music, documentation and well-motivated grouping of audio features such as the CEICES Feature Coding Scheme [102], standardised feature sets as provided by the openSMILE [72] and openEAR [73] toolkits or MPEG7-LLD standard, and machine learning frameworks [103]. Such standardised feature extraction and classification allows to evaluate the feature extraction and classification components of a recognition system separately. To further increase the reproducibility and comparability of results, well-defined evaluation settings should be employed, such as the ones provided by the named challenge events [12]. Finally, communication between system components in real-life applications requires standardisation of recognition results for application embedding. This can be achieved by mark-up languages such as EMMA [104], EmotionML [105], MIML [106] or VoiceXML, or the MIDI standard in music. Many further ones are, however, still needed. This holds in particular for general audio analysis.

13.3 Remaining Challenges

A number of challenges remain in order to reach full applicability of systems for all tasks such as those that were presented.

More robustness: Robustness issues in Intelligent Audio Analysis can be categorised into technical robustness on the one hand and security on the other hand [66]. Technical robustness refers to robustness against signal distortions including additive noise, e.g., environmental noise or interfering signals of similar type (e.g., 'cocktail party problem' in speech analysis) and reverberation, but also change of transmission medium, e.g., from air to liquid or oxygen to helium, and artefacts of transmission due to package loss and coding. Many of these issues have been extensively studied in the context of ASR leading to a multiplicity of solutions ready for application and transfer, as has been shown in this book. These include audio enhancement, robust feature extraction, model-based techniques (i.e., learning the distortions), and recognition architectures such as the BLSTM RNN. However, they have so far only been investigated in few of the application scenarios: Apart from ASR in speech analysis, there do exist a few studies on technical robustness of affect analysis, e.g., [15, 107, 108]—other speaker classification and paralinguistic analysis tasks are yet to follow. In particular for the processing of music or sound, however, experience is missing if the signal is captured via a (distant) microphone rather than accessed in ideal condition, e.g., from a storage device. On another level, Intelligent Audio Analysis systems are hardly ready to recognising malicious mis-use in the sense of attempted fraud. Examples for such fraud include feigning of a speaker's age (e.g., in an audio-based system for parental control), degree of intoxication (e.g., to ensure clearance for high risk operations despite alcoholisation), or emotion (e.g., by faking anger in an automated voice portal system in order to be redirected to a human operator). Studies in this direction are sparse including detection of feigned depression and sleepiness [109, 110]. This is in massive contrast to the research devoted to speaker verification, i.e., robustness of speaker recognition systems against feigning of another speaker's identity. Similar scenarios can be thought of in the processing of sound such as feigning an explosion sound to alert security for distraction with criminal intentions. Still, the majority of research in Intelligent Audio Analysis assumes laboratory conditions including the fact that data is mostly recorded without consideration of real users with potentially black-hearted intentions.

Blind separation and multi-task processing of real-life streams: Once going for broader analysis of audio, highly complex blends of speech, music, and sound need to be considered rather than more or less isolated and slightly corrupted signals. In fact, experience is almost entirely lacking on performance once several non-correlated audio sources shall be identified and characterised at a time. In an unknown real-life audio-stream all types of speech, music, and sound can, and in fact very likely will often be present simultaneously. If the sources or the recording device is moving, the task to analyse such a stream may become even more challenging.

Massive weakly supervised learning: For a start, semi-supervised learning of sound events has been shown to be beneficial in this book [45] and other Intelligent Audio Analysis [46]. However, the Internet and radio and television broadcast media streams provide the potential of almost infinite audio provision to systems that can learn partly or non-supervised by themselves. Once the previously mentioned step is mastered, i.e., once systems are capable to blindly separate and handle complex blends of audio as they occur in most real-life settings, such systems can start to improve by 'teaching themselves'. In combination with active learning, they can occasionally ask for human help once not sufficiently confident on decisions to be made. In the shown example on semi-supervised AEC (cf. Sect. 12.2), only the AM was learnt semi-supervisedly by the system. Yet, the parametrisation of the signal enhancement, the feature space, configuration of the classifier and other steps along the signal processing chain be auto-adapted in a similar fashion.

Evolutionary learning: Besides such adaptation of parametrisation, the overall learning algorithm's layout and architecture could be learnt by future Intelligent Audio Analysis systems. As an example, LSTM RNN could decide in which layers to provide memory and how many cells should ideally be coupled, where to provide bottle-necks, e.g., for feature de-correlation [40], etc.

Closing the gap between analysis and synthesis: This book focused entirely on the analysis side of Intelligent Audio Processing. In fact, the synthesis is mostly handled in according isolation. This is to regret, as closing this gap holds many promises as in the named example of synthesising training material for analysis [44].

Cross-cultural and cross-lingual widening: One of the barriers to overcome if audio information retrieval systems are to be widely employed is to enable their use across cultural and lingual borders [84]. Yet, cross-cultural effects usually make tasks even more challenging—examples include speaker states and trait analysis [111], music mood, metre or dance style determination or emotion evoked in listeners by sounds. Concerning speech, it is still an open question which speaker states and traits manifest consistently across cultures and languages. For example, emotion recognition, has been shown to depend strongly on the language being spoken [112–114], while non-linguistic vocalisation such as laughter and speaker identification [115, 116], seem comparably culture and language independent. However, generally little attention is paid to the more subtle effects of the cultural background and language variation. It might turn out, though, that cross-cultural Intelligent Audio Analysis is just another instance of learning correlated tasks.

Novel tasks: Many tasks have been shown in this book, and many more are covered in the literature. Yet, approaching human audio analysis abilities, several remain that may have been of lower interest so far from an application point of view. Examples could be the recognition of a speaking condition during eating, playing effects of under-researched instruments in music such as bending or overblow on a blues harp, or the recognition of specific sound source traits. Further, supra-human capabilities could be targeted such as by determining the heart rate or skin conductivity from the speech signal or assessing meta-information concerning the audio encoding, recording equipment or transmission. Several novel tasks can also be derived by transfer from one domain to the other [117], such as musical instrument trait assessment, e.g., neck length of a guitar, etc.

Further unification and transfer of methods: Substantial unification efforts were made by this book. However, in order to show further transfer potential of the research efforts often co-existing with limited exchange for the diverse tasks, further unification will be needed. This is in particular reasonable to reach the ability of genuine multi-task analysis as described.

Confidence measures: The performances chosen as illustrative examples in this book, and as will be summarised in Sect. 14.1 clearly show the need for additional information of a machine learner's confidence in its result for most real-life application. A number of approaches for the topic of confidence measures have been proposed in the domain of ASR. These can be roughly grouped into three categories [118]: In the first, a binary true or false classifier is built based on a combination of so-called predictor features (e.g., acoustic stability and LM scores) that are collected during the decoding procedure. Various classification models have been used in this respect such as a linear discriminant function [119], or a maximum entropy model [120]. In the second category, an approximation of the posterior probability in the standard MAP criterion approach is taken as the confidence measure. The posterior probability is typically estimated from the speech system lattices or 'N-best' lists [121]. Methods in the third category treat the confidence estimation problem as an utterance verification problem. They make use of the likelihood ratio between the null hypothesis (e.g., the word is correct) and the alternative hypothesis (e.g., the word is incorrect) as a confidence measure [122]. Unfortunately, the use of confidence measures for practically any other Intelligent Audio Analysis task seems to have never drawn comparable attention so far. Thus, the existing approaches are primarily designed for ASR systems, as most of them rely almost entirely on properties of HMMs, such as acoustic scores or the word graph, which are not typical components of all the Intelligent Audio Analysis systems as were introduced.

Distributed processing: Effort have constantly been made to integrate audio processing technology into Internet technology to facilitate the interface for human users, as well as to decrease computing resources on the client side [123] allowing, e.g., for access on mobile devices. A further advantage of distribution is that models stored on the server can be updated periodically on this side rather than by the end-user. This can be done, for example, by semi-supervised learning. Comparing to stand-alone analysis, distribution involves diverse technologies including data compression, network data transmission protocols, and distributed computing [123]. Furthermore, for economic reasons and efficiency the solution for distribution should be inexpensive to implement on the client side, the required data transmission bandwidth should remain at a low level, and the recognition accuracy should ideally be (at least) approximately equal to the state-of-the-art. To implement an accordingly distributed system, the first problem arising is how to distribute the components of the recogniser over the Internet. An obvious choice is the 'classic' client-server architecture, as adopted in the widely used ETSI standard for Distributed Speech Recognition (DSR) [124]. The recognition processing is then usually separated into two parts—the feature extraction and compression front-end is being executed on the client side and the recognition is processed on the remote back-end. By this, there is only the need to send the reduced parameterised representation of audio, which also provides a favourable data reduction in the light of access security. In fact, vector quantisation of the audio features and related techniques allow for further reduction of required network traffic. From a method point of view, this may lead to semi-supervised learning with compressed features on the server side if models shall be updated. So far, several speech-based Internet applications have been well explored and even applied in practice, such as DSR [125, 126], distributed speaker verification [127], and first research on distributed speech emotion recognition [123]. Yet, in contrast to the attention paid to these speech-based applications and some on-line music recognition services, there are few such efforts that deal with the manifold further Intelligent Audio Analysis tasks shown and discussed in this book in a distributed manner. This requires according research efforts to make all these analyses available on-line.

New research challenges: Despite the series of competitive evaluation campaigns in the fields of speaker state and trait analysis [12] and further existing ones such as by NIST, MediaEval[2] or MIREX, many white spots remain in the map of audio analysis tasks—in particular for the sound analysis domain. These will help to further consolidate the field and provide benchmarks and common findings for further improvement.

References

1. Schuller, B., Lehmann, A., Weninger, F., Eyben, F., Rigoll, G.: Blind enhancementof the rhythmic and harmonic sections by nmf: Does it help? In: Proceedings InternationalConference on Acoustics including the 35th German Annual Conference on Acoustics,NAG/DAGA 2009, pp. 361–364. DEGA, Rotterdam, March 2009
2. Weninger, F., Wöllmer, M., Schuller B.: Automatic assessment of singer traits in popular music: gender, age, height and race. In: Proceedings 12th International Society for Music Information Retrieval Conference, ISMIR 2011, pp. 37–42. ISMIR, Miami (2011)
3. Weninger, F., Durrieu, J.-L., Eyben, F., Richard, G., Schuller, B.: Combining monaural source separation with long short-term memory for increased robustness in vocalist gender recognition. In: Proceedings 36th IEEE International Conference on Acoustics, Speech, and Signal Processing (ICASSP 2011), pp. 2196–2199. IEEE, Prague, Czech Republic, May 2011
4. Weninger, F., Lehmann, A., Schuller, B.: Openblissart: design and evaluation of a research toolkit for blind source separation in audio recognition tasks. In: Proceedings 36th IEEE International Conference on Acoustics, Speech, and Signal Processing, ICASSP 2011, pp. 1625–1628. IEEE, Prague, May 2011
5. Weninger, F., Geiger, J., Wöllmer, M., Schuller, B., Rigoll, G.: The munich 2011 chime challenge contribution: Nmf-blstm speech enhancement and recognition for reverberated

[2] http://www.multimediaeval.org

multisource environments. In: Proceedings Machine Listening in Multisource Environments, CHiME 2011, Satellite Workshop of Interspeech 2011, pp. 24–29. ISCA, Florence, Sept 2011
6. Weninger, F., Wöllmer, M., Geiger, J., Schuller, B., Gemmeke, J., Hurmalainen, A., Virtanen, T., Rigoll, G.: Non-negative matrix factorization for highly noise-robust asr: to enhance or to recognize? In: Proceedings 37th IEEE International Conference on Acoustics, Speech, and Signal Processing, ICASSP 2012, pp. 4681–4684. IEEE, Kyoto, March 2012
7. Weninger, F., Feliu, J., Schuller, B.: Supervised and semi-supervised supression of background music in monaural speech recordings. In: Proceedings 37th IEEE International Conference on Acoustics, Speech, and Signal Processing, ICASSP 2012, pp. 61–64. IEEE, Kyoto, March 2012
8. Weninger, F., Amir, N., Amir, O., Ronen, I., Eyben, F., Schuller, B.: Robust feature extraction for automatic recognition of vibrato singing in recorded polyphonic music. In: Proceedings 37th IEEE International Conference on Acoustics, Speech, and Signal Processing, ICASSP 2012, pp. 85–88. IEEE, Kyoto, March 2012
9. Joder, C., Weninger, F., Eyben, F., Virette, D., Schuller, B.: Real-time speech separation by semi-supervised nonnegative matrix factorization. In: Theis, F.J., Cichocki, A., Yeredor, A., Zibulevsky, M. (eds.) Proceedings 10th International Conference on Latent Variable Analysis and Signal Separation (LVA/ICA 2012). Lecture Notes in Computer Science, vol. 7191, pp. 322–329. Springer, Tel Aviv (2012)
10. Batliner, A., Steidl, S., Schuller, B., Seppi, D., Laskowski, K., Vogt, T., Devillers, L., Vidrascu, L., Amir, N., Kessous, L., Aharonson, V.: Combining efforts for improving automatic classification of emotional user states. In: Proceedings 5th Slovenian and 1st International Language Technologies Conference, ISLTC 2006, pp. 240–245. Slovenian Language Technologies Society, Ljubljana, Oct 2006
11. Schuller, B., Wimmer, M., Mösenlechner, L., Kern, C., Arsić, D., Rigoll, G.: Brute-forcing hierarchical functionals for paralinguistics: a waste of feature space? In: Proceedings 33rd IEEE International Conference on Acoustics, Speech, and Signal Processing, ICASSP 2008, pp. 4501–4504. IEEE, Las Vegas, April 2008
12. Schuller, B.: The computational paralinguistics challenge. IEEE Signal Process. Mag. **29**(4), 97–101 (2012)
13. Schuller, B., Weninger, F., Wöllmer, M., Sun, Y., Rigoll, G.: Non-negative matrix factorization as noise-robust feature extractor for speech recognition. In: Proceedings 35th IEEE International Conference on Acoustics, Speech, and Signal Processing, ICASSP 2010, pp. 4562–4565. IEEE, Dallas, March 2010
14. Schuller, B., Weninger, F.: Discrimination of speech and non-linguistic vocalizations by non-negative matrix factorization. In: Proceedings 35th IEEE International Conference on Acoustics, Speech, and Signal Processing, ICASSP 2010, pp. 5054–5057. IEEE, Dallas, March 2010
15. Weninger, F., Schuller, B., Batliner, A., Steidl, S., Seppi, D.: Recognition of non-prototypical emotions in reverberated and noisy speech by non-negative matrix factorization. EURASIP J. Adv. Signal Process. **Article ID 838790**, 16 (2011). Special issue on emotion and mental state recognition from speech
16. Weninger, F., Schuller, B., Wöllmer, M., Rigoll, G.: Localization of non-linguistic events in spontaneous speech by non-negative matrix factorization and long short-term memory. In: Proceedings 36th IEEE International Conference on Acoustics, Speech, and Signal Processing, ICASSP 2011, pp. 5840–5843. IEEE, Prague, May 2011
17. Schuller, B., Gollan, B.: Music theoretic and perception-based features for audio key determination. J. New Music Res. **41**(2), 175–193 (2012)
18. Wöllmer, M., Eyben, F., Graves, A., Schuller, B., Rigoll, G.: A tandem blstm-dbn architecture for keyword spotting with enhanced context modeling. In: Proceedings ISCA Tutorial and Research Workshop on Non-Linear Speech Processing, NOLISP 2009, pp. 9. ISCA, Vic, June 2009
19. Wöllmer, M., Eyben, F., Schuller, B., Douglas-Cowie, E., Cowie, R.: Data-driven clustering in emotional space for affect recognition using discriminatively trained lstm networks. In:

Proceedings INTERSPEECH 2009, 10th Annual Conference of the International Speech Communication Association, pp. 1595–1598. ISCA, Brighton, Sept 2009
20. Eyben, F., Böck, S., Schuller, B., Graves, A.: Universal onset detection with bidirectional long-short term memory neural networks. In: Proceedings 11th International Society for Music Information Retrieval Conference, ISMIR 2010, pp. 589–594. ISMIR, Utrecht, Oct 2010
21. Böck, S., Eyben, F., Schuller, B.: Tempo detection with bidirectional long short-term memory neural networks. In: Proceedings Annual Meeting of the MIREX 2010 community as part of the 11th International Conference on Music Information Retrieval, pp. 3. ISMIR, Utrecht, August 2010
22. Böck, S., Eyben, F., Schuller, B.: Onset detection with bidirectional long short-term memory neural networks. In: Proceedings Annual Meeting of the MIREX 2010 community as part of the 11th International Conference on Music Information Retrieval, pp. 2. ISMIR, Utrecht, August 2010
23. Arsić, D., Wöllmer, M., Rigoll, G., Roalter, L., Kranz, M., Kaiser, M., Eyben, F., Schuller, B.: Automated 3d gesture recognition applying long short-term memory and contextual knowledge in a cave. In: Proceedings 1st Workshop on Multimodal Pervasive Video Analysis, MPVA 2010, held in conjunction with ACM Multimedia 2010, pp. 33–36. ACM, Florence, Oct 2010
24. M. Wöllmer, A. Metallinou, F. Eyben, B. Schuller, and S. Narayanan: Context-sensitive multimodal emotion recognition from speech and facial expression using bidirectional lstm modeling. In: Proceedings INTERSPEECH 2010, 11th Annual Conference of the International Speech Communication Association, pp. 2362–2365. ISCA, Makuhari, Sept 2010
25. Landsiedel, C., Edlund, J., Eyben, F., Neiberg, D., Schuller, B.: Syllabification of conversational speech using bidirectional long-short-term memory neural networks. In: Proceedings 36th IEEE International Conference on Acoustics, Speech, and Signal Processing, ICASSP 2011, pp. 5265–5268. IEEE, Prague, May 2011
26. Eyben, F., Petridis, S., Schuller, B., Tzimiropoulos, G., Zafeiriou, S., Pantic, M.: Audiovisual classification of vocal outbursts in human conversation using long-short-term memory networks. In: Proceedings 36th IEEE International Conference on Acoustics, Speech, and Signal Processing, ICASSP 2011, pp. 5844–5847. IEEE, Prague, May 2011
27. Wöllmer, M., Weninger, F., Eyben, F., Schuller, B.: Acoustic-linguistic recognition of interest in speech with bottleneck-blstm nets. In: Proceedings INTERSPEECH 2011, 12th Annual Conference of the International Speech Communication Association, pp. 3201–3204. ISCA, Florence, August 2011
28. Wöllmer, M., Blaschke, C., Schindl, T., Schuller, B., Färber, B., Mayer, S., Trefflich, B.: Online driver distraction detection using long short-term memory. IEEE Trans. Intell. Transp. Syst. **12**(2), 574–582 (2011)
29. Wöllmer, M., Metallinou, A., Katsamanis, N., Schuller, B., Narayanan, S.: Analyzing the memory of blstm neural networks for enhanced emotion classification in dyadic spoken interactions. In: Proceedings 37th IEEE International Conference on Acoustics, Speech, and Signal Processing, ICASSP 2012, pp. 4157–4160. IEEE, Kyoto, March 2012
30. Wöllmer, M., Kaiser, M., Eyben, F., Schuller, B., Rigoll, G.: Lstm-modeling of continuous emotions in an audiovisual affect recognition framework. Image and Vision Computing, Special Issue on Affect Analysis in Continuous Input, p. 16, 2012
31. Reiter, S., Schuller, B., Rigoll, G.: A combined lstm-rnn-hmm-approach for meeting event segmentation and recognition. In: Proceedings 31st IEEE International Conference on Acoustics, Speech, and Signal Processing, ICASSP 2006, vol. 2, pp. 393–396. IEEE, Toulouse, May 2006
32. Wöllmer, M., Eyben, F., Keshet, J., Graves, A., Schuller, B., Rigoll, G.: Robust discriminative keyword spotting for emotionally colored spontaneous speech using bidirectional lstm networks. In: Proceedings 34th IEEE International Conference on Acoustics, Speech, and Signal Processing, ICASSP 2009, pp. 3949–3952. IEEE, Taipei, April 2009
33. Wöllmer, M., Eyben, F., Schuller, B., Sun, Y., Moosmayr, T., Nguyen-Thien, N.: Robust in-car spelling recognition: a tandem blstm-hmm approach. In: Proceedings INTERSPEECH 2009, 10th Annual Conference of the International Speech Communication Association, pp. 1990–9772. ISCA, Brighton, Sept 2009

34. Wöllmer, M., Eyben, F., Graves, A., Schuller, B., Rigoll, G.: Bidirectional lstm networks for context-sensitive keyword detection in a cognitive virtual agent framework. Cogn. Comput. **2**(3), 180–190 (2010). Special issue on non-linear and non-conventional speech processing
35. Wöllmer, M., Eyben, F., Graves, A., Schuller, B., Rigoll, G.: Improving keyword spotting with a tandem blstm-dbn architecture. In: Sole-Casals, J., Zaiats, V. (eds.) Advances in Non-Linear Speech Processing: International Conference on Nonlinear Speech Processing, 25–27 June 2009 (NOLISP 2009). Revised Selected Papers, Lecture Notes on Computer Science (LNCS), vol. 5933/2010, pp. 68–75. Springer, Vic (2010)
36. Wöllmer, M., Schuller, B., Eyben, F., Rigoll, G.: Combining long short-term memory and dynamic Bayesian networks for incremental emotion-sensitive artificial listening. IEEE J. Sel. Top. Signal Proces. **4**(5), 867–881 (2010). Special issue on speech processing for natural interaction with intelligent environments
37. Wöllmer, M., Sun, Y., Eyben, F., Schuller, B.: Long short-term memory networks for noise robust speech recognition. In: Proceedings INTERSPEECH 2010, 11th Annual Conference of the International Speech Communication Association, pp. 2966–2969. ISCA, Makuhari, Sept 2010
38. Wöllmer, M., Eyben, F., Schuller, B., Rigoll, G.: Recognition of spontaneous conversational speech using long short-term memory phoneme predictions. In: Proceedings INTERSPEECH 2010, 11th Annual Conference of the International Speech Communication Association, pp. 1946–1949. ISCA, Makuhari, Sept 2010
39. Wöllmer, M., Marchi, E., Squartini, S., Schuller, B.: Multi-stream lstm-hmm decoding and histogram equalization for noise robust keyword spotting. Cogn. Neurodyn. **5**(3), 253–264 (2011)
40. Wöllmer, M., Schuller, B., Rigoll, G.: A novel bottleneck-blstm front-end for feature-level context modeling in conversational speech recognition. In: Proceedings 12th Biannual IEEE Automatic Speech Recognition and Understanding Workshop, ASRU 2011, pp. 36–41. IEEE, Big Island, Dec 2011
41. Wöllmer, M., Eyben, F., Schuller, B., Rigoll, G.: A multi-stream asr framework for blstm modeling of conversational speech. In: Proceedings 36th IEEE International Conference on Acoustics, Speech, and Signal Processing, ICASSP 2011, pp. 4860–4863. IEEE, Prague, May 2011
42. Wöllmer, M., Schuller, B.: Enhancing spontaneous speech recognition with blstm features. In: Travieso-González, C.M., Alonso-Hernández, J. (eds.) Advances in Nonlinear Speech Processing, 5th International Conference on Nonlinear Speech Processing, 7–9 Nov 2011 (NoLISP 2011). Proceedings, Lecture Notes in Computer Science (LNCS), vol. 7015/2011, pp. 17–24. Springer, Las Palmas de Gran Canaria (2011)
43. Schuller, B., Wöllmer, M., Moosmayr, T., Rigoll, G.: Recognition of noisy speech: a comparative survey of robust model architecture and feature enhancement. EURASIP J. Audio Speech Music Process. **Article ID 942617**, 17 (2009)
44. Schuller, B., Burkhardt, F.: Learning with synthesized speech for automatic emotion recognition. In: Proceedings 35th IEEE International Conference on Acoustics, Speech, and Signal Processing, ICASSP 2010, pp. 5150–515. IEEE, Dallas, March 2010
45. Zhang, Z., Schuller, B.: Semi-supervised learning helps in sound event classification. In: Proceedings 37th IEEE International Conference on Acoustics, Speech, and Signal Processing, ICASSP 2012, pp. 333–336. IEEE, Kyoto, March 2012
46. Zhang, Z., Weninger, F., Wöllmer, M., Schuller, B.: Unsupervised learning in cross-corpus acoustic emotion recognition. In: Proceedings 12th Biannual IEEE Automatic Speech Recognition and Understanding Workshop, ASRU 2011, pp. 523–528. IEEE, Big Island, Dec 2011
47. Schuller, B., Steidl, S., Batliner, A., Burkhardt, F., Devillers, L., Müller, C., Narayanan, S.: The interspeech 2010 paralinguistic challenge. In: Proceedings INTERSPEECH 2010, 11th Annual Conference of the International Speech Communication Association, pp. 2794–2797. ISCA, Makuhari, Sept 2010
48. Schuller, B., Wöllmer, M., Eyben, F., Rigoll, G., Arsić, D.: Semantic speech tagging: Towards combined analysis of speaker traits. In: Brandenburg, K., Sandler, M. (eds.) Proceedings AES 42nd International Conference, pp. 89–97. Audio Engineering Society, Ilmenau, July 2011

49. Schuller, B., Köhler, N., Müller, R., Rigoll, G.: Recognition of interest in human conversational speech. In: Proceedings INTERSPEECH 2006, 9th International Conference on Spoken Language Processing, ICSLP, pp. 793–796. ISCA, Pittsburgh, Sept 2006
50. Schuller, B., Müller, R., Hörnler, B., Höthker, A., Konosu, H., Rigoll, G.: Audiovisual recognition of spontaneous interest within conversations. In: Proceedings 9th ACM International Conference on Multimodal Interfaces, ICMI 2007, pp. 30–37. ACM, Nagoya, Nov 2007
51. Vlasenko, B., Schuller, B., Mengistu, K.T., Rigoll, G., Wendemuth, A.: Balancing spoken content adaptation and unit length in the recognition of emotion and interest. In: Proceedings INTERSPEECH 2008, 9th Annual Conference of the International Speech Communication Association, Incorporating 12th Australasian International Conference on Speech Science and Technology, SST 2008, pp. 805–808. ISCA/ASSTA, Brisbane, Sept 2008
52. Schuller, B., Rigoll, G.: Recognising interest in conversational speech: comparing bag of frames and supra-segmental features. In: Proceedings INTERSPEECH 2009, 10th Annual Conference of the International Speech Communication Association, pp. 1999–2002. ISCA, Brighton, Sept 2009
53. Schuller, B., Müller, R., Eyben, F., Gast, J., Hörnler, B., Wöllmer, M., Rigoll, G., Höthker, A., Konosu, H.: Being bored? recognising natural interest by extensive audiovisual integration for real-life application. Image Vis. Comput. **27**(12), 1760–1774 (November 2009). Special issue on visual and multimodal analysis of human spontaneous behavior
54. Wöllmer, M., Weninger, F., Eyben, F., Schuller, B.: Computational assessment of interest in speech: facing the real-life challenge. Künstliche Intelligenz (German J. Artif. Intell.) **25**(3), 227–236 (2011). Special issue on emotion and computing
55. Schuller, B., Batliner, A., Steidl, S., Schiel, F., Krajewski, J.: The interspeech 2011 speaker state challenge. In: Proceedings INTERSPEECH 2011, 12th Annual Conference of the International Speech Communication Association, pp. 3201–3204. ISCA, Florence, August 2011
56. Weninger, F., Schuller, B.: Fusing utterance-level classifiers for robust intoxication recognition from speech. In: Proceedings MMCogEmS Workshop (Inferring Cognitive and Emotional States from Multimodal Measures), Held in Conjunction with the 13th International Conference on Multimodal Interaction, Nov 2011 (ICMI 2011). ACM, Alicante (2011)
57. Krajewski, J., Schnieder, S., Sommer, D., Batliner, A., Schuller, B.: Applying multiple classifiers and non-linear dynamics features for detecting sleepiness from speech. Neurocomputing **84**, 65–75 (2012). Special issue from neuron to behavior: evidence from behavioral measurements
58. Schuller, B., Kozielski, C., Weninger, F., Eyben, F., Rigoll, G.: Vocalist gender recognition in recorded popular music. In: Proceedings 11th International Society for Music Information Retrieval Conference, ISMIR 2010, pp. 613–618. ISMIR, Utrecht, Oct 2010
59. Schuller, B., Eyben, F., Rigoll, G.: Fast and robust meter and tempo recognition for the automatic discrimination of ballroom dance styles. In: Proceedings 32nd IEEE International Conference on Acoustics, Speech, and Signal Processing, ICASSP 2007, vol. I, pp. 217–220. IEEE, Honolulu, April 2007
60. Eyben, F., Schuller, B., Reiter, S., Rigoll, G.: Wearable assistance for the ballroom-dance hobbyist: holistic rhythm analysis and dance-style classification. In: Proceedings 8th IEEE International Conference on Multimedia and Expo, ICME 2007, pp. 92–95. IEEE, Beijing, July 2007
61. Schuller, B., Eyben, F., Rigoll, G.: Tango or waltz?—putting ballroom dance style into tempo detection. EURASIP J. Audio Speech Music Process. **Article ID 846135**, 12 (2008). Special issue on intelligent audio, speech, and music processing applications
62. Schuller, B., Hantke, S., Weninger, F., Han, W., Zhang, Z., Narayanan, S.: Automatic recognition of emotion evoked by general sound events. In: Proceedings 37th IEEE International Conference on Acoustics, Speech, and Signal Processing, ICASSP 2012, pp. 341–344. IEEE, Kyoto, March 2012
63. Schuller, B., Steidl, S., Batliner, A.: The interspeech 2009 emotion challenge. In: Proceedings INTERSPEECH 2009, 10th Annual Conference of the International Speech Communication Association, pp. 312–315. ISCA, Brighton, Sept 2009

64. Schuller, B., Steidl, S., Batliner, A.: Introduction to the special issue on sensing emotion and affect: facing realism in speech processing. Speech Commun. **53**(9/10), 1059–1061 (2011). Special issue sensing emotion and affect: facing realism in speech processing
65. Schuller, B., Batliner, A., Steidl, S., Seppi, D.: Recognising realistic emotions and affect in speech: state of the art and lessons learnt from the first challenge. Speech Commun. **53**(9/10), 1062–1087 (2011). Special issue on sensing emotion and affect—facing realism in speech processing
66. Schuller, B., Steidl, S., Batliner, A., Burkhardt, F., Devillers, L., Müller, C., Narayanan, S.: Paralinguistics in speech and language: state-of-the-art and the challenge. Comput. Speech Lang. **27**(1), 4–39 (January 2013). Special issue on paralinguistics in naturalistic speech and language
67. Schuller, B., Steidl, S., Batliner, A., Nöth, E., Vinciarelli, A., Burkhardt, F., van Son, R., Weninger, F., Eyben, F., Bocklet, T., Mohammadi, G., Weiss, B.: The interspeech 2012 speaker trait challenge. In: Proceedings INTERSPEECH 2012, 13th Annual Conference of the International Speech Communication Association, p. 4. ISCA, Portland, Sept 2012
68. Schuller, B., Valstar, M., Cowie, R., Pantic, M. (eds.): In: Proceedings of the First International Audio/Visual Emotion Challenge and Workshop, AVEC, Oct 2011. Lecture Notes on Computer Science (lncs), Part II, vol. 6975. Springer, Memphis (2011)
69. Schuller, B., Valstar, M., Eyben, F., McKeown, G., Cowie, R., Pantic, M.: Avec 2011: the first international audio/visual emotion challenge. In: Schuller, B., Valstar, M., Cowie, R., Pantic, M. (eds.) Proceedings First International Audio/Visual Emotion Challenge and Workshop, Oct 2011 (AVEC 2011), Held in Conjunction with the International HUMAINE Association Conference on Affective Computing and Intelligent Interaction 2011 (ACII 2011), vol. II, pp. 415–424. Springer, Memphis (2011)
70. Schuller, B., Valstar, M., Eyben, F., Cowie, R., Pantic, M.: Avec 2012: the continuous audio/visual emotion challenge. In: Morency, L.-P., Bohus, D., Aghajan, H.K., Cassell, J., Nijholt, A., Epps, J. (eds.) Proceedings of the 14th ACM International Conference on Multimodal Interaction, ICMI, pp. 449–456. ACM, Santa Monica, Oct 2012
71. Schuller, B., Metze, F., Steidl, S., Batliner, A., Eyben, F., Polzehl, T.: Late fusion of individual engines for improved recognition of negative emotions in speech: learning versus democratic vote. In: Proceedings 35th IEEE International Conference on Acoustics, Speech, and Signal Processing, ICASSP 2010, pp. 5230–5233. IEEE, Dallas, March 2010
72. Eyben, F., Wöllmer, M., Schuller, B.: Opensmile: the munich versatile and fast open-source audio feature extractor. In: Proceedings of the 9th ACM International Conference on Multimedia, MM 2010, pp. 1459–1462. ACM, Florence, Oct 2010
73. Eyben, F., Wöllmer, M., Schuller, B.: Openear: introducing the munich open-source emotion and affect recognition toolkit. In: Proceedings 3rd International Conference on Affective Computing and Intelligent Interaction and Workshops, ACII 2009, vol. I, pp. 576–581. IEEE, Amsterdam, Sept 2009
74. Weninger, F., Schuller, B.: Optimization and parallelization of monaural source separation algorithms in the openblissart toolkit. J. Signal Process. Syst. **69**(3), 267–277 (2012)
75. Weninger, F., Schuller, B.: Audio recognition in the wild: Static and dynamic classification on a real-world database of animal vocalizations. In: Proceedings 36th IEEE International Conference on Acoustics, Speech, and Signal Processing, ICASSP 2011, pp. 337–340. IEEE, Prague, May 2011
76. Schuller, B., Knaup, T.: Learning and knowledge-based sentiment analysis in movie review key excerpts. In: Esposito, A., Esposito, A.M., Martone, R., Müller, V., Scarpetta, G. (eds.) Toward Autonomous, Adaptive, and Context-Aware Multimodal Interfaces: Theoretical and Practical Issues: Third COST 2102 International Training School, 15–19 March 2010, Caserta, Italy. Revised Selected Papers of Lecture Notes on Computer Science (LNCS), vol. 6456/2010, pp. 448–472, 1st edn. Springer, Heidelberg (2011)
77. Schuller, B., Dorfner, J., Rigoll, G.: Determination of non-prototypical valence and arousal in popular music: features and performances. EURASIP J. Audio Speech Music Process. **Article ID 735854**, 19 (2010). Special issue on scalable audio-content analysis

78. Eyben, F., Petridis, S., Schuller, B., Pantic, M.: Audiovisual vocal outburst classification in noisy acoustic conditions. In: Proceedings 37th IEEE International Conference on Acoustics, Speech, and Signal Processing, ICASSP 2012, pp. 5097–5100. IEEE, Kyoto, March 2012
79. Schuller, B., Wimmer, M., Arsić, D., Rigoll, G., Radig, B.: Audiovisual behavior modeling by combined feature spaces. In: Proceedings 32nd IEEE International Conference on Acoustics, Speech, and Signal Processing, ICASSP 2007, vol. II, pp. 733–736. IEEE, Honolulu, April 2007
80. Schröder, M., Bevacqua, E., Eyben, F., Gunes, H., Heylen, D., ter Maat, M., Pammi, S., Pantic, M., Pelachaud, C., Schuller, B., de Sevin, E., Valstar, M., Wöllmer, M.: A demonstration of audiovisual sensitive artificial listeners. In: Proceedings 3rd International Conference on Affective Computing and Intelligent Interaction and Workshops, ACII 2009, vol. I, pp. 263–264. IEEE, Amsterdam, Sept 2009
81. Schröder, M., Bevacqua, E., Cowie, R., Eyben, F., Gunes, H., Heylen, D., ter Maat, M., McKeown, G., Pammi, S., Pantic, M., Pelachaud, C., Schuller, B., de Sevin, E., Valstar, M., Wöllmer, M.: Building autonomous sensitive artificial listeners. IEEE Trans. Affect. Comput. **3**(2), 165–183 (2012)
82. Eyben, F. Wöllmer, M., Valstar, M., Gunes, H., Schuller, B., Pantic, M.: String-based audiovisual fusion of behavioural events for the assessment of dimensional affect. In: Proceedings International Workshop on Emotion Synthesis, Representation, and Analysis in Continuous Space, EmoSPACE 2011, Held in Conjunction with the 9th IEEE International Conference on Automatic Face & Gesture Recognition and Workshops, FG 2011, pp. 322–329. IEEE, Santa Barbara, March 2011
83. Metallinou, A., Wöllmer, M., Katsamanis, A., Eyben, F., Schuller, B., Narayanan, S.: Context-sensitive learning for enhanced audiovisual emotion classification. IEEE Trans. Affect. Comput. **3**(2), 184–198 (2012)
84. Schuller, B., Weninger, F.: Ten recent trends in computational paralinguistics. In: Esposito, A., Vinciarelli, A., Hoffmann, R., Müller, V.C. (eds.) 4th COST 2102 International Training School on Cognitive Behavioural Systems. Lecture Notes on Computer Science (LNCS), p. 15. Springer, Berlin (2012)
85. Schuller, B., Zhang, Z., Weninger, F., Rigoll, G.: Using multiple databases for training in emotion recognition: to unite or to vote? In: Proceedings INTERSPEECH 2011, 12th Annual Conference of the International Speech Communication Association, pp. 1553–1556. ISCA, Florence, August 2011
86. Stuhlsatz, A., Meyer, C., Eyben, F., Zielke, T., Meier, G., Schuller, B.: Deep neural networks for acoustic emotion recognition: Raising the benchmarks. In: Proceedings 36th IEEE International Conference on Acoustics, Speech, and Signal Processing, ICASSP 2011, pp. 5688–5691. IEEE, Prague, May 2011
87. Schuller, B., Vlasenko, B., Eyben, F., Wöllmer, M., Stuhlsatz, A., Wendemuth, A., Rigoll, G.: Cross-corpus acoustic emotion recognition: variances and strategies. IEEE Trans. Affect. Comput. **1**(2), 119–131 (2010)
88. Eyben, F., Batliner, A., Schuller, B., Seppi, D., Steidl, S.: Cross-corpus classification of realistic emotions: some pilot experiments. In: Devillers, L., Schuller, B., Cowie, R., Douglas-Cowie, E., Batliner, A. (eds.) Proceedings 3rd International Workshop on EMOTION: Corpora for Research on Emotion and Affect, Satellite of LREC 2010, pp. 77–82. European Language Resources Association, Valletta, May 2010
89. Jia, L., Chun, C., Jiajun, B., Mingyu, Y., Jianhua, T.: Speech emotion recognition using an enhanced co-training algorithm. In: Proceedings of the 2007 IEEE International Conference on Multimedia and Expo, ICME 2007, pp. 999–1002. IEEE, Beijing (2007)
90. Mahdhaoui, A., Chetouani, M.: A new approach for motherese detection using a semi-supervised algorithm. In: Machine Learning for Signal Processing XIX: Proceedings of the 2009 IEEE Signal Processing Society Workshop, MLSP 2009, pp. 1–6. IEEE, Grenoble (2009)
91. Yamada, M., Sugiyama, M., Matsui, T.: Semi-supervised speaker identification under covariate shift. Signal Process. **90**(8), 2353–2361 (2010)

92. Lee, K., Slaney, M.: Automatic chord recognition from audio using a supervised hmm trained with audio-from-symbolic data. In: Proceedings of the ACM Multimedia '06, Santa Barbara, USA, pp. 11–20. ACM, New York (2006)
93. Wu, S., Falk, T.H., Chan, W.: Automatic speech emotion recognition using modulation spectral features. Speech Commun. **53**(5), 768–785 (2011)
94. Mahdhaoui, A., Chetouani, M., Kessous, L.: Time-frequency features extraction for infant directed speech discrimination. Lecture Notes in Computer Science (including subseries Lecture Notes in Artificial Intelligence and Lecture Notes in Bioinformatics), vol. 5933 LNAI, pp. 120–127. Springer, Berlin Heidelberg (2010)
95. Ringeval, F., Chetouani, M.: A vowel based approach for acted emotion recognition. In: INTERSPEECH 2008: 9th Annual Conference of the International Speech Communication Association, pp. 2763–2766. ISCA, Brisbane (2008)
96. Reisenzein, R., Weber, H.: Personality and emotion. In: Corr, P.J., Matthews, G. (eds.) The Cambridge Handbook of Personality Psychology, pp. 54–71. Cambridge University Press, Cambridge (2009)
97. Provine, R.: Laughter punctuates speech: linguistic, social and gender contexts of laughter. Ethology **15**, 291–298 (1993)
98. Ververidis, D., Kotropoulos, C.: Automatic speech classification to five emotional states based on gender information. In: Proceedings of 12th European Signal Processing Conference, pp. 341–344, Vienna, 2004
99. Vogt, T., André, E.: Improving automatic emotion recognition from speech via gender differentiation. In: Proceedings of Language Resources and Evaluation Conference (LREC), Genoa, 2006
100. Stadermann, J., Koska, W., Rigoll, G.: Multi-task learning strategies for a recurrent neural net in a hybrid tied-posteriors acoustic mode. In: Proceedings of Interspeech 2005, pp. 2993–2996. ISCA, Lisbon (2005)
101. Byrd, D.: Relations of sex and dialect to reduction. Speech Commun. **15**(1–2), 39–54 (1994)
102. Batliner, A., Steidl, S., Schuller, B., Seppi, D., Vogt, T., Wagner, J., Devillers, L., Vidrascu, L., Aharonson, V., Kessous, L., Amir, N.: Whodunnit: searching for the most important feature types signalling emotion-related user states in speech. Comput. Speech Lang. **25**(1), 4–28 (2011). Special issue on affective speech in real-life interactions
103. Hall, M., Frank, E., Holmes, G., Pfahringer, B., Reutemann, P., Witten, I.H.: The WEKA data mining software: an update. SIGKDD Explor. **11**(1), 10–18 (2009)
104. Baggia, P., Burnett, D.C., Carter, J., Dahl, D.A., McCobb, G., Raggett, D.: EMMA: Extensible MultiModal Annotation Markup Language, World Wide Web Consortium, Recommendation REC-emma-20090210, M. Johnston (ed.), February 2009
105. Schröder, M., Devillers, L., Karpouzis, K., Martin, J.-C., Pelachaud, C., Peter, C., Pirker, H., Schuller, B., Tao, J., Wilson, I.: What should a generic emotion markup language be able to represent? In: Paiva, A., Picard, R.W., Prada, R. (eds.) Affective Computing and Intelligent Interaction: Second International Conference, Lisbon, Portugal, 12–14 Sept 2007 (ACII 2007). Proceedings, Lecture Notes on Computer Science (LNCS), vol. 4738/2007, pp. 440–451. Springer, Berlin (2007)
106. Mao, X., Li, Z., Bao, H.: An extension of MPML with emotion recognition functions attached. LNAI of Lecture Notes in Computer Science. Lecture Notes in Artificial Intelligence and Lecture Notes in Bioinformatics, vol. 5208. Springer, Berlin Heidelberg (2008)
107. Schuller, B.: Affective speaker state analysis in the presence of reverberation. Int. J. Speech Technol. **14**(2), 77–87 (2011)
108. Tabatabaei, T.S., Krishnan, S.: Towards robust speech-based emotion recognition. In: Proceeding of IEEE International Conference on Systems, Man and Cybernetics, pp. 608–611. IEEE, Istanbul (2010)
109. Cannizzaro, M., Reilly, N., Snyder, P.J.: Speech content analysis in feigned depression. J. Psycholinguist. Res. **33**(4), 289–301 (2004)
110. Reilly, N., Cannizzaro, M.S., Harel, B.T., Snyder, P.J.: Feigned depression and feigned sleepiness: a voice acoustical analysis. Brain Cogn. **55**(2), 383–386 (2004)

111. Boden, M.: Mind as Machine: A History of Cognitive Science, Chapter 9. Oxford University Press, New York (2008)
112. Shami, M., Verhelst, W.: Automatic classification of expressiveness in speech: a multi-corpus study. In: Müller, C. (ed.) Speaker Classification II. Lecture Notes in Computer Science/Artificial Intelligence, vol. 4441, pp. 43–56. Springer, Heidelberg (2007)
113. Chen, A.: Perception of paralinguistic intonational meaning in a second language. Lang. Learn. **59**(2), 367–409 (2009)
114. Esposito, A., Riviello, M.T.: The cross-modal and cross-cultural processing of affective information. In: Proceeding of the 2011 Conference on Neural Nets WIRN10: Proceedings of the 20th Italian Workshop on Neural Nets, vol. 226, pp. 301–310, 2011
115. Bellegarda, J.R.: Language-independent speaker classification over a far-field microphone. In: Mueller, C. (ed.) Speaker Classification II: Selected Projects, pp. 104–115. Springer, Berlin (2007)
116. Kleynhans, N.T., Barnard, E.: Language dependence in multilingual speaker verification. In: Proceedings of the 16th Annual Symposium of the Pattern Recognition Association of South Africa, pp. 117–122, Langebaan, Nov 2005
117. Weninger, F., Schuller, B., Liem, C., Kurth, F., Hanjalic, A.: Music information retrieval: An inspirational guide to transfer from related disciplines. In: Müller, M., Goto, M. (eds.) Multimodal Music Processing, volume Seminar 11041 of Dagstuhl Follow-UpsSchloss, pp. 195–215. Dagstuhl, Germany (2012)
118. Jiang, H.: Confidence measures for speech recognition: a survey. Speech Commun. **45**(4), 455–470 (2005)
119. Sukkar, R.: Rejection for connected digit recognition based on GPD segmental discrimination. In: IEEE International Conference on Acoustics, Speech, and Signal Processing, 1994 (ICASSP-94), vol. 1, pp. I-393–I-396
120. White, C., Droppo, J., Acero, A., Odell, J.: Maximum entropy confidence estimation for speech recognition. In: IEEE International Conference on Acoustics, Speech and Signal Processing, 2007 (ICASSP 2007), vol. 4, pp. 809–812
121. Wessel, F., Schluter, R., Macherey, K., Ney, H.: Confidence measures for large vocabulary continuous speech recognition. IEEE Trans. Speech Audio Process. **9**(3), 288–298 (2001)
122. Rahim, M., Lee, C., Juang, B.: Discriminative utterance verification for connected digits recognition. IEEE Trans. Speech Audio Process. **5**(3), 266–277 (1997)
123. Han, W., Zhang, Z., Deng, J., Wöllmer, M., Weninger, F., Schuller, B.: Towards distributed recognition of emotion in speech. In: Proceedings 5th International Symposium on Communications, Control, and Signal Processing (ISCCSP 2012), pp. 1–4. IEEE, Rome, May 2012
124. ETSI. ETSI ES 202 050 V1.1.5: Speech processing, transmission and quality aspects (STQ), distributed speech recognition, advanced front-end feature extraction algorithm, compression algorithms (2007)
125. Zhang, W., He, L., Chow, Y.L., Yang, R., Su, Y.: The study on distributed speech recognition system. In: Proceedings of ICASSP, pp. 1431–1434, Istanbul, 2000
126. Tsakalidis, S., Digalakis, V., Neumeyer, L.: Efficient speech recognition using subvector quantization and discrete-mixture hmms. In: Proceedings of ICASSP, pp. 569–572, Phoenix, 1999
127. Jain, A.K., Flynn, P.J., Ross, A.A.: Handbook of Biometrics. Springer, Heidelberg (2008)

Chapter 14
Vision

We can only see a short distance ahead, but we can see plenty there that needs to be done.

—Alan Turing

In this last chapter, a short summary of the exemplary results that were selected for application illustration is given that motivates the vision given in the concluding section of this book.

14.1 Summary of Results

Rather than repeating details from the individual sections and formerly derived best practice recommendations, this summary aims to provide an estimation of applicability of systems as they were shown in this book. By that, let us have a general overview on the results as were presented in Part III of this book in a 'less formal' way. In Table 14.1 results of regression are summarised and in Table 14.2 according results are presented for classification—the tasks drum-beat separation, onset detection, and structure analysis were left out due to their more specific testing conditions.

Looking at the results in Table 14.1, it becomes evident that improvement is needed in practically any of these, and that human performance level is only reached in some rare exceptions such as speaker intoxication classification [1]. On the other hand side, any of these tasks are highly significantly above chance level, and many are ready for first real-life application—given sufficient failure tolerance.

B. W. Schuller, *Intelligent Audio Analysis*, Signals and Communication Technology,
DOI: 10.1007/978-3-642-36806-6_14,

Table 14.1 Overview on presented regression results by CC and MLE for diverse intelligent speech, music, and sound analysis tasks

Task	Range	Database	# Train	# Test	Test method	CC [%]	MLE [–/cm]
Sentiment	[0, 100]	METACRITIC	49 698	52 924	T/T	0.570	14.1
Interest	[-1, +1]	TUM AVIC	2 673	1 207	T/D/T	0.421	0.146
Height	[144, 204]	TIMIT	4 620	1 680	T/T	0.296	7.05
Sound arousal	[-1, +1]	Emo findsounds	390	390	SCV	0.606	–
Sound valence	[-1, +1]	Emo findsounds	390	390	SCV	0.473	–

Given are the numeric range, the database with training (uniting training and development instances) and test instances and the test method, where T/(D/)T are train, (develop), and test and SCV is always ten-fold. The level of precision of the presented results depends on the number of test instances

14.2 Future Perspective

As an overall future vision, the field of Intelligent Audio Analysis can be expected to lead to genuine 'computer audition' or 'machine listening' capability in the sense of *Holistic Evolutionary Audio Analysis*. Once the different tasks described in this book are not handled in isolation, an exemplary future system output after analysis of a real-life audio-stream could thus be:

> At high confidence, the auditory scene overall seems very relaxing: There is a lowly arousing and pleasant sound of waves, two to three singing birds together with tranquil flute and strings music in duple metre and a likable calm voice of an aged mid-sized very open and conscientious male person of Asian origin talking to a younger female in English saying it was a lovely day. She—rather tall at about 175 cm yet light-weight and likely of European origin—seems interested in what he says and in a joyful, yet slightly tired state with low heart rate. This is also manifested by her mild speech-laughter replying that he was right, indeed.

Table 14.2 Overview on presented classification results by UA and WA for diverse intelligent speech, music, and sound analysis tasks

Task	# Classes	Database	# Train	# Test	Test method	UA [%]	WA [%]
Speech classes							
Digits	10	TI46	800	1 280	T/T	99.92	99.92
Spelling	26	TI46	2 080	3 328	T/T	93.09	93.09
Phonemes	41	COSINE	9.59 h	1.81 h	T/D/T	–	66.41
Words	4.8 k	COSINE	9.59 h	1.81 h	T/D/T	–	46.50
Non-linguistics	5	TUM AVIC	4 050	2 184	T/D/T	88.6	83.9
Sentiment	3	METACRITIC	49 698	52 924	T/T	53.99	53.71
Emotion	2	FAU AIBO EC	9 959	8 257	T/T	67.7	65.5
Age	4	aGender	53 076	17 332	T/D/T	48.91	46.24
Gender	3	aGender	53 076	17 332	T/D/T	81.21	84.81
Intoxication	2	ALC	9 360	3 ,000	T/D/T	65.9	66.4
Sleepiness	2	SLC	6 281	2 808	T/D/T	70.3	72.9

(continued)

Table 14.2 continued

Task	# Classes	Database	# Train	# Test	Test method	UA [%]	WA [%]
Music classes							
Metre	2	BRD	1 855	1 855	SCV	97.1	96.6
Dance style	9	BRD	1 855	1 855	SCV	88.9	89.1
Tempo	141	BRD	1 855	1 855	SCV	89.0	88.5
Key	12	KEY-ALL	521	521	SCV	–	77.3
Key	24	KEY-ALL	521	521	SCV	–	62.1
Chords	24	ChoRD	10 702	10 702	LOSO	–	60.13
Chords	36	ChoRD	10 702	10 702	LOSO	–	48.84
Mood–arousal	3	NTWICM	1 376	1 272	T/D/T	56.2	58.7
Mood–valence	3	NTWICM	1 376	1 272	T/D/T	61.2	61.0
Voice presence	2	UltarStar	326 527	97 144	T/D/T	75.77	75.81
Singer age	2	UltraStar	315 043	93 342	T/D/T	57.55	56.56
Singer gender	2	UltraStar	326 198	96 373	T/D/T	89.61	93.60
Singer height	2	UltraStar	280 714	80 962	T/D/T	72.07	78.26
Singer race	2	UltraStar	321 178	96 563	T/D/T	63.30	76.98
Sound classes							
Birds	2	HU-ASA	868	868	SCV	80.0	81.3
Animals	5	HU-ASA	1 063	1 063	SCV	49.5	64.0
Acoustic events	7	FindSounds	11 292	5 645	T/D/T	66.5	71.7

Given are the number of classes, the database with training (uniting training and development instances) and test instances and the test method, where T/(D/)T are train, (develop), and test and SCV is always ten-fold. The level of precision of the presented results depends on the number of test instances

A challenge remaining at that point will be the careful evaluation of ethical issues if machines can listen to and understand arbitrary audio including personal information and details.

Finally, given such holistic analysis capability basing on very efficient source separation and synergistic coupling of tasks, future audio analysis systems can start to train themselves in a massive way such as by crawling the Internet for audio, or listening to very general media broadcast potentially reaching supra-human capabilities in some of the alluded tasks.

Reference

1. Schiel, F.: Perception of alcoholic intoxication in speech. In Proceedings of Interspeech, pp. 3281–3284. Florence (2011)

Appendix
openSMILE Standardised Feature Sets

All's well that ends well.
—William Shakespeare.

In Table A.1 the LLDs and functionals and their frequency across the four openSMILE standard feature sets as were mentioned in this book are given.

LLDs are processed by simple moving average (SMA) low-pass filtering.

Delta regression coefficients are added per LLD. The total number of features is—in principle—obtained by multiplying the number of LLD times two times the number of functionals. However, for the two larger feature sets exceptions hold from this strict brute-forcing rule as are indicated. This prevents creation of nonsense features.

B. W. Schuller, *Intelligent Audio Analysis*, Signals and Communication Technology,
DOI: 10.1007/978-3-642-36806-6,

Table A.1 openSMILE standard features sets by LLDs and functionals

Feature	EC	PC	SSC	AVEC
Frequencies				
# LLDs	16	38	59	31
# Functionals	12	22	41	42
# Features	384	1 582	4 368	1 941
LLDs				
RMS energy	✓		✓	
Sum of auditory spectrum (loudness)		✓[a]	✓	✓
Sum of RASTA-sytle filtered auditory spectrum			✓	
ZCR	✓		✓	✓
Energy in bands from 250–650 Hz, 1–4 kHz			✓	✓
Spectral roll-off points 25 %, 50 %, 75 %, 90 %			✓	✓
Spectral flux			✓	✓
Spectral entropy			✓	✓
Spectral variance			✓	✓
Spectral skewness			✓	✓
Spectral kurtosis			✓	✓
Spectral slope			✓	
Psychoacousitc sharpness				✓
Harmonicity				✓
MFCC 0		✓		
MFCC 1–10	✓	✓	✓	✓
MFCC 11–12	✓	✓	✓	
MFCC 13–14		✓		
Log Mel frequency band 0–7		✓[a]		
LSP frequency 0–7		✓		
RASTA-style auditory spectrum bands 1–26 (0 – 8 kHz)			✓	
F_0 (ACF based)	✓			
F_0 (SHS based)		✓		
F_0 (SHS based followed by Viterbi smoothing)			✓	✓
F_0 envelope		✓		
Probability of voicing	✓	✓	✓	✓
Jitter		✓	✓	✓
Jitter (delta: 'jitter of jitter')		✓	✓	✓
Shimmer		✓	✓	✓
Logarithmic HNR				✓
Functionals				
Positive arithmetic mean				✓[d]
Arithmetic mean	✓	✓	✓	✓[d]
Root quadratic mean				✓
Contour centroid			✓	
Standard deviation	✓	✓	✓	✓
Flatness				✓

(continued)

Table A.1 (continued)

Feature	EC	PC	SSC	AVEC
Skewness	✓	✓	✓	✓
Kurtosis	✓	✓	✓	✓
Quartiles 1, 2, 3		✓[a]	✓	✓
Inter-quartile ranges 2-1, 3-2, 3-1		✓[a]	✓	✓
Percentile 1 %, 99 %		✓[a]	✓	✓
Percentile range 1–99 %		✓	✓	✓
% frames above minimum + 25%, 50% of range				✓
% frames above minimum + 75 % of range		✓[a]		
% frames above minimum + 90 % of range		✓[a]	✓	✓
% frames below minimum + 25 % of range			✓	
% frames rising			✓	✓
% frames falling			✓	
% frames left, right curvature			✓[f]	
% frames that are non-zero			✓[b]	
Linear regression offset	✓	✓[a]		
Linear regression slope	✓	✓[a]	✓	✓[c]
Linear regression approximation error (MAE)		✓[a]		✓[c]
Linear regression approximation error (MSE)	✓	✓[a]	✓	
Quadratic regression coefficient *a*			✓	✓[c]
Quadratic regression coefficient *b*			✓	
Quadratic regression approximation error (MAE)				✓[c]
Quadratic regression approximation error (MSE)			✓	
Maximum, minimum	✓			
Maximum–minimum (range)	✓			
Rising, falling slopes (min to max) mean, standard deviation				✓[c]
Inter maxima distances mean, standard deviation			✓	✓[c]
Amplitude mean of maxima relative to mean				✓[c]
Amplitude range of minima relative to mean				✓[c]
Amplitude range of maxima relative to mean				✓[c]
Relative position of maximum, minimum	✓	✓[a]		
LP gain			✓	✓[c,e]
LP coefficients 1–5			✓	✓[c,e]
Peak value arithmetic mean			✓	
Peak value arithmetic mean–arithmetic mean			✓	
Segment length mean, max, min, standard deviation			✓[b]	✓[e]
Input duration in seconds		✓[b]	✓[b]	

EC:INTERSPEECH 2009 Emotion Challenge, PC:INTERSPEECH 2010 Paralinguistic Challenge, SSC:INTERSPEECH 2011 Speaker Trait Challenge, AVEC:Audio/Visual Emotion Challenge 2011

[a]Only used for the TUM AVIC baseline (PC)

[b]Only applied to F_0

[c]Not applied to delta coefficient contours

[d]For delta coefficients the mean of only positive values is applied, otherwise the arithmetic mean is applied

[e]Not applied to voicing related LLDs

[f]Only applied to voicing related LLDs. For the PC feature set, the two additional features turn duration and number of voiced segments (F_0 onsets) were added

Index

B. W. Schuller, *Intelligent Audio Analysis*, Signals and Communication Technology,
DOI: 10.1007/978-3-642-36806-6,

MIX
Papier aus verantwortungsvollen Quellen
Paper from responsible sources
FSC® C105338

FSC
www.fsc.org

If you have any concerns about our products,
you can contact us on
ProductSafety@springernature.com

In case Publisher is established outside the EU,
the EU authorized representative is:
Springer Nature Customer Service Center GmbH
Europaplatz 3, 69115 Heidelberg, Germany

Printed by Libri Plureos GmbH
in Hamburg, Germany